Environmental Resources

A.S. Mather and K. Chapman

Longman
Scientific &
Technical

Longman Scientific & Technical
Longman Group Limited
Longman House, Burnt Mill, Harlow
Essex CM20 2JE, England
and Associated Companies throughout the world

Copublished in the United States with
John Wiley & Sons, Inc., 605 Third Avenue, New York
NY 10158

© Longman Group Limited 1995

First published 1995

British Library Cataloguing in Publication Data
A catalogue entry for this title is available from the British Library.

ISBN 0-582-10168-9

Library of Congress Cataloging-in-Publication data
Mather, Alexander S. (Alexander Smith)
    Environmental resources/A.S. Mather and K. Chapman.
      p. cm.
    Includes bibliographical references and index.
    ISBN 0-470-23491-1
    1. Natural resources.   2. Conservation of natural resources.
  3. Environmental protection.   I. Chapman, Keith.   II. Title.
S930.M158   1995
333.7–dc20   94-25284
                                    CIP

Set by 22 in 9/11pt Times
Produced by Longman Singapore Publishers (Pte) Ltd.
Printed in Singapore

Contents

Preface

The relationship between humans and their environment now attracts greater attention than at any time in recent history. Despite being better fed and more affluent than in the past, humans are increasingly anxious about their life-support systems and about the quality of the environments in which they live. In recent decades concern has broadened from a focus on prospects for food and energy resources to embrace wider issues about global warming and its consequences, about biodiversity and its accelerating loss, and about the overall state of the environment and the sustainable use of environmental resources. The side-effects of resource use are now far more widely acknowledged than a generation ago, as is the inter-relatedness of different parts of the world and of the environment and the economy. The state of the environment and the use of environmental resources have become political issues at local, national and international levels. An outpouring of books, pamphlets and papers on environmental themes threatens to swamp even the most avid student. Why, then, do we seek to add to the torrent?

The origins of most books are probably complex, and multiple rather than single factors probably have to be invoked in most cases. The primary reason in this case, however, is that in our experience of university teaching, few books on environmental resources have sought to bring together conceptual and empirical material in compact, textbook form. Many excellent volumes treat environmental philosophy, environmental economics, the tropical forest, problems of global food supply and of energy resources, and numerous other environmental fields. Even more numerous polemical tomes burden the bookshelves. Our experience, however, suggests that few publications prove suitable as texts which will encourage thoughtfulness and a deepening of the understanding of issues of environmental resources – and in a form and at a price suitable for a student readership.

Our book is aimed at an undergraduate audience (in geography, environmental science and related fields) seeking a knowledge of contemporary environmental resource issues and an understanding of how and why they have developed. It is assumed that the reader will have previously acquired a basic understanding of the ecology of environmental resources. Whilst we hope that we have achieved a balance between alarmism and complacency, we do not claim that what we say is 'objective'. Like prospective readers, we have our own values, although we hope that we have succeeded at least at times in offering different viewpoints. We share the conviction that the roots of environmental resource problems are social rather than environmental, and that they are not capable of solution by technical means alone. Our aim is to encourage the reader to *think* about resource issues, rather than to accept at face value simplistic formulations of resource problems and glib solutions to them.

The book is structured in three main sections. The first and third of these are essentially conceptual, while the middle section attempts to relate contemporary trends and issues in individual resource sectors to the wider conceptual framework. After basic concepts of environmental resources are introduced in the first part of Chapter 1, types of environmental attitudes are briefly reviewed and concepts such as sustainability (which underlie much of the ensuing material) are outlined. In Chapter 2, the economic systems and regimes under which environmental resources are used and managed are discussed. Chapter 3 considers the use of environmental resources in space and time, focusing in particular on the development of the world economy and the global integration of the areas of supply and demand that has characterised recent centuries. Chapters 4 to 10 deal with individual resource sectors – land and food, forests, biodiversity, energy, minerals, freshwater and the ocean. These sectoral reviews are necessarily selective and they attempt to demonstrate the relevance of concepts previously introduced and subsequently developed to an understanding of contemporary resource problems. References in the text point to more comprehensive sources of information on each of the resource sectors. The final section (Chapters 11 and 12) discusses the relationship between environmental resources and economic development before considering various scenarios of resource futures and tracing the evolution of the notion of sustainability.

This book would not have been produced without the encouragement of staff at Longman Higher Education. We acknowledge this encouragement with gratitude. We are grateful also to colleagues in the Department of Geography, University of Aberdeen for contributing to the congenial and stimulating atmosphere that has facilitated the writing of the book, and to numerous students on whom we have tried out our ideas over the years. The technical support of Sue Allen (word-processing), and Jenny Johnston, Susan Powell and Alison Sandison (cartography) has been much appreciated and is gratefully acknowledged.

Acknowledgements

We are grateful to the following for permission to reproduce copyright material:

Academic Press for Table 1.4 (Hall and Hall, 1984); Association of American Geographers for Fig. 1.1 (Whitaker, 1941); J R Coull for Plate 10.3; R Dunlap for Table 1.3 (Dunlap and van Liere, 1978); Economic Geography for Table 3.3 (Peet, 1969); Edward Arnold (Publishers) Ltd. for Fig. 11.1 (S R Eyre, *The real wealth of nations*, 1978); J Fairbairn for Plates 4.1 and 4.2; J R Flenley for Figs 11.4 and 11.5 (Bahn and Flenley, 1992); Food and Agriculture Organization of the United Nations for Fig. 5.7 (FAO, 1984); D B Grigg for Fig. 4.4 (Grigg, 1984); L Joels for Plate 8.2; J Livingstone for Plates 1.1a and 1.1b; Macmillan Press Ltd. for Fig. 1.4 (Gregory and Walford, 1989); the Mary Evans Picture Library for Plates 1.3 and 11.1; National Academy of Sciences for Fig. 7.6 (Hubbert, 1962); National Resources Journal for Fig. 5.6 (Kitibatake, 1992); Oxford University Press for Fig. 3.5 (M Healey and B Ilbery, *Location and change*, 1990); J Rees for Fig. 9.11 (Rees, 1985); Resources for the Future for Fig 11.2 (Barnett and Morse, 1963); Routledge for Figs 9.4 and 9.12 (Newson, 1992); Scientific American for Fig. 7.2 (Hubbert, 1971); Scientific Committee on Problems of the Environment (SCOPE) for Tables 3.1 and 3.2 (Wolman and Fournier, 1987); Shell International Petroleum Company, Shell Centre, York Road, London, SE1 7NA for Fig. 8.2 (SBS 4 1994, *Upstream Essentials*); World Resources Institute for Table 2.1 (Repetto *et al.*, 1989).

Whilst every effort has been made to trace the owners of copyright material, in a few cases this has proved impossible and we take this opportunity to offer our apologies to any copyright holders we may have unwittingly infringed.

CHAPTER 1

Introductory concepts

Some environmental resources are easily recognized. Most people would regard fertile farmland, for example, as such a resource, producing food for sale or for direct consumption, and many would see a fast-flowing river as a potential resource for producing hydro-electricity. Perhaps there might be more disagreement about whether a wilderness area or a barren mountain could be regarded as environmental resources. In the case of the fertile farmland and fast-flowing river, useful products such as food and electricity can be derived, but the value of the wilderness area or mountain cannot be expressed in similar material forms. This is not to say that they have no value, but their value is of a kind different from that of the farmland and river.

Natural resource is a term that has for long been applied to certain forms of land and water, and natural resources are, of course, the physical bases for human life. The term environmental resource is used here for a number of reasons, including a wish to embrace the wilderness or mountain that may be considered valuable even if it does not yield physical products or materials of economic value.

Environmental resources are parts of nature that humankind considers to be useful or valuable. What is meant by 'nature' can be debated endlessly, but for our present purposes it is simply the physical environment or the non-human world around us, including the land, sea and air and their plants and animals. 'Useful' here relates to the production of material benefits, such as food and electricity, while 'valuable' refers to human perceptions of an aesthetic or other non-material kind that may exist whether or not 'useful' products are derived. In the words of Holmes Rolston, 'valuable' means 'able to produce valued experiences' (1988: 272). He then goes on to list no fewer than 14 different types of value associated with nature, ranging from economic and life-support value to religious and character-building value. Alternatively and rather more prosaically, environmental resources can be defined as those parts of nature that can provide the goods and services sought by humans – including services such as opportunities for recreation, the appreciation of scenic beauty, or the disposal of wastes.

Three main groups of environmental resources can be recognized:

- One consists of raw materials and energy sources used by humans, usually as inputs into the economic system. These have traditionally been regarded as natural resources. Obvious examples include mineral ores, coal and oil.
- Another group comprises parts of the environment that can provide services rather than material goods – for example in outdoor recreation and appreciation of wildlife and scenery.
- Thirdly, the natural environment provides the essential life-support system for humans, including oxygen to breathe and water to drink as well as material goods such as food. It also provides the sump into which the waste products of the economic system and human life in general are put.

Resource management aims to provide goods and services, and to maintain essential life-support systems. It is a means towards an end, or a group of ends. The underlying objective is utility, which can take the form of production of materials such as food and wood, or less tangible forms such as pleasure or happiness. The goal may be survival, profit or capital accumulation, or the enjoyment of scenery or recreation. Resource management is concerned with the physical or biological functioning of part of the environment (for example a forest), but also with the allocation of resource products, within the frameworks of particular legal and cultural settings. It

Plate 1.1 Fertile lowlands with deep soils, gentle slopes and favourable climates have for many centuries been highly valued as resources for producing food. Uplands and mountainous areas, on the other hand, were until comparatively recently usually perceived negatively, as harsher environments in which food and other physical requirements were difficult to produce. By the end of the eighteenth century, however, more positive perceptions were being established. By this time, the scenic beauty of mountains was increasingly appreciated and tourism was beginning to develop. While the physical environment of such areas did not change appreciably, human appraisal did, and what had previously not been regarded as a valuable or useful part of the environment was now perceived as a resource. (a) Arable farmland in fertile lowland near Aberdeen, Scotland. (*Source:* J. Livingston)

Plate 1.1 (b) Wester Ross in the north-west Highlands of Scotland. (*Source:* J. Livingston)

therefore has three different dimensions. These were labelled by Firey (1960) as ecological, economic and ethnological (i.e. social or cultural). If an environmental resource is to be used, its use must be physically possible, economically viable and culturally acceptable.

Each of these dimensions is complex, and it is therefore not surprising that many texts on resource management concentrate on one or other of them. In practice, however, environmental resources have to be managed with regard to all three dimensions: the physical management of the resource has to take place within particular economic and cultural climates. Conversely, the management of an environmental resource in order to satisfy particular economic or social goals may have effects on the physical or biological nature of the resource.

Many of the problems encountered in resource management stem from conflicts that develop between different goals. For example, a quest for more food may lead to the intensification of agriculture. This in turn can lead to landscape change and a loss of scenic beauty as hedges and trees are removed, and also to the increased contamination of rivers through accelerated soil erosion, fertilizers and pesticide residues.

Resource definitions – material or functional?

'Resources' itself is a most elusive term, as Table 1.1 suggests. It is noticeable that there are different types or groups of meanings. On the one hand, resources may be 'stock that can be drawn on': environmental resources might include 'stocks' of coal or iron deposits, for example. On the other hand, there are several human-centred definitions, such as 'means of supplying a want' and 'skill in devising expedients'. These definitions imply or presuppose that a person is involved, and that resources are not simply 'stocks'.

These contrasting interpretations characterize the enormous body of writing that has been produced on natural or environmental resources. Some commentators regard resources as stocks of substances or materials found in nature or the environment, while others take a more human-centred view and see them as the result of human appraisal. The practical significance of this difference in viewpoint could scarcely be more profound. If environmental resources are simply stocks of substances found in nature, then they are inevitably fixed and limited in

Table 1.1 Selected dictionary meanings of 'resource' and 'resources'

1. 'means of supplying a want; stock that can be drawn on; country's collective means for support and defence; expedient device; skill in devising expedients, practical ingenuity, quick wit'
2. 'available means; something to which one has recourse in difficulty; capability or skill in meeting a situation'
3. 'source or possibility of help; an expedient means of support'
4. 'source or possibility of help; cleverness in finding a way round difficulties'
5. 'that to which one resorts, or on which one depends, for supply or support; funds, wealth, riches, available means'

Sources:
1. *Concise Oxford Dictionary*, 1964.
2. *Webster's Third New International Dictionary*, 1971.
3. *Chambers Twentieth Century Dictionary*, 1977.
4. *Longman Dictionary of Contemporary English*, 1978.
5. *Collins New English Dictionary*, 1956.

quantity. Limits to resource use must inevitably exist. If, on the other hand, resources reflect human appraisal, then the conclusion is quite different. In this case, their limits are not imposed by the non-human environment, but rather by human ingenuity in perceiving usefulness or value.

One of the most outstanding contributions to the theory of resources is that of Erich Zimmerman, who was an economics professor in the University of Texas. His classic *World Resources and Industries* was first published in 1933, and a revised edition followed in 1951. It sought essentially to work out a new synthesis between economics and geography, and in particular to develop a new view and new understanding of natural resources. Zimmerman clearly takes the 'appraisal' or functional view of resources: in his words ' "resource" does not refer to a thing or a substance but to a *function* which a thing or a substance may perform' (1951: 7) (italics added). More simply. 'Resources are not, they become' (p. 15). By this he means that no part of nature has intrinsic physical or chemical properties that make it a resource, but any part can become a resource when people perceive it as having utility or value. Resources are defined by humans, rather than by nature. In this sense, therefore, the term natural resource is not meaningful: no environmental material is a natural resource because of its intrinsic properties, and any material can become a natural resource.

Since the significance of this tension between the

material and perceptual view of resources is so fundamental, it is worth exploring further by means of specific examples. One case that exemplifies the perceptual view is that of snow on the Cairngorm Mountains in Scotland. During the 1960s this 'environmental resource' began to be utilized for skiing and as the basis of a winter tourist industry. The amount and composition of the snow did not alter significantly at that time, but the human perception of it did. What was previously not perceived as a 'useful or valuable part of nature' now became an environmental resource supporting an economic activity. To paraphrase Zimmerman, the snow was not previously a resource, but it now became one. Not everyone, however, regarded the snow as a resource, or as a useful or valuable part of the environment. Sheep farmers in the same area doubtlessly continued to view it negatively, as it caused them problems. In other words, the perception of the snow not only changed through time, but also differed between individuals and groups. It would seem that the 'material' view of resources, as stocks of substances, is untenable in this case.

A further and broader example of contrasts in perceptions between groups of people is quoted by Davidson (1990) in relation to the tropical rain forest in South-East Asia. Many Javanese people consider the forest to be a threat rather than a resource, because it harbours wild animals and pests which damage their homes and crops, and because they regard it as the domain of evil spirits. When they resettle in other parts of Indonesia, they seek to clear the forest back from their homes and cropland. Zimmerman uses the term 'resistance' to describe such negatively perceived parts of the environment. On the other hand many government officials and planners see the expanses of *Imperata* grassland that have replaced the forest there and in other parts of South-East Asia as waste, representing environmental degradation and a lost resource. Local people, however, may regard the grassland as far more valuable than the forest, offering a source of roof thatch, fodder for livestock and habitat, and hunting ground for edible wildlife. To them, the grassland may be more useful and valuable than closed forest. Again, it would seem that only the perceptual concept of environmental resources is tenable in this example.

Both Cairngorm snow and South-East Asian forest are environmental resources that are meaningful in perceptual rather than material terms. This does not necessarily indicate that 'material' definitions are meaningless. Only certain amounts of snow and certain areas of forest and grassland exist: for any particular perceptual definition of environmental resources, material definitions in terms of hectares, tonnes, cubic metres or barrels may apply. The two definitions to some extent complement rather than compete with each other. The material definition is meaningful if related to particular perceptual definitions. On its own, the material definition leads to what Zimmerman (1951: 7) regards as the 'false impression of resources as something static, [and] fixed, whereas actually they are as dynamic as civilization itself'. Equally, the perceptual definition on its own has limited meaning unless complemented by statements or measurements of defined environmental phenomena.

Creation and destruction of resources

Creation

The example of snow in the Cairngorms illustrates one way in which environmental resources can be created. At some point in time, a component of the environment that was previously not viewed as useful or valuable is perceived to be a resource. This change may follow from broad cultural or societal trends, related for example to affluence, free time or attitudes to nature. It may also stem from more sudden and abrupt changes in technology, allowing natural materials that previously seemed to have no great utility from the human viewpoint to become very useful. One example is the discovery in the nineteenth century of the process of vulcanization of rubber, whereby its durability and flexibility are increased. This led to the natural product of rubber being used in a whole variety of new applications, and in particular for the tyres of bicycles. Demand for rubber increased enormously. Rubber trees were now perceived positively, and valued as environmental resources. Another example is the afforestation of peatlands in Scotland. Until recent decades, they were regarded as unsuitable for commercial timber growing. Over the last 30 years, however, silvicultural techniques have been devised that permit acceptable timber growth rates to be achieved. What were previously negative areas for timber growing became positive areas.

Some environmental resources are suddenly and abruptly perceived as such, for example when a technological advance allows previously unusable materials to be utilized. More generally, the recognition of resources is a slower and more gradual process, involving a series of stages.

- Initially some resource potential is recognized, but the technical difficulty or cost of utilization means that the potential is not realized.
- Then with technological improvements and falling costs of production, perhaps combined with rising costs of producing alternative resources, it may become economically viable to utilize the resource. In other words, technical and economic criteria need to be satisfied before the potential resource becomes an actual resource.
- A third group of criteria is becoming increasingly significant: the side-effects and consequences of resource use have to be acceptable in environmental terms. Forms of mining, for example, that give rise to severely scarred landscapes and to badly polluted water may now be regarded as completely unacceptable. The potential resource would become an actual resource only if more acceptable extraction practices could be worked out.

The sequence of events leading to the recognition of some material resources (such as mineral deposits) is often quite clearly defined. The same is much less true of the recognition of aesthetic resources such as landscape. In this sector of environmental resources, the technological and economic criteria that are important in the perception of material resources are more indirect. Changing lifestyles and societal values are the main determinants of these resources. Lifestyle changes have economic and technological bases, but such changes are more difficult to identify and predict than those that are of a directly technical or economic nature.

Destruction

If resources can be created by changing perceptions, they can also be destroyed, or have their value reduced, in the same way. Environmental materials that are at one time regarded as resources can cease to be perceived as useful, and can fall into disuse. One example is oil-shale deposits in east-central Scotland. In the early part of the present century they were mined as sources of oil. With the growth in production of petroleum in other parts of the world, oil from the shale deposits could no longer compete in terms of costs, and the mines were closed. One resource product was simply substituted or replaced

by another product that was cheaper or in some other respect better. A similar example on a much grander scale is the natural (guano) nitrate deposits in Chile. At the end of the nineteenth century, it was feared that these would soon be exhausted, with dire consequences for agriculture (not least in Europe) because they were the main source of nitrogenous fertilizers. Just prior to the First World War, however, a process for producing nitrogen fertilizer from the atmosphere was devised in Germany. This technological advance not only changed the significance and resource value of the Chilean nitrate deposits, but also removed a limiting factor in agricultural productivity and hence in food production.

Environmental resources can therefore be created and in a sense destroyed by changing perceptions and changing technology. There are also other ways by which resources can be destroyed. They may simply be used up, and cease to exist: the resource may be physically exhausted. While theoretically possible, this rarely happens in an absolute sense. As the resource is depleted and exhaustion looms, the unit price of the material rises sharply, discouraging use and encouraging the search for further occurrences of the same resource or for substitutes that can fulfil the same function.

A more common form of resource destruction involves degradation of the resource. Resource use can be carried out in such a way that productivity is reduced. Cultivation practices can lead to soil erosion, which in turn leads to declining land productivity and crop yields. Similarly, material may be taken from the resource at a rate faster than it can be replenished by natural biological processes, for example through hunting or fishing. The destruction can in some cases be complete, leading to the extinction of a hunted species, but more usually it will be partial. The process of depletion is illustrated in Fig. 1.1(a), where 'zero base level' represents absolute exhaustion. As Fig. 1.1 indicates, the process can occur in a number of phases or stages. A change in technology or increased demand may lead to renewed pressures on the resource, and in turn to a stepped pattern of decline (b). A further variant is where pressures are transferred in sequence from one resource to another as depletion proceeds (c).

Resources can therefore cease to exist in at least three ways, some of which may be inter-related. They may cease to be perceived as useful or valuable, for example when cheaper or more widely available substitutes become available or when tastes and

Plate 1.2 Destruction of resources – severe gully erosion of agricultural land, near Pamplona, northern Spain. (*Source:* A. Mather)

values change. They may be exhausted; and they may be damaged or destroyed in the course of use and management.

Resources, resistances and hazards

The example of snow in the Cairngorms is a reminder that while some aspects of the environment are perceived positively, others are viewed negatively. Prior to the development of the skiing industry, Cairngorm snow was widely perceived in negative terms, as an environmental material disadvantageous to humans. Zimmerman described such parts of the environment as 'resistances'. As in the case of resources, the perception of resistances can vary through time. What is inimical to human activities at one period is not necessarily so at another, when culture and technology have changed.

Another category, which overlaps with resistances, is that of environmental hazards. These are discrete environmental events that are damaging to property or life. They include events such as floods, hurricanes, volcanic eruptions and earthquakes. In some cases their probability in terms of expected frequency of occurrence is known (for example a flood of a given magnitude may be expected to occur on average once in 100 years), but their timing is usually unpredictable. The incidence of some hazards can be reduced by human action. For example, flood control is often one objective in multi-purpose river-basin development schemes (Chapter 9). Another approach is to adopt 'defensive' strategies, such as, for example, special building techniques in earthquake-prone areas. A third approach, of course, is simply to avoid susceptible areas. Specific localities and geographical areas are sometimes recognized as prone to hazards, but their resource values may at the same time be high. For example, the slopes of a volcano may have fertile soils which attract human settlement, or a beach-front location may be an attractive site for a holiday home despite being at risk from attack by storm waves during hurricanes. In such cases a conscious trade-off can occur in the mind of the resource user, in which the risk of damage is set against the benefits of the location. In other cases, the resource users may be unaware of the hazard or have little option but to accept the risk. In extensive deltaic parts of Bangladesh, for example, the risk of catastrophic flooding is high but for millions of people there is little or no possibility of relocating to safer areas. For further discussion of environmental hazards, see Smith (1991).

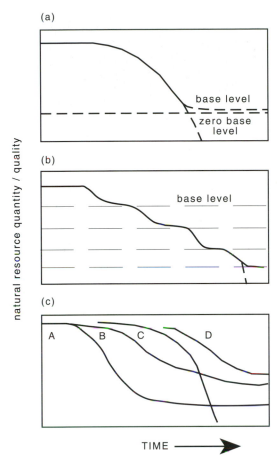

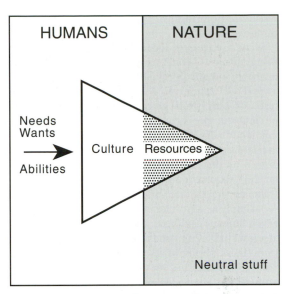

Fig. 1.2 Culture, nature and resources. (*Source:* extensively modified after Zimmerman, 1951.)

Fig. 1.1 Resource depletion curves: (a) the resource is depleted partially (to a base level above zero) or completely (i.e. to zero base level); (b) the resource is depleted in a series of steps, each reflecting new technological or economic conditions which give rise to a new phase of resource use. Depletion may continue until the resource is exhausted (i.e. zero base level is reached); (c) depletion of resources in a serial pattern, as pressures are transferred from A to B, C and D. Transfers may result from new resource discoveries or new technologies or transport systems, or from the increasing costs resulting from the progressive depletion of the first resource. Note that each resource may be depleted to a different level, and not necessarily to zero base level. (*Source:* modified after Whitaker, 1941.)

Culture and resources

Perceptions of resources vary through time and between different groups and individuals. These basic truisms lie at the root of many resource conflicts, and underline the fundamental role of culture in defining environmental resources. By culture is meant human attitudes and values, technical and organizational abilities, and social and political relationships. The interaction between culture and nature gives rise to environmental resources. Figure 1.2 depicts this interaction. It is inspired by Zimmerman's depiction of 'man, culture and nature', but has been radically modified and simplified.

Several important – and perhaps controversial – points can be drawn from Fig. 1.2. First, it is clear that only part of nature is perceived as constituting resources. The size of this part, however, can change through time. As culture develops, the proportion of 'nature' that is perceived to be 'resources' may increase (Chapter 3). In other words, new resources may be identified, and quantities or volumes of existing resources can change and even increase through time, even though we might intuitively expect them to become smaller as use proceeds. The classic example of radical changes in environmental perceptions is that of mountain environments. In both the West and China, a broad sequence can be discerned whereby awe, aversion and generally negative perceptions subsequently gave way to aesthetic appreciation and to the modern view of mountains as recreational resources (Tuan, 1974). Negative perceptions were replaced by positive perceptions: what was at one time not an environmental resource, and indeed was widely regarded as a resistance (for example, a barrier to travel),became a resource.

Second, nature is regarded in Fig. 1.2 as simply being 'neutral stuff'. Zimmerman's phrase was probably intended originally to convey the idea that a part of nature or the environment became a resource only when perceived as such by humans: no inherent property made it a resource. Perhaps unintentionally, however, the use of the phrase may indicate a particular attitude towards nature. Nature is seen as the raw material for natural or environmental resources, and as something which has no value or significance independent of humans. It is seen simply as a storehouse of potential resources, on which humans can draw. The perception of nature as 'neutral stuff' is an issue of fundamental significance from the viewpoint of resource use and conservation. Some commentators see the type of environmental attitude that is reflected in the phrase 'neutral stuff' as a source of many of today's environmental and resource problems.

Environmental attitudes and resources

Attitudes to nature

The way in which humans regard nature is of primary importance in conditioning their thinking about the use and management of environmental resources. Some people have a purely utilitarian view of nature, seeing it simply as a storehouse of potential resources on which we can draw. In other words, nature is of instrumental value, or of value as a means to an end (of human satisfaction). The use of environmental resources is largely a question of economics and technology: an anthropocentric (human-centred) or technocentric view of nature is taken. In the technocentric view, environmental resources and their management are essentially technical rather than moral issues. Implicit in this view is the belief that humans can use nature for their own purposes (i.e. anthropocentrism). Other people regard nature or the environment as having intrinsic value (or even rights), irrespective of practical usefulness to humans. In the case of the latter, the use of environmental resources becomes a moral and ethical issue: in this case there is a more biocentric or ecocentric view.

In passing, it may be noted that the definition of environmental resources is inherently anthropocentric, as usefulness and value are meaningful only in relation to humans. Nevertheless, while this is true in the strict sense of the term, different positions on an anthropocentric–biocentric continuum may be adopted by different individuals and groups. Alternatively, the continuum may be said to range from technocentrism to ecocentrism. For example, O'Riordan (1977, 1981) subdivides technocentrism into 'cornucopian' and 'environmental manager' groups, and ecocentrism into 'deep ecology' and 'self-reliant' groups.

Different positions are also taken up by different groups and societies at different times. Changes in attitudes to nature and its various parts have occurred in the West over several centuries (e.g. Thomas, 1983). Roderick Nash (1989) in *The Rights of Nature* refers to the recent emergence of the idea that the human–nature relationship should be treated as a moral issue, and regards it as one of the major developments in recent intellectual history. Certainly the role of morals, ethics and philosophy has become much more prominent in recent decades, especially since the plea by Aldo Leopold (1949) for a new 'land ethic' and the growth of the 'deep ecology' movement associated particularly with Arne Naess (e.g. Naess, 1990; Devall, 1980). Indeed Lester Milbrath (1985: 162) has claimed that 'Americans are undergoing a profound transformation of their basic beliefs about the proper relationship between humans and their environment'.

The idea of an ethical or moral dimension to the use of resources may be growing in prominence, but it is certainly not entirely new: it can be traced back over many centuries, for example to guidelines for land management in ancient Israel. There, cropland was to be rested every seventh year. Fallow fields were to be left open for wildlife, and crop fields left partly unharvested so that the poor could obtain food (e.g. Clark, 1990). The use of the land resource was therefore circumscribed by provisions for the welfare of humans and wildlife, even if these provisions were sometimes (or often) ignored. Perhaps what Nash refers to is thus the recent re-emergence, rather than emergence, of the moral dimension in the relationship between humans and nature. On a wider canvas, Worster (1977) sees the history of environmental attitudes as a struggle between rival views of the relationship between humans and nature: intrinsic value versus exploitation. Another – and probably correlated – dichotomy is between views of nature as an actual or potential commodity, which can be bought or sold on the market, and as an inheritance (from God, from previous generations, or from both) and as something to be stewarded and passed on.

Dominion and exploitation

The root of the exploitative, utilitarian view is traced by one school of thought to the idea of human domination of nature. This is one of three major views or perpectives identified in Table 1.2. It should be emphasized that the contents of the table are greatly simplified: an enormous range of shades of opinion exists, and the classification shown on the table is somewhat arbitrary.

The idea of domination is linked by some to Christianity and to the idea of humans 'having dominion' over nature. It is not unusual to find 'dominion' and 'domination' being equated (e.g. Tisdell, 1990). 'Domination' attitudes have been expressed especially strongly in the former Soviet Union and in China, during the days of Stalin and Mao-Tse-Tung respectively. In the case of the former, for example, there was a 'Great Plan for the Transformation of Nature'. It is often assumed that such outlooks of domination or conquest stemmed from Marxism, but the writings of Marx himself are more environmentally friendly (or at least ambivalent) than these later attitudes – influenced by Leninism and reaching an extreme in Stalinism – would suggest.

In a famous paper, Lynn White (1967) identified this notion of dominion as being disruptive of the relationship between humans and nature. In this view, he was anticipated by Robert Burns, the Scottish ploughman poet of the eighteenth century, when on seeing the disturbance he caused a field mouse, wrote:

I'm truly sorry man's dominion
Has broken nature's social union.

(Burns, 1789: 40)

In the view of White, humans considered themselves as licensed to exploit the Earth, and were not held back by attitudes of reverence or worship. The Earth was simply a storehouse of potential resources, and one that could be drawn upon freely and without moral or ethical constraint. It could be exploited to the full, with little or no regard for environmental damage. Some other commentators have held even stronger views. For example Ian McHarg, whose most influential book was *Design with Nature* (1969), referred to the dominion injunction as a declaration of war on nature.

If Christianity encouraged exploitative attitudes and lay at the root of environmental problems, the corollary was that it should be replaced by other religions with more benign attitudes towards nature

Table 1.2 Three human perspectives on nature

1. *Domination.* 'Dominion' becomes 'domination'. Nature is controlled and exploited in order to satisfy the needs and wants of humans. It is simply a storehouse of potential resources, with no intrinsic value. Nature and its fruits are commodities, which are subject to market forces. (Humans are 'apart from' nature.)

2. *Stewardship.* Human dominion over nature is conditioned and moderated by stewardship. Humans are the stewards, not the owners, of nature. Creation is good, and should not be defiled or abused by exploitative behaviour. As stewards, humans ought to be concerned with the equitable distribution of the fruits of nature, as well as with how these fruits are produced. (Humans are 'apart from' but 'part of' nature'.)

3. *Romanticism/deep ecology.* An intrinsic value is attached to all forms of life, and especially to wild nature unmodified by humans. A dichotomy between humans and the environment is rejected; in other words 'biospherical egalitarianism' is observed, with non-human species seen as having rights and being of intrinsic value. Nature may be worshipped (pantheism): it is not seen as something that should be exploited for human ends. A reduction in the size of the human population will be required if non-human life is to flourish. (Humans are 'part of' nature).

Note: In addition, a Marxist view of nature is sometimes recognized. From the scattered writings of Marx, it is difficult to distil a consistent and comprehensive view into clear and concise form. The same is true of the writings of Engels. What is apparent, however, is that in the former Soviet Union and in China around the mid-twentieth century, attitudes of domination or conquest of nature were expressed by Stalin and Mao-Tse-Tung, and were translated into projects of environmental transformation in some cases. In many such instances, severe environmental problems were the result.

and a more harmonious relationship between humans and their environment. Indeed this was advocated by writers such as Toynbee (1972). At one level, this theory could readily be shown to be flawed. If correct, then areas where other religions prevail should be free from the environmental problems found in Christendom, and should be characterized by more harmonious use of environmental resources. This is clearly not the case. Cultures and societies with reverent attitudes towards nature have despoiled their environments in the same way as have those with attitudes of 'dominion'. For example, environmental damage in the form of severe deforestation and soil erosion occurred in ancient China, where reverential attitudes predominated (Tuan, 1968).

In North America, there is substantial evidence that the landscape had been extensively modified prior to the arrival of Europeans and of Christianity. The composition and extent of the forest had been altered, wildlife had been affected by hunting and erosion accelerated in places (e.g. Denevan, 1992). Many extinctions of mammals may have been caused by hunting early in prehistory in North America, and almost certainly occurred in New Zealand. In neither of these areas was the relationship between humans and their environment altogether harmonious, and the use of environmental resources was probably not, in modern parlance, permanently sustainable. In both these cases the biblical concept of dominion was absent. There is therefore a mismatch in both space and time: environmental damage certainly was not confined to Christendom.

At another level, the 'dominion' theory has been criticized as mis-specified and imbalanced. Opponents of the theory have challenged the view or implication that dominion can be equated with domination, in the form of despotic control of nature. They point to biblical verses such as Genesis 2:15 'The Lord God took man and put him in the garden of Eden to dress it and keep it'. The nature of dominion is therefore balanced and qualified. Although humans have mastery over nature, this should be expressed as husbandry rather than as subjugation (Wilkinson, 1980). Furthermore, even if the biblical concept of dominion had been accurately interpreted by White, other writers such as Black (1970) and Attfield (1983) pointed to the parallel concept of stewardship: 'The Earth is the Lord's, and the fulness thereof' (Psalm 24: 1). Humans are not free to exploit the Earth as their own: they are stewards rather than owners, and dominion needs to be viewed in the contexts of stewardship and accountability. A remarkably similar concept occurs in Islam: humans act as 'vice-regents' exercising only delegated authority towards nature – as stewards rather than sovereigns (Palmer, 1990; Deen, 1990). In the view of Black (a professor of forestry at Edinburgh University), it is the fading of this concept of stewardship and its counterbalancing effect on dominion that has given rise to environmental problems, and not the concept of dominion itself.

Critics of the 'dominion' theory of environmental damage also contest the suggestion that Christianity encourages a low view of nature. From the Creation narrative in Genesis they conclude that God considered His Creation to be good, and that therefore nature ought to be respected and not abused or selfishly exploited. This respect would not stem from a pantheistic view that nature was the very essence of the Deity, as might be held for example in oriental religions, but would nevertheless militate against potentially damaging behaviour towards the environment, and against its being viewed merely as 'neutral stuff'.

Some critics also contend that 'dominion' should be seen in the context of a change in the relationship between humans and their environment that occurred early in human history. At the Fall of Man, Adam and Eve went against the Creator's plan, and thereby destroyed harmony and introduced disorder. 'Cursed is the ground because of you; in toil you shall eat of it all the days of your life; thorns and thistles it shall bring forth to you' (Genesis 3: 17–18): the fall of man led to the fall of nature (e.g. Bradley, 1990). Perhaps it was here that 'dominion', in the form of husbanding, became 'domination', in the form of exploitation and subjugation. In the words of Robert Moore, 'If humankind was created, as Genesis states, in the image of God, then our exploitative, battering and polluting behaviour towards nature is a corruption of our own status' (1990: 107). Humans became alienated from nature as they became alienated from God, and indeed also became alienated from each other. In other words social and environmental problems stem from a common root.

Environmental and social relations

Both Christian and non-Christian writers have referred to the association or relationship that frequently exists between environmental and social problems, and to common origins for these problems. One recent example is Carolyn Merchant, who observes that 'the domination of nature entails the domination of human beings along lines of race, class and gender' (1992: 1). Engel (1988: 35) refers to similar analyses, which claim, for example, the 'subjugation of women and the rape of nature are causally, as well as metaphorically, related', and that the root source is 'variously identified as human aggression, pride, ignorance, greed, free market capitalism, totalitarianism...'. Lewis Moncrief (1970) highlights the moral disparity that exists when an executive can suffer a prison sentence for embezzling company funds, but be applauded for increasing company profits by ignoring pollution standards – even though the cost to society at large may be far greater from the latter than from the former.

This distortion – and the degradation of the environment more generally – is blamed by some on capitalism (e.g. Johnston, 1989). The quest to make profits and accumulate capital may be rational in the economic sense, but may be less rational when judged by wider criteria – as well as being amoral. For example, some conclude that capitalism has reduced nature and people alike to the status of raw materials or inputs into systems of production, with scant regard for any values beyond the strictly utilitarian (e.g. Nash, 1989). Nature is 'neutral stuff': humans are units of labour or market. Similarly, Bookchin (1985) considers that the domination of nature by humans stems from the domination of humans by humans. Put another way, 'People can't change the way they use resources without changing their relations with each other' (Stretton, 1976: 3). The same inter-linkage between social and environmental domination is reflected in the words of C. S. Lewis (1947: 69): '... Man's power over Nature turns out to be a power exercised by some men over other men with Nature as its instrument.'

Various examples of an association between oppression and social injustice on the one hand and environmental problems on the other have been quoted. The degradation of humans and the degradation of the environment in slave-based systems of production in the United States, for instance, have been highlighted by Genovese (1965). His conclusions have not gone unchallenged, and there has been debate about whether slavery as an institution was responsible for soil erosion, or whether the trappings of slavery – such as a lack of motivation and ability on the part of the slave workers and the greed of the overseers – were responsible. In part of the American South, however, Trimble (1985) found that not only were slavery and erosion correlated, but also that erosive land use in 1860 and earlier could be predicted from slave densities.

In more recent times severe environmental problems have been experienced under oppressive regimes in many parts of the world. One example is South Africa, where under apartheid policies large numbers of Blacks were confined to small areas of the country, with consequent stresses on overused land, manifested in soil erosion and other symptoms of land degradation (Daniel, 1988; Durning, 1990). Symptoms of misuse of environmental resources are widespread both in areas dominated by capitalism (such as the United States (e.g. Petulla, 1977)) and in those under socialism (such as the former Soviet Union (e.g. Komarov, 1980)). Simple correlations involving broad political ideologies are probably no more generally applicable than those involving

religions. Nevertheless, blatantly unjust and oppressive regimes are often associated with severe resource problems. At a general level, it will be shown in the ensuing pages that many modern problems of environmental resources are primarily problems of distribution, and that they have social and political roots rather than simply ecological or technical causes. If this is so, resource management, in the technical sense, is unlikely to solve environmental resource problems. Ethics relating solely to the management of the environment (such as Leopold's land ethic) are likely to be similarly limited in effect if the wider social and political dimensions are excluded. A similar conclusion has been reached, for different reasons, by Kay (1985). She considers that no single land ethic is likely to be effective in ensuring a more harmonious relationship with nature. This is partly because of observed discrepancies between beliefs held and behaviour practised, and partly because of the widely differing environments and pressures on resources found around the world (see also Box 1.1).

Stewardship

In place of a specifically environmental ethic, some writers such as Attfield (1981) have advocated a more basic and wide-ranging concept of stewardship. This concept embraces ideas of responsibility to other humans presently living, and to future generations as well as to God. Stewardship is therefore a broad concept, and different emphases, and even different interpretations, are possible. Nevertheless, it reflects a very different view from that of 'neutral stuff' and of nature as merely a storehouse of potential resources. And it is difficult to disagree with Young (1982: 5) when he concludes that '... the idea of stewardship as a perspective on man–nature relations is making some headway against the entrenched view that natural resources exist solely to facilitate the pursuit of human welfare narrowly defined'. A similar view that an acceptable basis for an environmental ethic is now emerging, based on stewardship and the related concept of sustainability, is espoused by Berry (1992: 251). In support he points to the 1990 UK Government White Paper on the environment, and to the fact – remarkable for a government document, in his view – that it began with a moral principle: 'The starting policy for this Government is the ethical imperative of stewardship which must underlie all environmental policies.' A further indication of the recent emergence of the concept of stewardship was

Box 1.1 Religion and environmental ethics

The relationship between religion, ethics, environmental attitudes and how environmental resources are used is an interesting but confusing one. At one level, as has been demonstrated in the main body of the text, the reverential attitudes towards nature associated with some oriental religions have not always been matched by harmonious relations between humans and their environment.

Jeanne Kay (1985) identifies a number of important factors that complicate the relationship between environmental attitudes and how resources are used. One is the discrepancy that frequently exists between attitudes and behaviour. The other is that some environments are more fragile than others: the same type of behaviour might cause degradation in some environments but not in others. Furthermore, the pressures on resources, for example from population, are greater in some areas than in others. For these reasons, therefore, she concludes that no single land ethic can be guaranteed to ensure a harmonious relationship with the environment. Nevertheless, there is some evidence that different religious groups in the United States have had different environmental impacts.

Kay and Brown (1985) conclude that the beliefs of Mormons produced a conservation ethic in Utah, despite some harmful land-use practices resulting from pioneers settling in an unfamiliar and fragile environment. The Amish maintain a traditional pattern of resource use, rejecting mechanization and focusing on small-scale, energy-efficient agriculture that can be characterized as environmental nurture (Paterson, 1989). During the Dust Bowl episode of the 1930s, Mennonites were less likely to be forced to move west from Kansas, as they were less likely than other farmers to have been heavy borrowers and therefore less likely to suffer severe financial problems (Worster, 1979). On the other hand, their stewardship ethic in farming has not always been matched by conservative practices (Paterson, 1989) – a fact that has been explained by Paterson in terms of different interpretations of 'stewardship'.

the Brussels Conference in 1989, involving participants from the seven Economic Summit countries, the European Community and international organiza-

tions (Box 1.2). The scope for disagreement about environmental ethics is enormous: the relative roles of self, society, future generations and nature are all

Box 1.2 The ethics of stewardship

The emergence of 'bioethics' and stewardship as major issues is reflected in the subject of the Economic Summit Conference held in Brussels in 1989, and presented to the full Economic Summit meeting held in Paris later that year. Part of the conclusion reads as follows:

The way we think and feel about the world we live in, the sum total of our values, attitudes and predispositions, determines our behaviour and the choices we make. This means that it is urgent to agree upon a code of environmental ethics able to serve as a guideline for the behaviour of humankind in all activities which may have an environmental impact. Such an ethic should be a shared one to which people of different cultural and religious backgrounds can agree and should be based on a pragmatic approach.

Because man [sic] is both a part of and yet apart from nature, any environmental ethic should take account of the wellbeing of the earth's human inhabitants, of other life forms, as well as of the environment in general, including its nonliving components (e.g. the atmosphere, etc.). It should also be broad enough to cover the needs of future generations and include a global view of the planet as a life-supporting system.

Environmental ethics should not therefore be based on the image of the exploitative management of nature, but on the concept of human 'stewardship' of nature, i.e. its positive and responsible management. A steward does not own the property, nor can the latter be used solely for the steward's benefit. The steward has to manage wisely, to increase the stores if possible, and to give a good account of the use of the environment to those who come after. This leads to the concept of 'trusteeship' for future generations and to the issue of present discounting against future needs.

Source: Bourdeau, P., Fasella, P.M. and Teller, A. (eds) (1990) *Environmental ethics: man's relationship with nature, interactions with science.* Luxembourg: EEC, p. 3.

areas with great potential for discord. And as the debate triggered by White indicates, controversies can become heated. The fact that governments and supra-governmental institutions are now searching for ethical frameworks for environmental management is revealing: it is a tacit admission that technical solutions and solutions relying on economics have proved inadequate.

The concept of stewardship can include both the notion of human responsibility in the management of the environment and its resources in a technical sense, and also in the sense of sharing these resources in an equitable manner. In the words of the Second Vatican Council 'God destined the earth and all it contains for the use of every individual and all people', and 'It is manifestly unjust that a privileged few should continue to accumulate excess goods, squandering available resources, while masses of people are living in conditions of misery at the very lowest level of subsistence' (Pope John Paul II, 1990: 4). In other words, stewardship includes a social, distributional dimension as well as one relating directly to environmental management.

Environmental attitudes and the New Environmental Paradigm

White's theory that the dominion injunction lies at the root of environmental problems may be an over-simplification or a distortion. Nevertheless, it is true that 'dominating' attitudes towards nature are wide-spread and persistent, and that nature is seen by many people as no more than 'neutral stuff' – as a storehouse of potential resources that may be drawn upon or even ransacked. 'Dominating' or 'domineering' attitudes have been more widespread in some cultures and periods than in others, whether or not they can be linked to the biblical concept of dominion or to a waning and more recent waxing of the counterbalancing notion of stewardship.

Environmental attitudes are frequently found to vary with socio-economic factors. When levels of affluence rise above those required to meet basic needs of food and shelter, increasing concern is often expressed about environmental issues. 'Green' attitudes are usually found to be more prevalent at the upper end of the socio-economic spectrum than in the population at large. Whether it is significant that this group has greater control over resources and of influence in the formulation of policies about the environment within individual countries is debatable. On the other hand, women are usually found to express greater concern for the environment than men (e.g. Mohai, 1992): in general terms control over resources in most parts of the world is biased towards men. Attitudes in the well-fed developed world may appear to be 'greener' than in much of the developing world, where adequate supplies of food and other basic requirements are urgent priorities. One conse-quence is that some resource-producing or resource-processing activities that are not perceived as acceptable in the developed world are located in the developing world. Possible examples include metal smelters, oil refineries and perhaps intensive horti-culture employing large inputs of pesticides.

Interesting measures of environmental attitudes and dispositions have been obtained using the 'New Environmental Paradigm' (NEP) scale devised by Dunlap and van Liere (1978) (Table 1.3, Box 1.3). A paradigm is simply a kind of belief pattern – in the case of NEP of a belief that nature is intrinsically valuable, that humans should behave with compas-sion towards other humans, species and generations, and that limits to growth exist. Milbrath (1985) has discussed how the NEP has been challenging the 'Dominant Social Paradigm' (DSP) in recent years, notably in the United States but also in many other parts of the world. DSP, the prevailing set of beliefs, is characterized by a utilitarian view of nature, compassion only for those 'near and dear', and no limits to growth. NEP-based research has yielded interesting contrasts in mean scores between different population groups, as well as offering a means for charting more general trends in the long term (see Wilkerson and Edgell, 1993 for a useful review). It also, however, raises profound issues about the confused relationships that exist between attitudes and beliefs on the one hand, and actual behaviour on the other.

Nevertheless, the more ecocentric view of nature associated with NEP has potential significance from the viewpoint of the use of environmental resources, and especially from the viewpoint of the conservation of the environment. The ecocentric view is that nature should be conserved for its own sake, whereas more anthropocentric attitudes would see conservation in more utilitarian and pragmatic terms. Almost by definition, extreme anthropocentric and ecocentric outlooks are held by small minorities of the total population in most countries (although the balance between them may be shifting). Both are also untenable from the Christian viewpoint. Dominion or domination without stewardship is unbalanced and unbiblical. Nature is not just 'neutral stuff', but

Box 1.3 New Environmental Paradigm

The NEP scale consists of a series of statements with which respondents are asked whether they agree (Table 1.3). Scoring for the first eight items is on the basis of 4 for 'strongly agree' through to 1 for 'strongly disagree'. It is reversed for items 9–12, which relate particularly to the concept of dominion. High overall scores indicate agreement with NEP and a view that may be characterized as biocentric. Nature is regarded as something of intrinsic value, sometimes to the extent of being revered or worshipped.

As might be expected, members of environmental groups such as Greenpeace are usually characterized by high NEP scores. Perhaps less expected is the tendency in some areas for some groups of direct resource users, such as fishermen and farmers, to have lower scores than the general public or urban dwellers. One piece of research found that commercial fishers in British Columbia, for example, averaged 2.1 on the NEP scale, compared with 3.3 for the general public, while farmers in Iowa scored 2.9 against urban dwellers' 3.2 (Edgell and Nowell, 1989). It is not known how general or widespread these apparent differences between direct resource users and other citizens are. They may, however, reflect a basic difference that helps to explain many of the conflicts that have occurred in resource use in recent decades.

Table 1.3 The 'New Environmental Paradigm'

1. The balance of nature is very delicate and easily upset.
2. When humans interfere with nature it often produces disastrous consequences.
3. Humans must live in harmony with nature to survive.
4. Mankind is severely abusing the environment.
5. We are approaching the limit to the number of people that the earth can support.
6. The earth is like a spaceship with only limited room and resources.
7. There are limits to growth beyond which our industrialized society cannot expand.
8. To maintain a healthy economy we will have to develop a 'steady state' economy where industrial growth is controlled.
9. Mankind was created to rule over the rest of nature.
10. Humans have the right to modify the natural environment to suit their needs.
11. Plants and animals exist primarily to be used by humans.
12. Humans need not adapt to the environment because they can remake it to suit their needs.

Source: Dunlap and Van Liere (1978).
Scoring: for items 1–8, 4 for 'strongly agree', 3 for 'agree', 2 for 'disagree', 1 for 'strongly disagree'; for items 9–12, 1 for 'strongly agree', etc.

neither should it be worshipped as a god. It should be treated with respect as the creation of God, and it should be used in such as way as to provide for all the Earth's inhabitants rather than for the selfish ends of those claiming the ownership of particular resources. In other words, the use of environmental resources is not just a technical matter of applied environmental science or economics, but also a matter for observing moral and ethical principles in relation both to nature itself and to other humans.

Types of environmental resources

Environmental resources can be classified in various ways – whether they are scarce or abundant, whether they are widespread or localized in occurrence, and whether they are exhaustible or renewable. The simplest and most common division is into renewable and non-renewable resources, alternatively labelled flow and fund or stock resources. Solar energy and water power are examples of flow resources, which are permanently available (at least on the human time scale) in a continuous supply. Such resources are sometimes defined as perpetual, whereas some other flow resources (such as soil or forests) are regarded as only potentially renewable. The former is largely independent of modification by humans, while the latter group can be modified very easily in the course of use. The most obvious examples of fund or stock resources are the fossil fuels of coal, oil and gas. These materials are available in fixed and limited quantities, and from human perspectives are not renewable. In a sense they represent a flow resource – solar energy – that has been stored over geological time periods.

Between these simple examples of renewable and non-renewable or flow and fund resources lie other resource categories that are harder to classify. They highlight the limitations of this simple classification. How, for example, should metal ores be classified? At first sight they seem simply to be clear examples of non-renewable stock resources. They differ from the

TYPE	RENEWABILITY		EXAMPLE

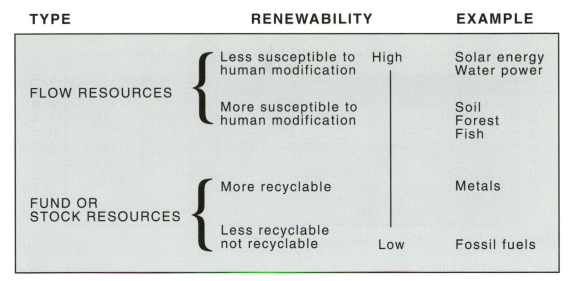

Fig. 1.3 Types of environmental resources.

fossil fuels, however, in being capable of re-use. The parent ores from which the metals are refined exist as stocks and can be mined only once, but the refined metals can thereafter be recycled as scrap. The resource products are renewable, even though their parent ores are not. In theory complete recycling could be carried out, with fixed quantities of the metal being maintained in circulation. In practice, however, this can rarely be achieved. The metals or other materials eventually become too dispersed and mixed with other substances to be usable. In other words, the law of entropy applies.

It is even more difficult to accommodate land in the simple framework of the flow–fund classification. Land can be useful or valuable in many different ways. It may offer sites for building houses and factories, and in this sense would seem to be a stock resource. Only a fixed amount of space is available, and in any particular area it can soon be fully used. The land, however, is not permanently consumed. The space cannot be destroyed, as, for example, coal is destroyed by being reduced to waste products of carbon dioxide and ash on combustion. In this sense land as space for building is a renewable, and indeed perpetual, resource. Land or soil can also give rise to a stream of products such as food and wood, and in this function seems to be a flow resource. Certain forms of land use, however, can result in the degradation of the land and in a reduction or loss of its ability to produce these goods. Thus it can become a non-renewable resource.

Land in this second function (as food-producing soil) is only a conditionally renewable resource. So also is the sea, as a fisheries resource. Over-fishing can threaten the renewability of the resource, and the result may be that in practice it resembles a fund resource. Too many fish are taken to allow for natural biological replenishment, and in a sense the resource is simply being 'mined'. If the use of these environmental resources exceeds a certain level, then the resources are not renewable, at least in the short term or on the human time scale. Flow resources have become fund or stock resources, subject to exhaustion (in some cases literally to the point of extinction) in the same way that a coal deposit, for example, can be worked out. This level or threshold of use, beyond which resource use is not sustainable, was defined by Ciriacy-Wantrup (1952) as the critical zone.

In hunting and fishing, the critical zone is defined simply by the level of use, in the sense of the number of fish, animals or birds taken. Beyond a certain rate of cropping the breeding stock is destroyed and hence a resource flow ceases. In agriculture and forestry, the type of use may be as significant as the level: some cultivation practices, for example, can lead to soil erosion which in turn makes the land less productive or completely unusable.

The apparently simple two-fold classification thus soon breaks down, and at the very least needs to be modified along the lines of Fig. 1.3. Here the two basic classes are subdivided on the basis of the degree of susceptibility to modification by humans, and the

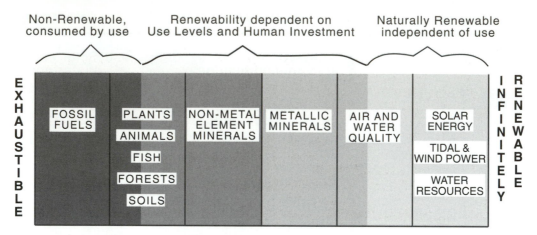

Fig. 1.4 The resource continuum. (*Source:* Rees, 1989.)

extent to which recycling is possible. Modification can of course take various forms. Flows can be increased by some forms of management (such as the application of fertilizers or other farming techniques) and depleted by others (such as over-cropping). A more radical modification suggested by Rees (1989, 1991a) is shown in Fig. 1.4. Here the idea of a resource continuum is presented as an alternative to the conventional classification into flow and fund resources. At one end there are resources such as the fossil fuels, which are exhaustible and non-renewable. They are available as fixed stocks, and are irretrievably changed in the course of use. At the other end of the continuum lie the 'infinitely renewable' resources such as solar energy and water resources. Perhaps some caution is appropriate in accepting solar power or water power as 'naturally' or 'infinitely' renewable. While it is true that the output of solar energy is independent of human activity, it is also true that the reception of solar energy can be affected by human activities that result in air pollution. Similarly, the amount of water power available can depend to some extent on land use within the river catchment, and on various human activities that can modify climate.

It is doubtful whether any classification of environmental resources can at the same time be fully comprehensive, logically robust and consistent, and comprise categories that are mutually exclusive. Aesthetic and recreational resources are especially difficult to accommodate in general resource frameworks. It may be possible in some cases to restore damaged or degraded landscapes or the quality of air or water, and hence such resources may be potentially renewable (as Fig. 1.4 suggests). In terms of perception, management and control, however, they

are different from material resources such as soil and minerals. Some authors suggest for them separate categories such as continuous resources or ambient resources.

Resource scarcity

The availability of environmental resources, in terms of scarcity or abundance, is a major and controversial issue. The symptoms of scarcity are sometimes all too stark, especially when expressed in the form of famine and starvation. Down through history such scarcity has been experienced by countless people, and many today are short of basic resource products such as food and fuelwood. History shows several different responses to scarcity. One is obviously to go without – in some cases simply to starve. Environmental deterioration, frontier expansion, migration and annexation, and the introduction of new technologies are others (e.g. Boserup, 1976; Boughey, 1980). Scarcities are seen by some as the driving force for major innovations leading to dramatic changes in the perception and use of resources. One well-known exponent of such views is Richard Wilkinson. He regards the Industrial Revolution in England as having stemmed from resource scarcities:

The ecological roots of the English industrial revolution are not difficult to find. The initial stimulus to change came directly from resource shortages and other ecological effects of an economic system expanding to meet the needs of a population growing within a limited area. As the traditional resources became scarce, new ones were substituted which usually involved more processing, used more productive labour and frequently resulted in what was regarded as an inferior product. As these initial changes were felt in the economic system, the preexisting technical consistency was disturbed and various secondary changes set in motion.

(1973: 112).

The initial scarcities to which he refers were of land (or at least food) in the face of a growing population, which in turn led to agricultural reorganization, and of wood, which eventually led to the substitution of coal as fuel (and in turn to a coal–iron industry) as well as to imports of timber. Substitution of inorganic for organic sources of supply occurred, and at a stroke the limits imposed by available levels of organic supplies were raised (Wrigley, 1962).

In one sense it is true that scarcity has been the habitual human condition and that there has never been sufficient abundance to satisfy all human wants (Ophuls, 1977). Serious resource scarcities, however, have usually been restricted to local or regional areas. During the present century, fears have been expressed about potential global shortages of basic resources, and about the limits that such scarcities may impose on economic growth. These fears are usually based on the assumption that nature is niggardly at the global and local scales alike – that the physical potential of the environment is inadequate in relation to the demands made on it, and that substitutes are not available. In fact resource scarcity is a much more complicated concept than this simple assumption would suggest. It is usually associated with quantities or stocks, expressed in units such as tonnes or barrels. While these physical measures may be meaningful for stock resources such as minerals and fossil fuels, they are less easily applied to renewable resources. This is one of the reasons why Tilton and Skinner (1987) suggest that resource scarcity or availability is best measured by the costs of acquiring additional supplies.

Types of resource scarcity

Many different forms of resource scarcity exist. At one level, a distinction can be made between absolute and relative scarcities (e.g. Barbier, 1989; Miller, 1990). An absolute scarcity exists when insufficient physical quantities of the resource are available to meet the demand for it. Relative scarcity, on the other

hand, exists when the physical quantities of the resource are sufficient to meet demand, but problems arise over quality of supplies. Poorer grades of the resource may have to be exploited in order to meet demand, and this is likely to mean that greater effort is required, for example in supplying fertilizers on poorer land or in refining leaner grades of ore. This in turn will mean that costs of the resource products rise. Relative scarcity can also apply to resource products that are subject to limits of control and distribution. A company or country may deliberately restrict production of a resource in order to raise prices, or for political reasons. Oil was relatively scarce during the 1970s, for reasons that were tied up with politics. Production was not a problem in physical terms, but as is explained subsequently the main exporting countries restricted supplies to some of the main importing countries.

The distinction between absolute and relative scarcity is sometimes expressed in the terms Malthusian and Ricardian scarcity (Fig. 1.5, Table 1.4). Thomas Robert Malthus (1766–1834) considered that a fixed amount of land (a basic resource) existed and that therefore definite environmental limits existed. Costs of production may remain constant up to the time when the resource is fully used (Fig. 1.5(a)). David Ricardo (1772–1823), on the other hand, focused on the quality (rather than quantity) of land. With growth in population and the economy, progressively poorer grades of land have to be brought into use. The costs of production are likely to rise as poorer qualities of the resource are used (Fig. 1.5(b)). Increasing scarcity in this sense is therefore likely to be signalled by rising prices. In practice the two concepts of scarcity are perhaps more profitably seen as complementary rather than as alternatives. Clearly the land area of the Earth is limited, but it is also of variable quality for agriculture and other uses.

Another classification of resource scarcity is into the three distinct divisions of physical, economic, and geopolitical scarcity (Rees 1989, 1991). (She also recognizes a fourth category, which she labels 'renewable and environmental resources scarcity'.)

Physical scarcity

The most obvious type of scarcity is physical scarcity. Environmental materials such as land (and its mineral contents) and water are limited in quantity: the earth is finite and limited in content. Whether a particular resource is physically scarce

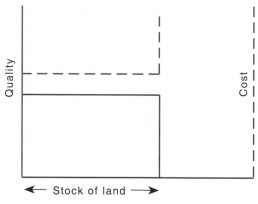

(a) Malthusian scarcity

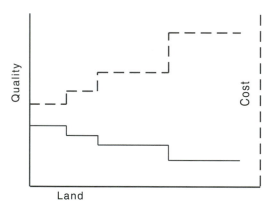

(b) Ricardian scarcity

Fig. 1.5 (a) The resource (land) is assumed to be available in fixed quantity and uniform quality. The cost of using it is constant. (b) There are small quantities of high-quality land, which can be brought into production cheaply. After all the high-quality land is used, further quantities of poorer land can be brought into production, but the cost increases as the quality decreases. .

Table 1.4 Types of resource scarcity

Malthusian stock scarcity (MSS). The resource is fixed in absolute size and extraction costs are constant.

Malthusian flow scarcity (MFS). The resource is fixed in absolute size but extraction costs increase with rate of extraction.

Ricardian stock scarcity (RSS). No constraint from absolute size of resource but extraction costs rise with rate of extraction and with the proportion of the resource extracted.

Ricardian flow scarcity (RFS). No constraint from absolute size of resource, but extraction costs increase with the rate of extraction.

Source: Hall and Hall (1984).

physical composition. For example around 8 per cent of the Earth's crust consists of aluminium, and in practical terms this imposes an upper limit on resource

Plate 1.3 Thomas Robert Malthus (1766–1834). Malthus was the originator of the theory that population growth *tends* to outstrip growth in food production. Scarcities of resources have been expected by some commentators ever since his day, and 'neo-Malthusian' views became very common during the 1970s in particular. (*Source:* Illustration courtesy of the Mary Evans Picture Library.)

depends on the quantity of the material that exists within the Earth (or for that matter within other planets from which production could take place) and on the level of consumption or demand. One resource might exist only in small quantities, but if demand is very low need not necessarily be scarce. Conversely, another resource existing in greater quantities might be scarcer, if demand were higher.

Physical scarcity (and hence relative abundance) can fluctuate through time, at least within certain parameters. These may be defined by the Earth's

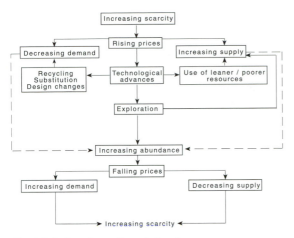

Fig. 1.6 Cycles of scarcity and abundance.

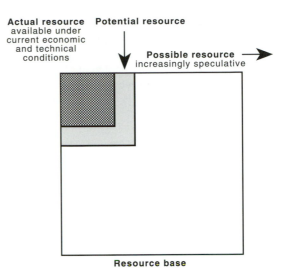

Fig. 1.7 Actual, potential and possible resource quantities. A small part of the resource base constitutes an actual resource which can be utilized under existing economic and technical conditions. It is possible that changing economic and technical conditions may mean that increasing proportions of the resource base will be used in the future.

availability. At least until this limit is approached, however, cycles of scarcity and abundance are likely to exist. When the known reserves of bauxite (Chapter 8) are largely used up, scarcity will loom and prices will rise. This will have two effects, both of which will tend to reduce scarcity. Demand is likely to fall, as aluminium is used more economically, as more recycling takes place, and as substitutes are found for some applications. At the same time deposits that previously could not be economically worked may become viable as the price rises (or as production techniques are improved), and efforts to find new deposits will be stepped up. In other words, feedback loops come into operation, and scarcity may be averted (Fig. 1.6).

Scarcity can be averted in this way, however, only in relation to resources that have economic value. Those that are completely outside the market system do not have these in-built checks. Furthermore, as has been indicated, this process cannot continue indefinitely. Eventually the Earth's crustal-content limit will be approached. This defines what may be termed the resource base. At any particular time only part of this resource base is regarded as a resource that is technically and economically usable. In the context of mineral resources, this part is known as the reserve. Figure 1.7 shows the relationship between the reserve and the resource base, and the zone between them. The inner part of this zone represents what could be worked with existing technology and might be worked if a price rise occurred. Thereafter, increasing uncertainty exists about if and when the material could be worked. The interaction of economic and technological factors will determine whether it will be worked. The resource base is usually very large compared to the actual resource or reserve. For example the resource base for copper has been estimated at 1.5×10^{15} tonnes, sufficient to last for millions of years at current rates of production. In contrast, world reserves of copper, at 5.1×10^8 tonnes, have been variously estimated to last for approximately 36 to 60 years (Tilton and Skinner, 1987).

The concept of the resource base is applied most easily to metals such as aluminium and to other mineral resources such as the fossil fuels. It can, however, also be applied to other resources. For example, the hypothetical resource base for agricultural land might be regarded as the entire land area of the Earth. With current technology and prices, only a very small part of this resource base actually constitutes a land resource at present. Similarly, the forest resource base is much larger than the timber resource: timber cannot be economically produced from much of the resource base because extraction costs would be higher than timber prices. The resource base for flow or renewable resources could also be conceptualized in terms of reception of solar energy, which is not only the ultimate determinant of biological processes in agriculture and forestry, but is also the driving force for resources such as wind and water power.

Economic scarcity

While physical scarcity can have an economic dimension, and may be reflected in rising prices, economic scarcity is rather different. Under some conditions (such as those prevailing in most of the Western world) the market is an important factor in relation to demand, supply and scarcity. If the price of a resource product such as food or wood rises, two consequences are likely. Demand slackens, and production increases (if, for example, farmers are stimulated by the prospect of higher prices to increase their production). Supplies thus tend to stabilize, and scarcity is averted. In areas of great poverty such as those in parts of the developing world, people simply cannot afford to pay higher prices for food, wood or water. The need may be great, but demand in the economic sense does not increase, and farmers (or other resource controllers) lack the stimulus of higher prices that would lead them to increase production and hence supplies. The system is therefore not self-stabilizing, but slides into a vicious spiral of decline with rising prices and growing need.

Geopolitical scarcity

Geopolitical scarcity came into prominence during the 1970s. During the Arab–Israeli dispute in 1973, Arab oil producers cut production and placed embargoes on sales to the United States and some other Western countries. Panic buying allowed the Organization of Petroleum Exporting Countries (OPEC) to double oil prices (see Chapter 7). It seemed at the time that similar developments could happen in other resource sectors, as Western countries in general and the United States in particular were apparently becoming increasingly dependent on imports from Third-World and Soviet-bloc countries. Rapid increases in the prices of various other minerals such as phosphates and bauxite seemed to support the view that a change was occurring in the balance of power between the importing industrialized countries and the exporting countries of the Third World. In the event, the increased prices initially commanded by the resources did not endure, for a number of reasons that are explored in subsequent chapters. In theory, however, if a resource is of localized occurrence and concentrated in a small number of countries, it is possible for these producing countries to create geopolitical scarcity by restricting output and exports, at least for a time. This geopolitical scarcity has little to do with the physical quantities of the resource that are available, but is related to the spatial patterns and control of the resource.

Other forms of scarcity

This three-fold classification of scarcity does not readily accommodate resources such as attractive landscape and wildlife, or even essential resources such as clean air and water. The notion of physical scarcity is applicable to some extent, but not wholly. Rees (1991) therefore adds her fourth category of scarcity, termed 'renewable and environmental resources scarcity'. The label reflects a more abstract concept than that underlying the other three types of scarcity. Essentially she is referring to scarce qualities, rather than to scarce quantities of substances or material. These qualities may be aesthetic, as for example in the case of attractive landscape, or they may be more physical, as for example in the ability of the atmosphere to absorb waste gases derived from burning fossil fuels. Scarcity in this sense of pristine, unimpaired qualities may well increase through time, as the side-effects of exploitative use of environmental resources take their toll. In contrast with commodity resources such as oil and minerals, market forces and feedback loops do not respond to increasing scarcity in this sense and sphere, and other measures are required if scarcity is to be avoided. This is true also in relation to the ecological scarcity as defined by Ophuls (1977). He uses this term to include not only physical scarcities of food and other physical resource products stemming from the limited capacity of the Earth, but also the limited ability of the Earth's natural systems to absorb waste and pollution and to continue to support life.

The significance of resource scarcity

The significance of resource scarcity varies greatly, as do responses to it. At one level, the scarcity of food can obviously be of the utmost significance: it can simply be fatal. At the opposite extreme, many other resource scarcities have far less dire consequences, and, as has been indicated, some may even have (or may lead to) positive benefits.

The shortage of food has been a threat throughout most of human history, and famines are, unhappily, still with us. In the past the consequences of regional crop failure, through the occurrence of pest outbreaks, natural hazards or adverse weather were extremely serious. Today, with much improved transport and better developed distribution systems, the consequences should be less deadly. Unfortunately, economic scarcity, and scarcity arising from political reasons, are still all too common.

The combination of a growing population and fixed areas of land available for food production might be expected to give rise, inevitably, to scarcity and to resource problems. In practice, however, growing pressure on land resources may be the driving force that leads to technical improvements in agriculture and to more intensive use of land (Boserup, 1965). Similar sequences exist in other resource sectors, as is demonstrated in subsequent chapters. Actual or impending scarcity, therefore, can lead to more productive use of resources.

Scarcity arising from resource depletion can also lead to better conservation of the resource. This is a theme associated particularly with Ernst Friedrich, a German geographer of the early twentieth century (Friedrich, 1904). 'Destructive exploitation' (or *Raubwirtschaft*) was seen by him to be an important step leading to more careful use of the environment. If the resource is abundant, there is no obvious need to husband it carefully. In the early settlement of the United States, for example, land was cheap and plentifully available, especially compared with the European homelands of most of the immigrants. Resources seemed so plentiful that it was almost inconceivable that they could run out. If damaged or depleted in fertility, it could simply be abandoned and new land taken into cultivation. As Lester Milbrath (1985: 162) put it, 'Wastefulness seems to accompany a sense of riches'. When limits begin to appear, and when no new land is available, however, habits of 'cut and run' are seen to be improvident and ultimately untenable. Better conservation then begins to be practised. In other words, scarcity leads to changes in resource management, and the perception of impending scarcity may mean that actual scarcity is averted. Indeed in some cases 'melioration' takes place, with the downward resource curve of Fig. 1.1 giving way to rising curves – for example of the forest area (Whitaker, 1941). The experience of the United States gives some support for the notion that conservation or better resource management can follow destructive exploitation.

The problem with the theory that destruction leads to conservation or re-creation of resources is that it ignores the possibility that the resource may be completely exhausted, so that 'melioration' is impossible. In addition, it can be shown that specific resource scarcities have not always been overcome, but have seriously handicapped the societies that experienced them. For example, a shortage of timber

is suggested by Braudel (1979) as a reason for the failure of Islam to gain further ground on Christendom during the medieval period. At a different scale and in a very different setting, the exhaustion of forest resources in Easter Island in the Pacific may have led to impoverishment and population decline (Bahn and Flenley, 1992). Population declines were certainly recorded in ancient civilizations, and in at least some of them may have involved resource problems. For example in the Tigris–Euphrates area, severe salinization of soil resulted from some of the attempts at irrigation, and some areas had to be abandoned completely. The use of environmental resources, in short, was accompanied by impacts that may have been partly responsible for population decline (Whitmore *et al.*, 1990). In effect, problems of resources scarcities were encountered.

Some resource scarcities have led to better and more productive use of environmental resources. Others have given rise to new trading alliances and, where diplomacy has failed, to the use of force to obtain the scarce resource. Others again have remained unsolved, to the serious detriment of the societies that experienced them.

Scarcities and limits

There has been endless debate and speculation over the last 200 years about the outlook for resources and their adequacy. On the one hand, there is the view that resources are likely to be scarce and that they will severely limit human activities. On the other hand, there is the view that resources will be adequate or plentiful.

One of the best-known and most influential contributions to the debate was that of Malthus, around 200 years ago. In short, his thesis was that the human population tends to grow more rapidly than the means of subsistence (see Chapter 5). The result of this contrasting growth pattern is dire: the human population will tend to outstrip its physical means of support, and famine and distress are the likely consequences. More recently, 'neo-Malthusians' have revived and extended his gloomy views, applying them to the supply of other resource materials as well as to food. Some fear that the rapid growth of the world population will inevitably lead to resource scarcities of a new order and severity. They point to famine and other forms of human misery as evidence that the population is tending to outstrip its resources. Others, from the time of Marx onwards, have interpreted scarcities in a rather different light, seeing them as the

Table 1.5 Limits to growth: comparative perceptions.
'There are limits to growth beyond which our industrialized society cannot expand.'

	Strongly disagree (%)						Strongly agree (%)		
	N	−3	−2	−1	0	1	2	3	Mean
General public									
United States	1513	11	11	12	17	18	15	16	5.55
United Kingdom	725	7	5	9	15	21	18	24	4.90
Germany	1088	2	3	4	10	14	20	47	5.78
Australia	390	7	7	7	18	15	18	28	4.93
Business leaders									
United States	223	22	19	14	9	14	11	10	3.47
United Kingdom	261	10	13	15	13	15	16	18	4.32
Germany	130	5	4	13	9	21	20	28	5.12

Source: compiled from data in Milbrath (1983).
Mean score is based on scale positions coded 1–7 from left to right.
N is number in sample: column data are percentages of total group sample.
No data for 'business leaders' for Australia.

products of unjust social and political systems. In other words, the scarcities are due to mal-distribution rather than to physical shortages or physical environmental problems.

In the early 1970s, neo-Malthusian views were boosted by the publication of *Limits to Growth* (Meadows, 1972), an immensely influential book reporting the results of attempts to model trends in population and resource use. These results indicated, essentially, economic collapse in the twenty-first century, and widespread starvation resulting from over-population, depletion of raw materials and increasing levels of pollution. In short, the message was one of severe resource scarcity. As is discussed later (Chapter 11), the bases for the modelling were both limited and flimsy, but the impact of the computer-based prediction was tremendous, especially at a time when an environmental movement was burgeoning in America and some other parts of the world and when some resources appeared to be becoming scarcer. Basically, the message of the book was that the availability of environmental resources imposed definite and rigid limits on continuing economic growth of the kind that had been experienced during the present century.

Within a few years, LTG was rendered largely irrelevant as rates of economic growth – and hence of growth in demand for resources – turned out to be much slower than those that had been assumed. Prospects of severe shortages or limits have, for a number of reasons, been more remote in recent years

than they were during the 1970s. Nevertheless, as many as half of the people in the United States believe that such limits exist, compared with 32 per cent who deny their existence (the remainder are neutral or undecided)(Milbrath, 1985). In an earlier study, Milbrath found some interesting patterns of belief in the existence of limits to growth. As Table 1.5 indicates, the belief appears to be stronger among the general public in Germany and the United Kingdom than in the United States. Milbrath (1983) speculates that this may reflect the lower population density and the richness of resources in the latter, but on the other hand the belief is also relatively strong in Australia, which might be expected to resemble the United States rather than the United Kingdom in this respect. Table 1.5 also shows that belief in the existence of limits is weaker among 'business leaders' than among the general public. Clearly their attitudes towards the use of environmental resources are of greater practical significance than those of the general public.

Acceptance of the concept of limits to growth is one of the most prominent aspects of the New Environmental Paradigm. The extent to which it is attributable to, or has followed from, the publication of *Limits to Growth* is debatable, but there is no doubt that the book had a huge impact. It would perhaps be an exaggeration to suggest that the authors of *Limits* began with the initial assumption that resources were available only in fixed quantities, and that their conclusions were therefore not surprising. Nevertheless, it is true to say that many neo-Malthusians assume that resources can be equated simply with certain materials and substances, and remain impervious to the functional view of resources associated with Zimmerman.

The corollary of limits is that of the concept of over-population and of definite carrying capacities. This latter concept is notoriously difficult to define in precise terms when applied to human populations, although the basic notion is quite simple. Complications arise not only over the role of trade but also over assumptions about the availability of various resources that can relax the limiting factor. For example, food production can usually be increased if more fertilizer is applied, but this requires inputs of resources such as energy or minerals. In addition, food supply has not been the only factor limiting human populations. Ester Boserup (1976) suggests that although wet tropical environments may allow abundant plant growth and hence potential food production, they also offer excellent conditions for the bacteria and parasites that have badly affected human populations throughout history. Thus environments in cooler or drier areas might support larger populations, although food production is more difficult. This view is a salutary reminder that environmental resources embrace more than the production of food and other materials: the life-support function must also be borne in mind.

Sustainability

Some of the trends of the 1950s and 1960s, which lay behind the Limits project, failed to persist into the late 1970s and beyond, and some of the apparent scarcities turned out to be relative or geopolitical rather than physical. Nevertheless, many resource problems persisted: population growth continued at high rates; severe fuelwood shortages hit many parts of the developing world; and land degradation, especially through soil erosion, affected parts of the developed and developing worlds alike. Pollution of the atmosphere, seas and rivers continued. During the 1980s, the concept of sustainable use of resources attracted as widespread attention as that of 'limits to growth' had done in the 1970s. First gaining prominence through the World Conservation Strategy (IUCN *et al.*, 1980), sustainability soon entered the everyday vocabulary of resource managers and politicians. It was greatly boosted by the report of the World Commission on Environment and Develop-

ment (1987) entitled *Our Common Future*, in which 'sustainable development' was a key theme. On the surface, sustainable use or sustainability is a simple idea: the use of resources should be sustainable, and not based on short-term exploitation. On closer examination, however, the meaning becomes more problematic, especially when applied to stock (non-renewable) resources. Much has been written on the subject, and numerous distinctions of varying degrees of subtlety have been drawn between terms such as sustainable use, sustainable growth, sustainable development and sustainability. These will be further explored and applied to particular sectors in the ensuing chapters.

In the meantime, it can be concluded that the general concept of sustainability has provided a focus around which those with different attitudes to the environment and to the use of environmental resources can come together. In this respect its vagueness and ambiguity is an asset, even though it is difficult to translate into practical terms that could serve as operational guidelines for resource users.

This first chapter has sought to set out some fundamental concepts about environmental resources. The most basic of these are the nature of environmental resources (what is a resource?), human attitudes to the environment and the extent to which nature can be viewed as 'neutral stuff', and the existence, type and significance of resource scarcities. These issues underlie many of the trends and problems that characterize individual resource sectors at present, and further reference will be made to them in the context of the individual resource sectors that are discussed in the ensuing chapters. Before turning to these sectors, however, it is important to consider basic questions of the control and management of environmental resources.

Further reading

Bourdeau, Ph., Fasella, P.M. and Teller, A. (eds) (1990) *Environmental ethics: man's relationship with nature: interactions with nature*. Brussels and Luxembourg: EEC.
MacLaren, D.J. and Skinner, B.J. (eds) *Resources and world development*. Chichester: Wiley.
O'Riordan, T. (1981) *Environmentalism*. London: Pion.
Rees, J. (1990) *Natural resources: allocation, economics and policy*. London: Routledge.
Zimmerman, E.W. (1951) *World resources and industries*. New York: Harper & Row.

CHAPTER 2

The political economy of environmental resources

The condition and availability of environmental resources depends not only on the environment and on our attitudes towards it, but also on how the resources are controlled and managed. Types of ownership and political climate (at various scales from the local to the international) are of fundamental importance, as are the economic motives which underlie many decisions about resources. In this chapter, some basic economic concepts about the use of environmental resources are introduced, and the characteristics of different resource regimes are reviewed. It is important to emphasize that behind the technical and legal issues of the economics of environmental resources and of their control frameworks lie fundamental philosophical questions. To what extent should environmental resources be regarded merely as commodities that are bought and sold? To what extent should they be regarded as an inheritance passed down from generation to generation, and in which all humankind has, or should have, a share? These basic questions clearly relate back to the attitudes to the environment discussed in Chapter 1.

Whatever our view of these questions may be, more and more parts of the environment have in practice become commodities during recent centuries, and the use of more and more environmental resources has been motivated by a wish to make a profit or to accumulate capital. Some people welcome this trend, since they consider that this goal, and the framework of private property that often accompanies it, leads to maximum overall benefit to humankind. Others deplore it, concluding that under capitalism there is a tendency to siphon off profits from resource use as quickly as possible, even if fertility is neglected or the resource degraded, and then to re-invest in other industries offering more attractive returns (e.g Perelman, 1975, paraphrasing Karl Marx). Strict govern-ment regulation or outright ownership is usually the remedy advocated by those holding such views, in the belief or hope that resources would then be managed for the long-term benefit of all and not just for the short-term benefit of the few.

The evidence from actual resource-management experiences around the world supports neither of these views very convincingly, and it is perhaps significant that many environmental resources – in many parts of the world – are used within frameworks that combine private ownership with state regulation.

Economics and environmental resources

All economic activity is based ultimately on environmental resources. Even human labour, which produces goods and services, depends on the natural resources that sustain life. All commodities can therefore be traced back to these resources (Dasgupta, 1990).

Since the importance of environmental resources to economic activity is so vital, it is not surprising that they have attracted the interest of economists. This interest has burgeoned in recent years, extending far beyond long-standing concerns with costs of production of basic resource products and analyses of demand–supply relationships in activities such as agriculture. More recently, much attention has been focused on the economics of wider environmental and resource management, including issues such as the maintenance of the quality of the atmosphere and of rivers. The extent to which economic analysis can be applied to issues of resource management, and in particular the relationship between economic analysis and the environmental ethics discussed in Chapter 1, is a matter of continuing debate. Environmental resources can be regarded as useful or valued parts of the environment, but value can have different meanings. In the words of Berry (1990: 203), it 'suffers from the arrogance of economists, obfuscation by philosophers, and rhetoric in politics'. He goes on to identify at least four different meanings: cost in the market–place; usefulness for individuals or society; intrinsic worth; and symbol or concept (e.g. national flag or liberty). To find a meaningful common yardstick, such as the pound, dollar or other monetary unit, is problematic to say the least.

Economics can be defined as the 'study of the allocation of scarce resources for the satisfaction of human wants, and the problems of choice that this involves' (Norton, 1984: 2). (As such, it is a technical discipline concerned with means rather than ends. Rightly or wrongly, however, it is often associated with the ends such as the maximization of profit or of monetary value.) 'Resources' to the economist traditionally were land (i.e. natural resources), labour and capital. This reminds us of the many meanings that 'resources' can have, as well as sometimes giving rise to confusion in discussing the economics of natural resources. 'Scarce' simply implies that supply is not unlimited, and the significance of 'human wants' and 'choice' in the definition is self-evident.

Economic systems

An economic system is the means used to decide how to produce and allocate or distribute goods and services. There are four main economic systems: traditional, market or commercial, command or centrally planned, and mixed. To some extent they correlate with the systems of ownership and control discussed later in this chapter.

Traditional systems

Under traditional systems, people produce enough goods for their basic survival. Production is geared largely to subsistence, with little or no surplus left over for sale or exchange. Market and money are unimportant in exchange, and customs and traditions are major influences on what is produced and how it is distributed and used. Peoples involved in traditional systems of resource use are usually able to see the consequences of their management at close quarters. This does not necessarily preclude signifi-

cant environmental change and resource depletion, as was mentioned in Chapter 1. On the other hand, many groups have devised customs and practices that have the effect of conserving resources, whether or not that was the initial intention. Serious problems of resource scarcity and depletion sometimes occur when the framework of custom and tradition breaks down, for example when a different economic system is established. It is sometimes suggested that the arrival of a money economy not only alters social relations among people but also affects their attitudes to nature and natural resources, encouraging the perception of natural resources as objects for exploitation (Omari, 1990).

Market systems

Traditional economic systems have now been widely replaced by market systems. Under market (or capitalist) systems, decisions are governed by interactions of demand, supply and price. These interactions are depicted in their simplest form in Fig. 2.1. Demand for a product such as oil tends to fall when the price rises and to rise when the price falls. Conversely, supply will increase when the price rises and decrease when the price drops. The relationship between price and quantities of demand and supply can be represented by demand and supply curves. Demand and supply curves intersect at the point where the quantity which suppliers are willing to sell is the same as that which consumers are willing to buy. This intersection is the market equilibrium point (Fig. 2.1(a)).

An increase in demand, with the supply remaining constant in the short term, will result in the market equilibrium point moving to a new position (from 1 to 2 in Fig. 2.1(b)). The higher price (b) will encourage an increased supply. This in turn will lead to a new supply curve and new market equilibrium point (3), and the price will then drop again, i.e. move back towards (a). Changes in demand can result from changes in the number of buyers and in their affluence and tastes. Changes in supply, which could have the same kind of effect, could result from changes in levels and costs of production. Usually both curves shift together. If demand increases, then prices rise and suppliers are likely to respond by producing more, which in turn leads to prices falling back until equilibrium is achieved once more.

It is obvious that only some environmental resources are used or exchanged within this system. Markets for some environmental resources (and

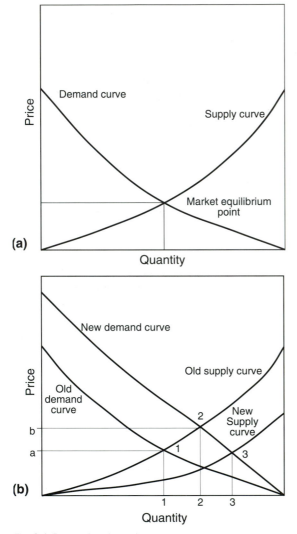

Fig. 2.1 Demand and supply curves.

resource problems. Furthermore, in the longer term, strong demand and high prices may encourage increases in production and supply, to the point that the resource is depleted beyond the critical zone where it can be self-renewed. Nevertheless, some people view the market system, perhaps with modifications to remove flaws, as the most efficient means of producing and distributing resources. Two hundred years ago, the economist Adam Smith wrote of the 'invisible hand' of the market, which would guide towards maximum economic well-being – at least if certain basic conditions such as the existence of free competitive markets were fulfilled. Self-interested, rational behaviour of individuals seeking to maximize their own profits or economic well-being could also achieve maximum well-being for society as a whole.

Not everyone shares this rosy view of the market system, especially as it applies to environmental resources. One particular problem is that of external-ities, or environmental side-effects such as air pollu-tion or water pollution from a petrochemicals plant or pulp mill. Part of the 'cost' of production is borne not by the plant operator, but by neighbouring residents or by society more generally. Such costs are often not reflected in the market price of the product and are usually difficult to incorporate in that price although governments may attempt to do so by means of levying taxes or 'selling' licences. Some flaws and imperfec-tions of the market are common to a number of sectors of economic activity. For example the emergence of monopolies or cartels can mean that supply is manipulated in order to push up prices. At least in theory, governments can take action to control or curb this problem. Some other perceived flaws are more fundamental and less amenable to treatment. Markets in the normal sense of the term simply do not exist for some environmental resources (such as, for example, clean air and beautiful views). The ensuing problems arising from reliance on a market system are not merely those of market failure, in response to which remedial government action could be taken, but are more fundamental. This issue is further discussed subsequently. Another problem is that the market system may separate the consumers of a resource product, and indeed also those who process it, from both the area and the consequences of production (e.g. Tisdell, 1990). Feelings of responsibility for these consequences may therefore be weakened. Similarly, shareholders individually may feel little responsibility for the environmental effects and consequences of their joint-stock companies, while managers may

especially those producing services rather than goods) simply do not exist. And problems can arise even in resource sectors where material goods are produced. For example the long cycles of production of some crops (and especially of timber and tree crops) and animals can mean that rapid adjustment between demand and supply is not possible. Gluts and shortages may thus follow in quick succession. At a time of severe shortage or famine, food prices may well rise, but rising prices will not succeed in stimulating an immediate increase in supply. Instead of supply rising to meet demand, demand is simply reduced by the drastic means of starvation. In such situations, market forces can do little to relieve

Plate 2.1 A landscape degraded by mining and industry, near St Helens, England. The costs of this degradation, in the form of air pollution, water pollution and loss of visual amenity, were not 'internalized' by the mine owners and industrialists in their production costs, but were 'externalities' borne by society at large. (This area has now been rehabilitated). (*Source:* K. Chapman)

consider that their primary responsibility is to make profits. In short, responsibility is both divided and weakened.

Overall, it is difficult to disagree with the conclusion reached in a World Bank review: 'The free market is a good servant but a bad master; since environmental problems often cannot be resolved in an efficient or equitable manner by unregulated market mechanisms, there is no alternative but some form of public intervention' (Schramm and Warford, 1989: 16).

Command systems

Under command systems, government intervention is wholehearted. Decisions are simply made by government, rather than through the market. The market system is viewed by some as badly flawed and seriously inefficient and inadequate as a means of producing and distributing resources. Hence governments assume responsibility for decisions of what and how much to produce, and how to distribute the products. In theory such a system, in which private profit is no longer the driving force, should be more just and less likely to degrade resources through excessive use. But neither in theory nor in practice is

there an obvious mechanism to link demand and supply for different environmental products. In the parts of the world where command systems have held sway for much of the present century, there have been many and serious resource problems, including severe famines and extensive environmental damage. In other words, if market failure is only too apparent, it has to be acknowledged that government failure may also occur (e.g. Dudley, 1990). It may do so for a variety of reasons, including not only lack of signals from prices but also fundamental questions about the nature of states and of governments. Whom do they represent, and who will guard the guardians?

Mixed systems

Mixed economic systems, combining elements of market and command economies, now exist around most of the world. There is a wide range of variation in the relative emphasis on market and command, but most mixed systems are essentially capitalist in character. In these mixed systems, governments have intervened to a greater or lesser extent to modify the workings of the market economy. Governments sometimes intervene in a relatively minor way to prevent monopolies being formed and to ensure free competition. Many have intervened to exert an influence on prices of agricultural products, rather than simply leaving them to be determined by market forces. They may also adopt strategies of regulation, in which they seek to restrict the activities of private resource users or to encourage particular activities by offering incentives such as grants (for example for tree planting). They may also set themselves up as the direct operating authorities for certain resources, such as oil or timber, with or without a monopoly over these resources.

In general terms, intervention usually reflects an awareness that markets have functioned imperfectly or have simply failed as a system for determining resource use (i.e. they are not maximizing collective welfare, however well they may have performed for certain individuals.)

Reasons for government intervention

Endless argument rages about the desirable extent of government intervention, but there is a large measure of agreement about why pure market systems and the efficient use of environmental resources are often incompatible. In general terms, what is optimal in economic terms for the individual resource user is not always optimal for society as a whole.

One major reason is that markets simply do not exist for some environmental resources. Many environmental resources are associated with public goods, such as clean air and water, scenery and wildlife. While some resource products are commodities or private goods (such as kilograms of potatoes and meat) that can be bought and sold in the market place, public goods cannot be divided and sold in units in this way. The distinction between commodity resources and amenity resources is important from the viewpoint of economics, as indeed is the relationship between the two when they are 'produced' in the same area. Amenity resources such as clean water and beautiful views may indeed have economic value (which is reflected in the higher prices that people are willing to pay for houses in attractive settings), but direct markets do not exist for them. Demand and supply cannot be directly linked by price.

A second major reason involves externalities. Essentially, these are the side-effects of resource use which impinge on persons other than the user. External costs are simply passed on to other people, to society as a whole or even to future generations. For

example, a particular form of agriculture might give rise to soil erosion and flooding. Persons downstream from the agricultural area might have to endure muddy rivers, impaired angling, and even damage to property through flooding and silting. Some people take the view that such issues are merely matters to be resolved privately between the resource user and those affected by the side-effects of the resource use. For example the former might simply compensate the latter. In practice, however, the group of 'those affected' may be impossible to define. For example, if a plant or animal species becomes extinct, arguably all humankind, including future generations, is adversely affected and compensation would be an undefinable and meaningless concept. Once driven to extinction, the species cannot be restored: the loss is simply irreversible. Indeed the concept of externality in its literal sense breaks down at larger scales: the loss of the species is not strictly external to the resource user. Another example would be changes in global climate consequent on deforestation or river diversion.

A third reason is connected with time scales. A personal or corporate resource user might consider it desirable to exploit an environmental resource at a rapid rate, and then invest the accumulated profits in other enterprises (such as, for example, other environmental resources or manufacturing). Under some circumstances it could be economically rational for the individual or company to deplete or degrade the resource completely, whether or not it is rational or optimal for society as a whole. This might happen if, for example, demand for a resource product such as wood or fish was keen, and could be satisfied only by exploiting the resource beyond the critical zone (Chapter 1). Similarly, in theory it might be economically rational for one generation to decide to use resources rapidly, regardless of the possible effects on future generations. Although many governments tend to operate on limited time scales, not least through a perceived need to be re-elected every few years, at least in theory they should be able to take a longer view.

There are several other reasons for government intervention. One is a desire to promote social justice and equity within a country, by providing sufficient access to resources to ensure that citizens are able to meet their basic needs. A related reason for intervention is to assist certain groups within a country, such as farmers or fishermen. These groups may have encountered particular problems, because of the peculiar economic nature of some environmental resources or for other reasons.

Intervention may also be motivated by considerations of national security. As has been indicated, environmental resources are essential for life and for economic activity. If a market system fails to deliver adequate supplies of basic materials such as food and wood, then governments may intervene to increase supplies, for strategic reasons of a military or economic nature.

Government intervention in issues of environmental resources takes place within the wider context of the nature and role of the state. Why and how the state should act in relation to environmental resources are political questions. One view is that the state is simply an arbiter, adjudicating between different claims, pressures, concerns and needs arising within the society it represents. Another view is that the state and capitalism are inextricably linked, and that the former promotes the interests of the latter. This view can lead to rather gloomy conclusions, such as those of Johnston (1989: 189): '... the creation of environmental problems is a product of the dominant mode of production in the world today, and the solution of those problems is difficult because the only institutions within which the necessary collective action could be mobilised exist to promote the interests of that mode of production'. Whether capitalism is chiefly to blame for environmental problems, and whether the state's chief aim is to promote its interests, are for the reader to decide. What is clear is that political economy is of fundamental importance in the use and management of environmental resources.

Economics and environmental resources: key issues

The role of economics and of economic analyses in the management and use of environmental resources is also the subject of some controversy. The roots of this controversy lie in the fundamentally different environmental attitudes discussed in Chapter 1. On the one hand, there is the view that the environment is simply a potential storehouse of resources, the use of which should be dictated by economic imperatives. The contrasting view is that nature has a value independent of human needs and wishes, and which cannot usefully or meaningfully be expressed in economic terms. Between these views is an extensive middle ground which recognizes the actual and potential usefulness of economic analysis in resol-

ving issues about the use of environmental resources, but which also is aware of certain fundamental difficulties. Three issues in particular encapsulate many of the theoretical and practical difficulties encountered in attempts to subject resource problems to economic analysis. These issues, which are partly related, are environmental accounting, cost–benefit analysis, and discounting. They in turn raise even more general questions. One is the extent to which resource values can be measured and expressed in terms of money, while the other is the question of what is meant by sustainability and how it relates to economics.

Environmental accounting

In recent years there has been a growing interest in environmental or natural resource accounting. This interest has stemmed in part from the inadequacy of traditional accounting measures such as gross national product (GNP) in relation to environmental resources. The GNP of a country would increase if its forests were cut down and converted to marketed timber, despite the obvious fact that its assets – and their ability to yield continuing streams of income – had been reduced. GNP measures do not show what is happening to the stock of capital assets in the form of environmental resources. The costs of resource depletion are simply not reflected by GNP, and a false sense of security may be engendered. What is required is some form of measurement of increase or decrease, improvement or deterioration, of the stock of environmental resources. Measurements of income flows can then be adjusted for changes in the value of assets or stocks of resources, or at the very least GNP (or gross domestic product (GDP)) can be viewed in the context of these changes in asset values. For example, Warford (1987) estimates that the costs of forest depletion in the 1980s in the major tropical hardwood exporting countries ranged from 4 to 6 per cent of GNP, largely offsetting any economic growth that might have been achieved. In the Philippines, for example, losses from deforestation averaged 3.3 per cent of GDP between 1970 and 1987. In Indonesia, the 4 per cent annual depreciation of soil fertility was roughly the same as the annual increase in farm production during the period between 1977 and 1984 (Repetto, 1992).

In principle, this concept of accounting is very simple, but in practice it is very difficult. Two stages are involved. A physical inventory of resources is required, so that year-to-year changes in assets can be identified, and then an evaluation of the changes needs to be carried out. In some countries, attempts have been limited to physical accounting.

Physical inventories of resources may be difficult to carry out, especially on a regular basis. For example accurate and reliable data on the extent of forest, arable land, degraded land and so on are difficult to assemble, and there is also the problem of common units of measurement. Norway has sought to adopt physical accounting, and under its system resources are divided into two categories, material and environmental (e.g. forest land and atmospheric quality, respectively). Inventories or accounts are easier to prepare for the former than for the latter.

Financial accounting suffers from the extra dimension of difficulty arising from the need to assign money values to environmental resources. An Indonesian example is illustrated in Table 2.1, where the gross domestic product is adjusted for changes in three resources – petroleum, forests and soils. The table may be deceptively simple, as questions of how changes in resources such as soils should be evaluated are not highlighted. Although only a few material resources are included, it is clear that after being adjusted for these resource changes, the net figure for domestic product is considerably lower than the gross figure. Even if not an exact science, environmental accounting of this type is useful in focusing thought on trends in the resource stock. It can thus help to counterbalance the overly sanguine interpretations of trends in GNP or GDP.

Cost–benefit analysis

Cost–benefit analysis has a long history of application in the field of environmental resources. It was used early in the present century by the Army Corps of Engineers in assessing the benefits of public expenditure in river and harbour projects in the United States. Its use has increased greatly during the second half of the century, especially in recent decades. In essence, cost–benefit analysis is a technique used to compare the costs and benefits of particular projects and policies, expressed in monetary units. The aim is to establish whether the benefits exceed the costs, and if so by how much. It may be used in various ways, for example to stimulate general awareness of issues and perhaps to influence opinion, as a basic criterion for decisions, and in making trade-offs between different courses of action. It can also be used at different levels, for example in relation to environmental policies, or in relation to individual projects.

Table 2.1 Environmental accounting: comparison of Indonesia's gross and net domestic products

Year	GDP	Net change in natural resources				NDP
		Petroleum	Forestry	Soils	Total	
1971	5 545	1 527	−312	−89	1 126	6 671
1973	6 753	407	−591	−95	−279	6 474
1975	7 631	−787	−249	−85	−1 121	6 510
1977	8 882	−1 225	−405	−81	−1 711	7 171
1979	10 165	−1 200	−946	−73	−2 219	7 946
1981	12 055	−1 552	−596	−68	−2 215	9 840
1983	12 842	−1 825	−974	−71	−2 870	9 972
Average annual growth 1971–84						
	7.1%					4%

Source: compiled from Repetto *et al.* (1989).
GDP, gross domestic product; NDP, net domestic product (millions of rupiahs).

In some applications, cost–benefit analysis can be carried out relatively easily. For example, the financial analysis carried out by a private company in considering a project is likely to involve only the direct financial costs and benefits affecting that company. Social and environmental costs incurred outside the company are likely to be regarded as irrelevant and to be ignored. On the other hand, the use of the technique by the environmental ministry of a national government or by a development agency is likely to be much more difficult. Some of the reasons for this have already been mentioned: the difficulty of attributing money values to resources for which there is no market or which have intrinsic value; the problem of externalities; and the question of time scales and of how to treat future costs and benefits. One danger is that the results of cost–benefit analysis may be accepted without adequate consideration of the assumptions and procedures on which it was based, and of its scope or range of factors included. Also the technique can be abused by manipulating the numerical or monetary values of essentially unquantifiable costs or benefits in order to produce a desired result. In other words, quasi- or pseudo-objectivity can be a problem (see Chapman, 1981).

Much effort has recently gone into ways of creating shadow prices to represent the value of environmental resources for which there is no direct market. A variety of contingent valuation methods exists, including willingness to pay (WTP) and willingness to accept (WTA) compensation for loss of environmental quality. These methods basically ask people what they are willing to pay for a benefit and what

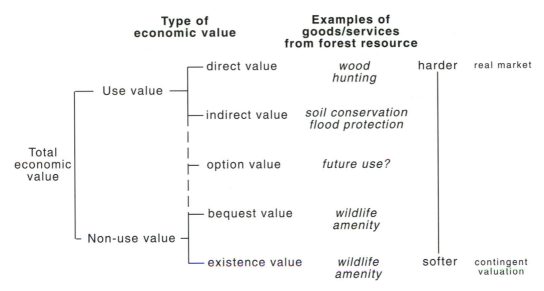

Fig. 2.2 Types of economic value.

they are willing to accept as compensation for a cost. Individual, personal valuations, contingent on a hypothetical market, are sought. These valuations may, and often do, include various components (Fig. 2.2). User values or benefits are the most obvious component. They would include the benefits enjoyed by an angler, hunter or walker who directly 'uses' environmental resources and benefits from them. In addition, however, there may be option values. These indicate the potential, as opposed to actual, benefits, that may exist in the option to make use of the resource at some time in the future. A third category is existence value. People may value the existence of a particular part of nature, such as a whale or grizzly bear, although no user or option values apply. For many types of environmental resources, it seems that existence values are large in relation to user or option values.

Another method of estimating the value of amenity resources is through travel costs, it being assumed that the 'value' of a recreational site, for example, is at least as great as the sum of the costs incurred in reaching it. This approach is simple in practice, but the results that it yields depend heavily on operational decisions and assumptions. For example, should the costs simply be those of transport, or should the traveller's time also be incorporated, and if so at what rate?

Box 2.1 outlines simple examples of how economic analysis might be carried out and applied in local resource-management issues, and also suggests some of the problems that can be encountered – especially when different methods yield different results.

It is theoretically very useful and helpful to set out the costs of conservation and of agricultural production 'forgone', and the benefits of recreational use and wildlife conservation in the way indicated in the second example in Box 2.1. In practice, however, great care is required in interpreting the results, as the contrasting figures derived from different approaches often remind us. It is an obvious truism that the results are meaningful only in the context of the assumptions and methods from which they have flowed. In the Netherlands example shown in Box 2.1, for example, the government ministry that sponsored the study does not use the results directly, because they would not be accepted by other ministries and other interested parties (Barde and Pearce, 1991). Nevertheless, the results may have an indirect influence on these ministries and parties, for example by making more modest figures from other studies of environmental damage more acceptable. A related issue is that accuracy and 'hardness' are, to a large extent, mutually exclusive (Barde and Pearce, 1991). Analyses employing 'hard' figures based on real markets are likely to under-estimate costs, while 'softer' figures based on imaginary markets may be more comprehensive but less readily defensible.

Some commentators also have more fundamental criticisms, considering that cost–benefit analysis is inherently biased against environmental concerns, because of difficulties of valuation, the problem of

---

Box 2.1 Economic evaluations in environmental issues

The first example relates to agriculture and amenity resources in the Yorkshire Dales in the north of England. The environment of the Yorkshire Dales is useful and valuable from a number of viewpoints, including agriculture, nature conservation and recreation. Management to safeguard these values involves both costs and trade-offs: for example ranger services may be necessary, and the safeguarding of the conservation value could have a cost in terms of agricultural output forgone. Estimates of the valuations placed on attractive landscape, recreation and nature conservation could help to inform policy formulation and management decisions.

An attempt was made to assess the value placed on the landscape by both visitors and local residents. An extensive questionnaire survey revealed a total 'willingness to pay' of around £42 million (Willis and Garrod, 1991). The meaning and significance of such a result could be debated endlessly. One conclusion, however, was that the benefits of National Park status and other policies designed to ensure landscape conservation far outweighed their costs, by a factor of about four. Although no market existed for these amenity resources, their management could be supported on the economic grounds that the benefits greatly exceeded the costs.

It is not surprising that different valuation methods often yield very different results, as they have such different bases and starting assumptions. One example, which incidentally also shows the type of application, is work carried out by Willis *et al.* (1988) in an upland farming area of value for conservation at Upper Teesdale in north-

east England. Here farmers are compensated, at a rate working out at £222 per hectare of land per year, for managing the land in such a way that the wildlife interest is protected. This means that agriculture has not been intensified, which in turn means that there is a 'social' cost (of production forgone) of £154. The site is visited both by wildlife enthusiasts and by general recreationists. By travel-cost methods, the 'value' of such use is in the range of £6–34 or £46–251 per hectare, depending on whether only wildlife enthusiasts or all visitors were included (the latter might continue to visit the site even if the wildlife were not conserved). Contingent valuation was also applied to 'experts', sampled from the membership of appropriate bodies: the result was a value of £25 per hectare per year.

An example of a different kind of application relates to severe damage to forests and heather moorland in the Netherlands which it is assumed will take place if current levels of air pollution are not reduced (Barde and Pearce, 1991). A study completed in 1988 used 'willingness to pay' methods and a random sample of households. Responding households were willing to pay, on average, 23 Dutch florins per month, representing an annual aggregate value to Dutch society of 1.45 billion florins. In comparison, the physical loss of timber production amounted to only 13 million per year: the amenity or 'environmental' value of the resource thus would appear to be much greater than its physical or material value. At least in theory, the cost of ameliorative measures could be compared with the aggregate value as indicated by the WTP survey.

---

irreversibility (as in species extinctions) and of discounting (e.g. Goodland *et al.*, 1989). Although these problems all have technical aspects, they also have philosophical and moral components. Some of these relate back to the environmental attitudes reviewed in Chapter 1. Deep ecologists, for example, and probably many holding 'stewardship' views, would reject the view that nature is merely a commodity and that its values can meaningfully be expressed in monetary terms. One question is whether it is appropriate to use 'willingness to pay' as a measure of value. Mark Sagoff (1988: 68) is especially sceptical, to say the least: 'The things we cherish, admire or respect are not always the things we are willing to pay for. Indeed, they may be cheapened by being associated with money.'

In their *Blueprint for a Green Economy*, David Pearce and his co-authors (1989) attempt to deal with the objection that parts of environments (for example animal species) are priceless. One interpretation of 'priceless' is 'of infinite value', but Pearce and his colleagues point out that 'priceless' paintings are merely those of great but not infinite value. (Whether the value of a painting or other work of art is measured simply by its price is another, but related, matter.) The other objection to which they refer is that valuation in money terms is simply impossible – for example in the case of human life. In practice, they claim, society sets definite limits to the level and kind of expenditure it is willing to undertake to save human life, even if it is 'priceless'. They conclude that money is a satisfactory 'measuring rod' for values in environmental resources.

Be that as it may, Sagoff levels another equally basic criticism. He sees a distinction between the judgements people make as citizens of a country and the preferences they hold as consumers. Environmental goals such as clean air and water and wilderness preservation are not to be construed, in his view, as personal wants or preferences, and are not to be 'priced' by markets or by cost–benefit analysis. Instead they are based on beliefs and values held by a people as a whole, stemming from the character of a people as forged through shared histories.

On the other hand, economists such as Pearce contend that there are several good reasons for using monetary measures. One is simply to indicate degrees of concern about environmental issues. Another is to lend support to arguments for environmental quality by employing (money) units which are familiar and meaningful to politicians and civil servants. A third is to permit comparison with other benefits which could arise from alternative uses of funds expended on environmental management. In their view, cost–benefit analysis allows the systematic weighing up of the advantages and disadvantages of an action. Others, in contrast, would hold the view that employing monetary measures simply legitimizes the perception of nature as a commodity, irrespective of any tactical benefits that might accrue from their use.

Future benefits and discounting

Even if we agree that the values of environmental resources can be expressed in monetary terms, we are faced with the problem of dealing with costs and benefits that will arise in the future. People usually prefer a benefit today or this year to one tomorrow or next year: a social time preference exists, for the present over the future. The future benefit is valued less than the present benefit. The weighting of the present over the future is known as discounting.

Interest rates help to explain discounting. With an interest rate of 10 per cent, £1 in year 1 will grow to £1.10 in year 2. In year 3, it would become £1.21 (£1.00 + 0.10 + 0.11), and so on. Conversely, with a discount rate of 10 per cent, a benefit of £1 one year from now would be worth £0.90 at present, and with a rate of 5 per cent the same future benefit would have a present worth of £0.95. The present value of £x in year t is given by $x/(1 + i)^t$, where i is the discount rate. Over long periods of 50 or 100 years, discounting can reduce very large sums to very small amounts (Fig. 2.3). One example quoted by Pearce *et al.* (1989) is that of a hypothetical environmental disaster

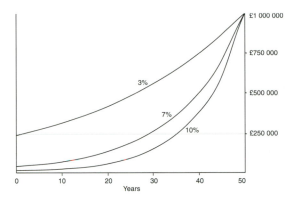

Fig. 2.3 Discounting and present values.

involving soil contamination 100 years from now. Their disaster is estimated to cost, at today's prices, £100 million, and it is considered that there is a 50–50 chance of its happening. The cost around AD 2090 could therefore be taken as £50 million. With a 10 per cent discount rate, this is equivalent to a present cost of only £36, and with a 2 per cent rate it is roughly £69 000. The difference between these sums is large. The higher the discount rate, the more the future is discounted – that is the more we prefer present to future benefits. Put another way, 'The higher the discount rate, the greater will be the discrimination against future generations' (Pearce and Turner, 1990: 221). What is even clearer from the example, however, is that discounting over long periods reduces very large sums to very small amounts in terms of present values. In practice, long-term environmental costs or benefits that may not arise or accrue until 20 years or more into the future are rendered almost costless by discounting in a typical cost–benefit analysis (e.g. Goodland *et al.*, 1989).

Many resources can be used over long periods, and difficult and important questions arise about optimal rates of use and about comparing costs and benefits that occur at different times. One of the factors on which answers to these questions depend is the discount rate employed in the analysis. With high discount rates, it may seem rational in economic terms to develop and use resources rapidly, rather than to save them for future use. To this extent, high discount rates may favour depletion of environmental resources, and are unlikely to favour long-term projects such as afforestation schemes where the benefits accrue some decades after the costs have been incurred. On the other hand, with high discount rates, fewer major long-term projects (such as hydro-electricity schemes) are likely to be viable, and hence fewer wilderness areas (for example) may come under threat.

'Discounting the future' also raises the question of how irreversible changes are to be treated and how the preferences of future generations are to be viewed. How much is a plant or animal species worth? How great is the loss if it becomes extinct? If a species is hunted to extinction, or if a mineral resource is completely exhausted, then an irreversible change has occurred. Options are simply closed off completely. The environmental preferences of individuals in the future will depend on what is then available. What they will want will depend on what earlier generations leave them, and there may be an impoverishment that cannot be expressed in monetary terms, irrespective of whatever discount rates are employed in the analyses. In Sagoff's striking words, 'Future generations might not complain. A pack of yahoos will like a junkyard environment. This is the problem. That kind of future is [economically] efficient' (1988: 63). In other words, 'the present stands as a dictator over the future' (Bromley, 1991: 87).

Even if species are not extinguished or resources are not totally exhausted, future generations will have a more or less a depleted resource base. For this reason, it is sometimes suggested that the future should not be discounted at all in analyses relating to environmental resources, or if it is, that only low discount rates should be employed. Some commentators consider that discounting is immoral because of the way in which it 'discriminates against future generations', and because it seems to be inconsistent with the concept of sustainability.

Sustainability

By now it is probably apparent that there are limits to the extent to which economics and the use of monetary units can or should provide answers to questions about the use of environmental resources. Philosophical, moral and ethical judgements must – or at least should – transcend the outcomes of economic analyses. The significance of such judgements is highlighted in particular in the concept of sustainability. As already indicated, this is a rather loose notion which is capable of an almost infinite variety of definitions. One view of it is that each generation should hand on to the next capital at least as great as it inherited. Major disagreement arises, however, between those who consider that this rule should apply to 'natural' capital alone (i.e. to environmental resources), and those who think that it should apply to natural and human capital combined. (Human capital includes forms of wealth such as money, factories and buildings, as well as knowledge.) The focus of disagreement is whether future generations can be compensated for depleted natural capital by enriched human capital (the potential usefulness of the environmental accounting discussed previously becomes obvious here).

Those holding technocentric views might argue that losses of natural capital are offset by gains in human capital, and in particular by advances in technology. Clearly this view is favoured by those seeking rapid growth and rapid use of resources. It is perhaps supported by those who argue that it is impossible to define what will constitute environmental resources in the future (i.e. what will be useful or valued parts of the environment), and that it is better to use them rapidly and convert the natural capital they represent into human capital. The same argument, however, can equally well be used in the opposite direction: since we do not know what will be an environmental resource in the future, we should in fairness to future generations (inter-generational equity, in the jargon) pass on undepleted natural capital.

Sustainability and maximum sustained yield

In some renewable resources, such as forests and fish stocks, the stock (i.e. the weight or biomass of wood or fish) may increase through time in a pattern that approximates to a S-shaped or logistic curve (Fig. 2.4(a)). The levelling off in the upper part of the curve occurs as the carrying capacity is approached. For example the carrying capacity of a fish population might be set by the amount of food available: beyond a certain level further growth in population could not occur. The growth pattern shown in Fig. 2.4(a) can be redrawn in another form (Fig. 2.4(b)). Here we can clearly see that the growth rate of the resource reaches a maximum, and thereafter slows down. This maximum can be regarded as the maximum sustainable yield (MSY). It should be emphasized that MSY relates to the maximum growth rate of the resource, and not necessarily to the maximum stock.

In a biological sense, production or harvesting at the level of MSY seems optimal. Concepts of 'sustained yield' are especially prominent in the literature on the use of resources such as forests and fisheries, and 'sustained yield' is often suggested as a

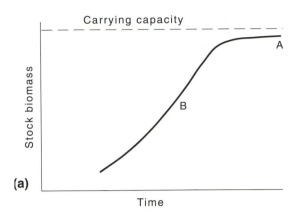

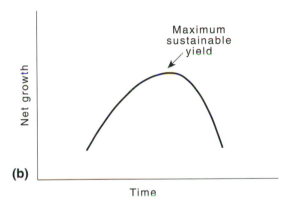

(a)

(b)

Fig. 2.4 Carrying capacity and maximum sustainable yield.

Table 2.2 Costs and benefits (£000s) of different levels of fishing

Number of fishing trips	Total catch	Additional catch*	Cost per trip	Average catch per trip
10	1 000	100	75	100
12	1 200	100	75	100
13	1 280	80	75	98.5
14	1 350	70	75	96.4
15	1 380	30	75	92.0
16	1 400	20	75	87.5
17	1 415	15	75	83.2
18	1 425	10	75	79.2
19	1 430	5	75	75.3
20	1 430	0	75	71.5

Source: modified after Norton (1984).
* Arising from last trip.

fundamental goal of resource management. Various problems, however, may arise when attempts are made to put the concept into practice. The basic problem is that what is optimal in physical or biological terms is not necessarily optimal in terms of economics.

There are two major economic issues. The first concerns the amount of effort required in relation to the level of harvest, and the related costs. This issue is illustrated in Table 2.2. As the number of fishing trips increases, the additional or marginal catch per trip is at first constant, but then begins to decrease. In the example illustrated in Table 2.2, this happens on the thirteenth trip. (For purposes of simplicity, it is assumed that the cost per trip is constant, although in practice this would not necessarily be so because of overhead costs.) Beyond 13 trips, the additional benefit per trip is less than the cost. In simple economic terms, the thirteenth trip is the optimal level of effort. This level, of course, does not

necessarily coincide with the optimal biological level of maximum sustainable yield. Another complication arises from the relationship between the biological growth rate of the resource and prevailing interest or discount rates. If a whale population, for example, grew at a rate of 7 per cent and commercial interest rates were 10 per cent, the best policy in terms of maximizing present value might be to catch as many whales as possible, even although this would not be sustainable in the long term. The revenue might simply be re-invested in some other enterprise which would yield a better future return than whaling. With high interest or discount rates, the present value of potential future harvests is relatively low. Economic imperatives, therefore, could encourage over-exploitation of the resource.

A further problem is that of access to the resource. In the example shown in Table 2.2, it was initially assumed that the fishery was under the control of one operator, who could behave in an economically rational manner. In practice, of course, most fisheries are not under the control of single operators. Instead, they may be regarded as a common-property resource, or they may have open access. The optimal level of fishing activity may depend on the nature of the control regime. The right-hand column in Table 2.2 indicates the value of the catch per trip. As long as this value exceeds the cost of the trip, it will be economically worth while for each operator to continue, although the extra catch is small or non-existent. The nineteenth trip may be worth while (if only just), with the benefit exceeding the cost by a small amount. On the twentieth trip, the costs would exceed the benefits.

This higher level of effort and greater catch make sense from the viewpoint of individual operators. If it is the case that the apparent economic optimum does not necessarily coincide with the biological maximum sustained yield, it is also true that the apparent economic optimum may depend on the nature of the resource regime. Under private ownership, for example, the goal may simply be to maximize personal profit, while under state ownership the objective may be to maximize social – as opposed to personal – profit. Before proceeding to consider the whole question of the ownership and control of resources, it is appropriate to refer back to the beginning of this chapter. Here reference was made to the continuing debate about the role of economics in the management of environmental resources.

This is a subject that engenders strong feelings. On the one hand, there is the view that economics offers a rational basis for the allocation of scarce resources, and that there are no fundamental, intrinsic reasons why the principles of economic analysis should not be applied to environmental resources. In the words of Redclift (1988: 54), 'Some economists clearly do not see the environment as a problem for economists...'.

But in the same sentence he continues '...even if economics is a problem for the environment'. Some other commentators are even more scathing about the role of economics in resource management, seeing it as a dangerous instrument leading to environmental degradation. For example Coddington (1970: 596) went so far as to suggest that in relation to environmental issues 'the greatest service economists can render to posterity is to remain silent'.

Such extreme views are not very helpful, and it is encouraging that environmentalists and economists have recognized some common ground in recent years. Some environmentalists have accepted that economic signals and incentives can sometimes have a part to play in protecting resources, even if they have also caused environmental damage in other settings. Conversely, many economists have begun to consider issues such as the value of environmental resources for which no market exists, and relationships between trends in GNP and in stocks or quality of resources. The use of environmental resources cannot be divorced from economics, and environmental accounting and other forms of economic analysis may be useful tools in the management of a wide range of environmental resources. On the other hand, economics is best seen as a tool or instrument, to be deployed within a defined philosophical and moral framework: what is rational in the narrow sense of economics will not necessarily be rational within that wider framework. It is difficult to disagree with Helm and Pearce (1990) when they suggest that the role of the state is to regulate through command and control procedures, for example in setting maximum pollution levels, while the role of the market (and by implication of economics) is to find the best means of achieving these ends. Equally, it is difficult to envisage practical alternatives to the market system for distributing basic environmental-resource products such as wood and minerals.

The ownership and control of resources

'Wherever a society has needed a natural resource – whether medieval grazing lands in England or wild beaver in Canada – rules for its orderly use have been worked out' (Berkes and Farvar, 1989: 10). It is debatable whether this statement is true, but there is no doubt that ownership and control are key factors in the use and management of resources. Humans, individually or collectively, seek to own or control resources for various reasons, ranging from a basic wish to ensure security of supply of food and other essential resources to a desire to make a profit and to accumulate capital, or in some cases to wield power over other humans.

Different regimes of ownership and control have been established at different times and in different places and different resource sectors. Regimes in this context are simply social and political structures or frameworks within which people manage their environmental resources. These frameworks relate to rights of access to the resource and powers to exclude others; to rights of withdrawal or taking of resources (for example catching fish); to rights to manage the resource in a physical or technical sense, and to sell or dispose of the resource. These rights need not all be held by a single individual or group. For example, the right to make use of a resource is not necessarily accompanied by a right to transfer it to another individual or group. Furthermore, the rights of an individual or group to use or manage a resource may be limited or regulated by the state – for example to ensure animal welfare, or to avoid severe environmental pollution or damage to the resource. In recent years the rights of individual states have in a similar way begun to be regulated by supra-national or international frameworks.

A tremendous diversity exists in the extent and

fullness of control systems or regimes, ranging from *laissez-faire* at one extreme to central planning and socialism at the other. They may develop spontaneously; they may be negotiated by the parties with interests in the resource, or they may simply be imposed on these parties by an external authority such as the state (e.g. Young, 1982).

In some settings, ownership and control are rather meaningless concepts. It is not at present meaningful, for example, to talk of ownership of the atmosphere. Nevertheless, in the course of time, ownership and control can become very significant issues in areas of resources where they were previously not appropriate. One example is that of coastal seas, as is discussed further in Chapter 10. In the past, the concept of ownership of land was equally meaningless, and to some extent it is still an alien idea in some parts of the world. In much of the world, however, land ownership by individuals or companies has become firmly established and enshrined in law.

These general examples remind us that ownership and control are dynamic rather than static. Major changes have occurred in the ownership and control of environmental resources, especially in the last few hundred years, and resource regimes are still evolving. In particular there has been continuing tension between individual and group objectives – whether the individual is a person, a company or a country, and whether the group is the tribe or larger units of society such as national states or even the world community.

Various classifications or typologies of resource regimes have been suggested: a developmental sequence of change through time is often either implied or made explicit. That sequence may be related to population densities: regimes that are appropriate at very low densities are not always appropriate when densities increase. Two dimensions of variation can be distinguished in most typologies:

one involves definitions of those individuals or groups with rights of access to the resource, and the other relates to limitations on extraction rates, or in other words the level of use of the resource by those holding access rights (Fig. 2.5). In practice a distinction is usually made between state, private and common-property regimes. State and common-property regimes may be similar in theory but usually differ in practice. Here, a fourth category – open-access 'regimes' – is recognized as separate from the common-property type with which it is often confused. It should be emphasized at the outset that the classification represents a simplification: in practice many intermediate or transitional types occur. It should also be emphasized that different types of regime may overlap in the same geographical area. For example, the fish in the river may be subject to one regime, the water to another, and land through which the river flows to a third.

Open-access resources

Many environmental resources are used initially on an open-access basis. This was probably true of hunting and gathering during early prehistory, and more recently applies to the use of parts of the New World as well as to that of the ocean and the atmosphere. Under open-access regimes (if regime can be used to described such circumstances), individual and groups take whatever resources they seek from a particular environment. No organized regulation of their activities is imposed, and no mechanisms exist for allocating resource products. Although some open-access regimes may have certain limitations on users, more typically this type of regime is associated with completely open access and unlimited extraction, with 'members only/ extraction unlimited' regimes representing a category transitional to common-property resources. Perhaps

	Private property	Common ownership	Open access
Access or 'right' to extract resource product	owner/occupiers	group members	anyone
Limitation on level of extraction by	owner's decision	group rules	unlimited

Fig. 2.5 Resource regimes: access to resources and limitations on extraction.

the variant of 'unlimited access/extraction limited' should also be recognized. In some hunting/gathering cultures, the landscape is divided into territorial blocks not in order to define zones of use exclusive to particular holders, but rather to regulate the use of dispersed resources over a common range. The 'owners' are simply the custodians of the part of the world that 'belongs' to them. What is possessed is custodianship: the notion of excluding others from the resource, and of the sale or disposal of the land, is quite alien (Ingold, 1986).

Open-access use can perhaps be sustained if the pressures and demands are very modest, but as soon as these increase various problems emerge. These include both the questions of allocation of resource products to the various individuals and groups seeking to make use of them, and the depletion of the resource.

Common-property resources

Common-property and open-access regimes differ in that in the former the use of the resource is restricted to defined individuals or groups, whereas under the latter there is, in effect, a 'free for all'. An evolutionary sequence may exist for some resources, whereby group control is imposed on areas previously subject to open access. For example, a hunting area or farmland may be controlled by a tribal group. Control may include provision for excluding non-members from using the resource, and for regulating the activities of members.

Common-property regimes are not necessarily incompatible with individual use. The ownership of farmland, for example, may be vested in a group, and the group leaders then allocate use-rights on portions of land to individual persons or families. Jurisdiction remains with the group: the individual is not free to pass on the land to others. Maori land in New Zealand, for example, was held by tribes with use rights being allocated to kinship groups according to need (Pawson and Cant, 1992). For other resources involving more extensive foraging and larger tracts of land, the clan or village may be the effective unit rather than the family. Examples of such resource sectors might include fish and fuelwood (Gadgil and Iyer, 1989).

The extent to which commercial elements feature in resource use under such regimes is often limited, although it is not necessarily so. Under common-property arrangements, land is usually not viewed as a commodity that can be sold on the market, but rather as an inheritance that should be passed on to future generations. On the other hand, it is possible that the resource products arising from such regimes are sold on the market (i.e. that the resource is used commercially), and it is possible for the land (or other resource) to be sold by the group to a private purchaser. More usually, however, private purchasers, perhaps aided and abetted by governments, have failed to recognize the rights of the traditional communal owners or have simply ignored them.

Common-property regimes persist in various resource sectors in various parts of the world, and indeed have seen a degree of recent resurgence in some areas. In parts of Australia, for example, the land rights of native peoples have increasingly been recognized in recent years, and now extend over 12 per cent of the country. In general, indigenous peoples in countries such as Australia and Canada were forced back to what was perceived as the worst land and harshest environments in the face of European settlement and its associated system of land rights (Chapter 3), although valuable resources such as minerals have subsequently been found in some of the areas in question (e.g. Peters, 1992). Common-property regimes, however, are by no means confined to such settings, occurring as they also do under some circumstances in countries such as Switzerland and Japan.

There are at present signs that common-property regimes are beginning to evolve at the global scale, in relation to oceanic and atmospheric resources. Up until now, the oceans and the atmosphere have in practice been open-access resources. With expanded resource perceptions and increasing concern about pollution there is now a widespread (though not yet universal) seeking after international cooperation and regulation in the use of these global commons.

Private-property resources

Common-property regimes have in numerous instances given way to those involving private ownership, by individuals or companies. The concept of communal ownership may alter in some settings, so that ownership is gradually identified with the tribal chief, initially on behalf of his group and ultimately as an individual. From there it is but a short step to simple private ownership. Bromley (1991: 25) has reworked Proudhon's famous saying: 'private property is not necessarily theft ... but a good deal of theft has ended up as private property'. He quotes in particular the European colonizers over-

Plate 2.2 Fence lines in the High Plains of Montana, USA. The concept of private land and of linear boundaries was introduced to many parts of the New World by European settlers in the nineteenth century. This concept conflicted with the common-property perceptions of many indigenous peoples. In some parts of the world, serious conflicts developed between the migratory habits of hunter-gatherers and the new settlers, fuelled by mutual incomprehension of attitudes, values and resource perceptions. (*Source:* K. Chapman)

seas, who imposed new regimes and acquired land previously held communally by the indigenous population. Europeans sought to take advantage of the absence of pre-existing systems of private property, ignoring or failing to recognize the traditional systems of common-property under which land and its resources were held. In North America, for example, European newcomers perceived land as property, and were therefore concerned with the boundaries of property, while the indigenous peoples perceived it as a communal environment where boundaries fluctuated with season and habitat (Golley, 1992). It is not surprising that such contrasts led to tension and conflict.

The arrival of the new concept of resource ownership as private property has often been traumatic, both for the resource and for people. Since the concept of private ownership was alien to many indigenous peoples, the significance of transfer of ownership was not necessarily understood by all those involved in the common-property use of the resource. Furthermore, the perceptions of land boundaries could be problematic. These were not always perceived in terms of defined blocks of land,

and indeed were not always physically demarcated. Control of the resource was achieved through social checks rather than the defence of territorial boundaries (e.g. Usher *et al.*, 1992), and geometric boundaries were as alien a concept to the previous communal users as private ownership itself. New concepts of territorial boundaries arrived with the colonists in some areas at the same time as new resource perceptions, and in particular new perceptions of resources as commodities for a market rather than for direct consumption. The significance of these fundamental changes for long-term resource use and for the environment is reviewed for New England in a classic study by Cronon (1983).

Furthermore, various cultural and religious practices may have helped to regulate the use of the resource and thus helped to conserve it, whether or not they were intended to do so directly. With the advent of private property, these checks and balances on resource use sometimes simply disappeared.

When land or other resources pass into private ownership, it may follow that ownership becomes concentrated in a few hands. Especially in the case of localized resources, relative scarcity and favourable prices can then readily be created. In some instances, governments may step in and seek to regulate the activities of the private owners, or to take over the resource directly.

State ownership

The simple developmental sequence from open-access through common-property to private-property resources is complicated by state ownership. State ownership can become established at various stages and for various reasons. In much of the developing world, post-independence state ownership of land and other resources has followed the earlier appropriation by colonial powers. Both colonial powers and modern governments have failed to recognize the communal patterns of 'ownership' and control by indigenous peoples, and have simply declared themselves to be the owner of the resource. A classic – but by no means unique – example is that of the United States, where much of the country was declared to be in the ownership of the federal government (the rights of native Americans were frequently disregarded). From this form of state ownership, the land could then easily be passed into private hands through leasing or sale, or alternatively the state (in the American case the federal agencies) could seek to control and manage the land directly.

Another means by which state ownership can become established is as a response to dissatisfaction with the established private pattern of ownership. This may stem from social and political reasons: for example the extensive land holdings of a few large land holders may be acquired by the state, and subdivided into small farms, or the oil reserves in the hands of private corporations may be taken over by the state in order that its citizens (at least in theory) can enjoy more of the benefits arising from the use of the resource. State ownership may also become established if private ownership proves unable to manage the resource in a sustainable manner. In Britain and some other European countries, for example, private ownership proved incapable of maintaining the degree of forest cover considered by the governments to be satisfactory, and state bodies were set up to acquire land and establish forests.

While large tracts of land have passed into state ownership around the world during the present century, in recent times the tendency has been for resources to be transferred from state to private ownership. Nevertheless, the role of the state as an owner of environmental resources remains a major

one. The United States (to take one example) is not usually associated with extensive state ownership, yet approximately one-third of the land, one-half of the standing stock of softwood and almost all the wild animals in the country belong to the state (Young, 1981). Most of Alaska, for example is federal land, but this has not prevented heated controversies over management issues relating to timber and oil production and to wilderness preservation. Around the world, certain resources are characterized by extensive state ownership: probably more than two-thirds of the world forest area is (at least nominally) in state hands (Chapter 5). In some parts of the tropical forest, in areas such as Amazonia and Borneo in particular, major conflicts and problems of management have arisen from (or been exacerbated by) the nature of the resource regime.

Regime performance

Although evolutionary sequences can be demonstrated in many resource regimes, it does not necessarily follow that later, more highly evolved regimes perform better than those at earlier stages of development. The standard or quality of performance, of course, depends on the criteria used for evaluation.

Three characteristics of successful regimes are suggested by Bromley (1991). The first criterion of success is that natural resources are not squandered (although it may be observed in passing that 'squandering' may be difficult to define precisely). Second, some level of investment in the resource should occur. Third, the co-owners (or those with interests in the resource) should not be in a permanent state of anarchy. An alternative set of criteria is suggested by Gibbs and Bromley (1989). These were originally suggested in the context of common-property resources, but to some degree can be applied to all regimes (Table 2.3). These criteria can be summarized as efficiency, stability, resilience and equitability.

It is argued by some that state ownership is desirable if problems of environmental resources are to be avoided or minimized. Under such a regime, the resource should be used and managed in the interests of the citizens of the state, and the problems arising from the pursuit of personal profit and from market failures would be avoided. It is clear, however, that state ownership does not always meet Bromley's

Table 2.3 Characteristics of well-functioning resource regimes

1. A minimum (or absence) of disputes and limited effort necessary to maintain compliance: the regime will be efficient.
2. A capacity to cope with progressive changes through adaptation, such as the arrival of new production techniques: the regime will be stable.
3. A capacity to accommodate surprise or sudden shocks: the regime will be resilient.
4. A shared perception of fairness: the regime will be equitable.

Source: compiled from Gibbs and Bromley (1989).

criteria for successful resource regimes. In many parts of the developing world, state ownership of land, forests and other resources has been declared by independent countries, following the pattern established by the colonial powers that preceded them. In at least some cases, state ownership has not been accompanied by effective state management of resources, because of other priorities and lack of financial and technical means and capability. At present, many tropical countries grant concessions for the logging of parts of the forests nominally in state hands. In theory the logging should be carried out under strict conditions, but enforcement often proves difficult. Similarly, in the nineteenth century large tracts of state (i.e. Crown) land in Australia and New Zealand were leased to private graziers on short-term bases. Again, while state control of resource use was theoretically possible, over-grazing and depletion of the resources were common.

A further problem is that the imposition of state ownership can lead to the breakdown of existing forms of management. State ownership is not necessarily accompanied by effective management, and the alienation of the traditional users of the resource can lead to the complete breakdown of local regulatory mechanisms. In practice, the imposition of state management in this way simply leads to a shift from communal management to open access, or a 'free for all'.

The result in many cases is that the resource is squandered, little investment occurs and near-anarchy exists among the resource users. In short, the resource regime has failed on all three of Bromley's criteria. Similarly, in parts of Eastern Europe and the former Soviet Union, state ownership of land and other environmental resources has not ensured efficient resource management, as is demonstrated by eroded land and polluted waters. In parts

of the West, state agencies involved in resource management have sometimes functioned as state capitalists, behaving little differently from private corporations in their pursuit of profit. Examples included forest services in many countries (at least until recently), and some state mining corporations such as the National Coal Board in Britain. Perhaps at least part of the problem lies in the nature and role of the state. As has already been indicated, this is a controversial subject in which many and strong views are held. One widely held view is that the state is simply not a purposive actor seeking to maximize in a particular direction (as, for example, a private company might seek to maximize profit). Instead, it is an institution whose function is (or should be) to 'aggregate the diverse preferences of individuals and groups over a wide range of issues into collective choices for the society at large' (Young, 1981: 182). This concept of the state is perhaps a happier one than the 'promoter of capitalism' referred to previously in this chapter, but it carries its own limitations from the viewpoint of resource management. If this concept is valid, then it is not surprising that state regimes may be characterized by inconsistency of objectives through time, and that economic efficiency is not always a primary objective.

The undistinguished performance of resource regimes involving state ownership does not necessarily mean that other regimes perform better. Private ownership has all too often resulted in the degradation of agricultural land, the scarring of landscapes and the pollution of air and water as a result of mining, and various other symptoms of sub-optimal performance. In many countries the state has had to intervene, not as owner but as regulator of private regimes and to seek to ensure that the worst malpractices are avoided.

In recent years much attention has been focused on common-property regimes, and their advantages and disadvantages. The controversy was triggered mainly by the publication in 1968 of a paper entitled 'The tragedy of the commons', by Garrett Hardin.

The tragedy of the commons

Hardin, a professor of biology in the University of California, considered that the growth of the human population in a world that was at least for practical purposes finite was a major problem, and one that had no technical solution. He advocated that curbs should be imposed on population growth, and opined that freedom to breed was intolerable under some circumstances. 'Mutual coercion, mutually agreed upon' is therefore in his view necessary. As a holder of such views, he is a representative of a school of authoritarian commentators on resource issues (see Chapter 12). One of two 'solutions' is often advocated as essential, if problems allegedly associated with common-property resources are to be avoided. The first is coercion by strong, authoritarian government action, if necessary riding roughshod over individual human rights. The second is simply conversion to private-property regimes.

Hardin is best known for his analogy of a common pasture on which each herdsman seeks to keep as many cattle as possible. If the numbers of herdsmen and animals are kept below the carrying capacity of the land (i.e. the level of use beyond which the resource will be depleted), all is well. When carrying capacity is reached, however, problems are likely to occur. Each herdsman seeks to maximize his advantage. If he adds one more animal to his herd, he will receive all the additional proceeds. The effects of over-grazing will be shared by all the herdsmen using the commons, and each individual will suffer only a small fraction of these costs (at least initially). In Hardin's terms, the positive utility to the herdsman of each decision to add an animal is nearly $+1$, whereas the negative utility is only a fraction of -1. In other words, the herdsman enjoys all the benefits but suffers only a small fraction of the costs of his decision to increase the stocking rate. The inevitable consequence is that the pasture is depleted, and eventually becomes useless for all. Although Hardin used this analogy to convey his concern about the rate of human reproduction in a finite world, he also applied it to resource areas such as the oceans and to cattle grazing and national parks in the American west. He considered that the 'tragedy of the commons' could be averted, at least for productive resources such as land, by establishing private-property regimes.

Such regimes are often assumed to perform better in several ways. There is more incentive for investment to occur in the resource, and more disincentive for damaging forms of resource management. It is therefore argued that the replacement of a common-property regime by one based on private property is desirable. It has even been suggested that the 'first economic revolution', when the transition was made from hunting and gathering to agriculture, depended on such a replacement (North and Thomas, 1977). With increasing population pressure on the resources of hunting and gathering, individual groups of humans began to exclude outsiders, and to become sedentary. There was now more incentive to improve the productivity of the resource base, since the additional fruits would not be freely available to all. Under what North and Thomas describe as common-property rights, there was little incentive to invest time and effort in developing new forms of resource use.

This interpretation of common-property rights, which has been shared by numerous other commentators, is confusing. Open-access and common-property resources are not identical. This was pointed out by Ciriacy-Wantrup and Bishop (1975) and numerous other commentators more recently (e.g. Schlager and Ostrom, 1992), but confusion still persists. Under open-access conditions, there are no limitations on access: a 'free for all' exists. In common-property regimes, the group can impose controls and limits on access. Hardin's analogy of the herdsmen and pasture, therefore, was really an example of an open-access resource. He ignored the possibility that social controls might be effective means of regulating levels of use. This criticism is epitomized in the words of Arthur McEvoy (1988: 226): 'A shortcoming of the tragic myth of the commons is its strongly unidimensional picture of human nature. The farmers in Hardin's pasture, for example, do not seem to talk to one another. As individuals, they are alienated, rational, utility maximising automatons and little else. The sum total of their social life is the grim...struggle of each against all and all together against the pasture in which they are trapped.' In short, culture and community are disregarded.

Other basic criticisms of Hardin's 'tragedy of the commons' have also been made. It has been observed by Berkes (1985) that three conditions must be fulfilled if the model is to apply. First, the users must be selfish and must be able to pursue private gain against the best interests of the community or group. Second, the environment must be limited and the rate of exploitation must exceed the rate of replenishment. Third, the resource must be freely open to any user. If these three conditions are fulfilled, the problem is self-evidently insoluble and 'tragedy' is inevitable. The fact that many common-property resources, and in particular fisheries, have not suffered tragedy is a reminder that the conditions do not necessarily apply. Social checks on individuals, means of controlling access and means of regulating rates of exploitation can all be parts of common-property regimes. Nevertheless, there is a widespread belief that resources that are held in common will be over-exploited and hence depleted. The concept of the tragedy of the commons has been used to explain over-exploitation of various resources including fisheries, forests and groundwater, and processes such as wildlife decline and the extinction of species have also been attributed to it. In many if not most of these cases, however, the problem lies in the breakdown or mal-functioning of the common-property regime, rather than in the regime itself. Some of the possible causes of breakdown are listed in Table 2.4. Those regimes that function well are usually characterized by clarity of definition, both of group membership, and of how compliance is to be achieved and disputes resolved (Gibbs and Bromley, 1989). By implication, when clarity in these areas is lost, and when sudden changes affect the common-property management system, problems arise.

Numerous examples have been reported of sustainable use under such regimes. Three from North America are reported by Berkes *et al.* (1989), including beaver hunting around James Bay in Quebec and lobster and fish management off the east coast of the United States. Other examples of succesful performance include grazing on Swiss alpine meadows (Nesting, 1986), Japanese mountain forests and meadows (McKean, 1986), and in irrigation systems around Valencia and Murcia in eastern Spain and in the Philippines (Ostrom, 1990).

Ostrom also describes failures of common-property regimes, in settings as diverse as Californian groundwater and Sri Lankan fisheries. On the basis of comparisons between different performances, she concludes that success is associated with a number of conditions. These include in particular the existence of clear definitions, both of the resource itself and of those having rights to use it, of capability to monitor use, of mechanisms for the resolution of conflicts and of systems of graduated sanctions on those who violate the rules (Table 2.5). The cases she regards as having failed to perform successfully, or to be in fragile states, generally are lacking in one or more of these conditions (Table 2.6). The examples quoted by Ostrom and others clearly demonstrate that common-property regimes are by no means invariably associated with unsatisfactory performance or with tragic degradation of the resource. Much depends on the social institutions that have evolved to regulate the behaviour of the resource users. Hardin apparently assumed that systems of social checks and balances did not exist: perhaps as a biologist the social aspect of human behaviour was

Table 2.4 Possible causes of breakdown of traditional common-property systems

1. Increased involvement of market economies, encouraging the over-exploitation of resources that were previously harvested only for local subsistence use.
2. Breakdown of traditional value systems, which often directly or indirectly encouraged conservation of environmental resources.
3. Population growth, leading to over-exploitation of resources in order to meet needs.
4. Technological change, often making it physically easier to over-exploit natural resources.
5. Increasing centralization of power, and inappropriate pricing policies, subsidies or other government incentives.

Source: modified after Goodland *et al.* (1989).

Table 2.5 Characteristics of long-enduring common-property regimes

1. *Clearly defined boundaries.* Clear definitions of families or households with use rights, and clearly defined boundaries of the common-property resource itself.
2. *Congruence.* Agreement and consistency between management rules and rights and local conditions.
3. *Collective-choice arrangements.* Individuals affected by operational rules can participate in modifying them.
4. *Monitoring.* Active auditing of conditions and behaviour of users: monitors are accountable to the users or are users themselves.
5. *Graduated sanctions.* Sanctions appropriate to the offence can be imposed on users violating rules.
6. *Conflict-resolution mechanisms.* Easy access to means of resolving conflicts among users.
7. *Recognition of rights to organize.* Rights of users to devise their own institutions are not challenged by external government authorities.

Source: modified after Ostrom (1990).

not at the forefront of his thinking. To be sure, some of the social checks may be crude: for example Berkes *et al.* (1989) suggest that in the Maine lobster fishery a would-be lobster fisherman has first to be accepted by the community. Once accepted, he can fish in the territory held by the community, but those who are not accepted are discouraged, if necessary by violence.

If common-property regimes can function satisfactorily when robust social institutions are in place, it follows that they may begin to break down if these institutions are weakened. Bromley (1991:10) suggests that 'the real and lasting "tragedy of the commons" is the gradual breakdown of institutional arrangements' that he considers has occurred in

Table 2.6 Common-property resources: characteristics and performance

Area	Resource type	Characteristic 1 2 3 4 5 6 7	Institutional performance
Törbel, Switzerland	Alpine meadow	p p p p p p p	Robust
Japanese mountains	Forest & meadow	p p p p p p p	Robust
Valencia, Spain	Irrigation systems	p p p p p p p	Robust
Bacarra, Philippines	Irrigation systems	p p p p p p p	Robust
Alanya, Turkey	Fishery	a p w p p w w	Fragile
Gal Oya, Sri Lanka	Fishery	p p p p – w w	Fragile
Port Lameron, Canada	Fishery	p p w p p p a	Fragile
Izmir & Bodrum, Turkey	Fishery	a a a a a a w	Failure
Mawelle, Sri Lanka	Fishery	a p a p p a a	Failure
Mojave, California	Groundwater	a a p a a p p	Failure

Source: modified after Ostrom (1990).
Code numbers refer to characteristics defined in Table 2.5.
p, present; w, present only in weak form; a, absent; –, missing information.

particular in much of the developing world. Under colonial administration, and subsequently since independence, traditional communal rights have been de-legitimized and property rights have been located in alien sources of authority, either in European colonial capitals or in post-independence national capitals. Declarations of state ownership, and the ensuing attempts to manage resources from afar, all too often leave a local managerial vacuum and the decay of the age-old mechanisms for regulating use. With this breakdown, a 'free for all' situation develops. The common-property resource has become, in practice, an open-access resource whatever the official legal status may be. In parts of India, for example, forests were used in a sustainable manner as common-property resources until the colonial conquest (Gadgil and Iyer, 1989). Under British rule, communal organization was disrupted, traditional checks on access and extraction broke down and were not effectively replaced, and the result was that the forests were used in unsustainable ways – despite some areas being declared as protected or reserved forests. This is an example of the type of change arising from the encroachment of the world economy (Chapter 3). In the context of this encroachment, Richards (1990: 176) concludes that the 'decline of common property generally has been a harbinger for land degradation'.

In a few instances, the evident failure of supposedly more advanced forms of resource regimes has been acknowledged, and a return made to common-property regimes. One example is the forests of Nepal. The Nepalese government reversed its earlier policy in 1978, and began to return the previously nationalized forests to local village control. During the period of nationalization, a kind of 'free for all' had occurred as the villagers used the forest resource under a state-ownership regime characterized by poverty of local control (Ostrom, 1990; Bromley, 1991). Nominal ownership and effective control proved to be very different things.

The process of resource degradation may be intensified by the tendency for the better, more productive and fertile areas of land to be first privatized, so that common-property regimes persist longest on the poorer, and possibly more fragile, areas. There the resilience of the resource in the face of human pressures may be lower and its susceptibility to damage greater. It is not surprising, therefore, that severe degradation of (former) common-property resources can occur when cultural and social institutions break down, for example as the result of the sudden impact of foreign political regimes, economies and technologies. On the other hand, Berkes (1989) suggests that change that is more gradual may allow common-property systems to evolve in such a way that successful resource management can continue. He concludes that the health of resources and the health of social systems go hand in hand.

Even on straightforward economic criteria, private-property regimes may have little advantage over common-property regimes. In comparing privately and communally owned alpine grazings in Switzerland, Stevenson (1991) concluded that the problems and costs of private management of remote areas could mean that rents might in any case be higher under common-property regimes.

This points to the conclusion that different regimes may be optimal for different resources. It is sometimes suggested that common-property regimes are very appropriate for certain kinds of resources and in particular for ubiquitous resources such as the atmosphere and for fugitive (mobile) resources such as fish, where the concept of private ownership is not easily applied (e.g. Ciriacy-Wantrup and Bishop, 1975). The long-term co-existence of common-property and private regimes in areas such as the Swiss Alps further suggests that different regimes may be appropriate and optimal for different types of resources (such as alpine meadows and cultivated land), and that an evolutionary chronological sequence in regimes is by no means inevitable.

Resource regimes and global commons

Several points about resource regimes in general and about common-property regimes in particular may be emphasized in conclusion. First, regimes are dynamic, and are likely to evolve in response to changing perceptions and definitions of resources, and to changing power relationships between different social groups. But perhaps they have not been sufficiently dynamic in some cases: perhaps the development of new regimes for the use of resources such as the atmosphere and the oceans has not kept pace with the growing demands placed on them (e.g. Pearse, 1991). Second, it is unlikely that any particular type of regime is best or worst in a general sense. Different regimes are likely to be appropriate at different times. Third, regimes are increasingly overlapping, as more and more attention is focused on the management of environmental resources at the national and global scales. At the former, states are increasingly imposing general frameworks of control or regulation on private resource users operating within their boundaries, for example in terms of ways in which land can be used and discharges made to rivers, seas or the atmosphere. At the international or global scale, attempts are being made to devise supra-national frameworks for the use of resources such as wildlife, the oceans and the atmosphere through agreements, conventions and treaties. Some forms of wildlife, for example, are widely recognized as being of value to humankind in general, and not just to the citizens of any one country. Hence attempts have been made to safeguard such wildlife by means of international conventions which, among other objectives, seek to safeguard certain habitats and to outlaw trade in certain wildlife products (Chapter 6).

Similarly, attempts have been made to define the oceans as the permanent common property of all, with some form of equity to be achieved in the distribution of the benefits of using oceanic resources (Chapter 10).

What in effect is a form of (state) privatization or an enclosure movement in the coastal seas has accompanied these attempts. That is, national sovereignty has been extended over territorial waters, even if the oceans themselves are regarded as common property whose use is subject to agreed controls and restrictions. With this national sovereignty comes exploitation of resources, either directly by the state itself or indirectly through licensing procedures and other means. Furthermore, a number of powerful states, including notably the United States, Britain and Germany, have declined to become signatories to international agreements on the recognition of the oceans as a common-property resource. The success of management of the oceans (and of the atmosphere) as a common-property resource depends on the willingness of states to surrender parts of their sovereignty to international institutions (e.g. Ross, 1971). The failure to secure the agreement of some economically strong states to do this, and the fact that most states are prepared to enter international agreements only when it is in their interests (or at least not against their interests), have given rise to widespread pessimism. Commentators such as Johnston (1992) observe gloomily that anarchy in the use of the oceanic resource is favoured by the economically powerful states. He goes on to conclude that if the 'tragedy of the commons' is to be averted in the oceans then a coercive superstate, which could impose regulation, is the only hope.

Equally gloomy conclusions about 'global commons' have been reached by many other commentators. At the level of the commons within a country, there is at least the possibility that the state can make binding and authoritative decisions about environmental resources and provide a framework within which the competing parties must operate.

With the possible exception of Antarctica, where an international treaty signed in 1961 has provided protection against resource exploitation by individual corporations and states, such frameworks do not yet exist at the global scale, and each state seeks to protect its own sovereignty in relation to issues such as the use of resources. The voluntary relinquishing of such sovereignty to some global organization has not and will not come easily. Observers such as Ophuls (1977: 218) consider that 'the world community as presently constituted is simply incapable of coping with the challenges of ecological scarcity, at least within any reasonable time': each state functions like Hardin's herdsmen.

If cooperation does not occur, the alternatives as envisaged by the pessimists are either direct international conflict over resources, or the imposition of some form of world government to regulate the use of the global commons.

This chapter has sought to review the economic systems under which environmental resources are managed, the role of environmental economics in that management, and the frameworks of control (i.e. regimes under which management takes place). Types of economic systems and of resource regimes are of fundamental importance to the management and use of environmental resources. Like the role of environmental economics, they are controversial topics, views and depend as much on social, political and philosophical factors as on technical assessments. In a sense they form the vital link between technical processes of resource management on the one hand, and human values and ethical and philosophical factors on the other.

Further reading

Berkes, F. (ed.) (1989) *Common property resources: ecology and community-based sustainable development*. London: Belhaven.

Bromley, D.W. (1991) *Environment and economy: property rights and public policy*. Oxford: Blackwell.

Mitchell, B. (1989) *Geography and resource analysis*. Harlow: Longman.

Pearce, D.W. and Turner, R.K. (1990) *Economics of natural resources and the environment*. Hemel Hempstead: Harvester Wheatsheaf.

Pearce, D.W., Markandya, A. and Barbier, E.B. (1989) *Blueprint for a green economy*. London: Earthscan.

Repetto, R. (1992) Accounting for environmental resources. *Scientific American*, **266** (June), 64–70.

Young, O.R. (1982) *Resource regimes; natural resources and social institutions*. Berkeley: University of California Press.

Young, O.R. (1989) *International co-operation: building regimes for natural resources and the environment*. Ithaca and London: Cornell UP.

CHAPTER 3

Environmental resources in space and time

The control of environmental resources and the means of exchange of resource products were discussed in Chapter 2. This chapter reviews the equally important questions of how resources are distributed in space and how they have been perceived and utilized through time. It is concerned with the wheres and whens of environmental resources, and in particular with how these are inter-related.

To state the obvious, the global physical environment is far from uniform. Climate, soil type and terrain all vary greatly over the Earth's surface. The distribution of the human population, and of the demand for environmental resources, is also highly variable. In addition, as Chapter 1 suggests, different attitudes to the environment exist in different parts of the world. This chapter shows that these factors have changed over time, at least in many parts of the world. It also highlights the significance of what has been described as the 'conquest of space': easier and cheaper forms of transport have combined with political and economic imperatives to transform the spatial scale of resource use. In the past, people could usually look only to local environments for resources, but today world-wide systems of supply exist for many resource products. Whereas we were once 'ecosystem people', dependent on what could be obtained from the local ecosystem, now we are 'biosphere people', drawing resources from around the world (Dasmann, 1988). The use of environmental resources is therefore carried out within a changing framework of space and time. An understanding of how this framework has evolved is essential to the understanding of modern issues and problems in the management of environmental resources.

Spatial characteristics of environmental resources

A spatial classification of resources

Some environmental resources are strongly localized, while others are widespread or even ubiquitous. The atmosphere and its oxygen are literally ubiquitous, while the ocean is very widespread and land is extensive. These spatial characteristics provide the basis for a simple classification of resources. Ubiquities occur everywhere, while commonalities occur in many and widespread areas. The atmosphere is an example of the former, and arable land or forests of the latter. On the other hand, rarities are strongly localized, occurring only in a few (usually small) areas. Some mineral deposits are among the best examples of resource rarities. In a few cases, they may amount to uniquities, occurring at only one point.

Like most classifications, this one tends to over-simplify. One complicating factor is that the Earth's surface is almost continuously variable. Some mineral deposits, for example, may be rarities, but the metallic elements they contain may be quite common and very widely distributed. What is rare, therefore, is not so much the presence of the material, but the degree of concentration of the metallic element in the deposit, and hence its ease of working. Similarly, a commonality such as arable land or forest is unlikely to be homogeneous throughout its extensive occurrence. Some areas are likely to be more fertile and productive than others. One point of profound significance in relation to resource problems is that certain parts of the world today suffer from multiple resource problems, involving shortages of food, fuel and water. These problems have in many instances been exacerbated by social and institutional factors, but in general terms basic resource products such as food, wood and water have been harder to obtain in an area such as the Sahel than in Western Europe or eastern North America.

Most of the Earth's surface is too cold, too hot, too wet or too dry for crop production (Table 3.1). Arable land is broadly defined by limits of temperature, rainfall, soil and terrain, but within these broad limits there is a 'highly productive' optimal or core area surrounded by a marginal zone of less productive land (Table 3.1, Fig.3.1). In the optimal area, the costs of production are relatively low and the yields (and income) high. Costs may rise towards the margins, and yields fall. Eventually an economic limit is reached, beyond which production is no longer profitable. The absolute or physical limit of growth may lie some distance beyond this point.

Economic limits depend on the prevailing costs and prices of production and can fluctuate considerably. Absolute limits are more permanent, but they are not necessarily completely unchangeable. Both natural and human-induced change may occur. Climatic change of a magnitude significant for crop growth may have occurred in some parts of the world during historical times. Drainage and irrigation schemes can push the limits back into country that is too wet or too dry, and in theory additional warmth can be provided in glass houses. These extensions of limits will almost invariably be costly to achieve, and normally would be considered only when the economic limits are pressing against the absolute limits. Less spectacular extensions may also be achieved through the breeding of special strains of crops to allow them to be grown in otherwise unsuitable conditions – for example in areas with short growing seasons. In Canada, for instance, the introduction of the Marquis variety of wheat in 1909 helped to push the boundary of wheat cultivation some 300 km northwards (Andraea, 1981).

In a sense resource rarities simply represent the strongly localized occurrence of 'optimal' conditions. Most metallic elements are quite widely distributed in the Earth's crust, but usually in very low concentrations. At a few localities, much higher concentrations occur, and at these points costs of extraction may be relatively low. If product prices rise, however, then extraction may become economically feasible in other areas, so that the rarities may become more numerous. The same effect can result from improvements in technology, which result in reductions in

Table 3.1 The earth's land surface: crop production constraints

	Area	
	Mha	%
Ice-covered	1 490	10
Too cold	2 235	15
Too dry	2 533	17
Too steep	2 682	18
Too shallow	1 341	9
Too wet	596	4
Too poor	745	5
Total non-productive	11 622	78
Weakly productive	1 937	13
Moderately productive	894	6
Highly productive	447	3
Total productive	3 278	22

Source: compiled from data in Buringh and Dudal (1987).

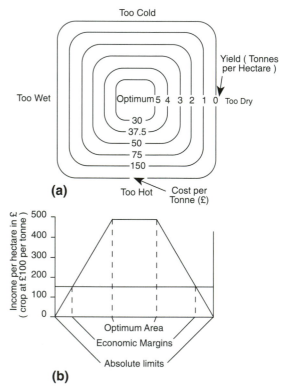

Fig. 3.1 Optimum and marginal areas for crop production. (*Source:* Grigg, 1985.)

costs of extraction or refining. In the same way, the economic limits are likely to contract during periods of low prices, so that the commonalities become less

(a)

(b)

Plate 3.1 Terracing in the Mediterranean world. Areas of steep slopes and thin soils are not normally valued highly for purposes of arable farming. The constraints imposed by these steep slopes and thin soils can, however, be overcome to some extent by terracing, if labour is available and if there is a driving force such as population pressure. (a) Near Villafran-cha, eastern Sicily. Note the bare and rocky hill slopes in the background, and the partly degraded terraces in the middle distance. If the maintenance of terraces is neglected, for example with the ageing of the population, out-migration or the growth of non-agricultural employment, the terraces can begin to break down, leading in some instances to accelerated downslope movement of soils. (*Source:* A. Mather) (b) Actively maintained terraces used for growing vegetables and citrus fruits, Estellencs, Mallorca. (*Source:* K. Chapman) common or widespread and the rarities more or less numerous.

Location and resources

Commonality and rarity are therefore not necessarily constant, unchanging characteristics of resources. A further complication is that place or location can be of primary significance in determining whether part of the natural environment is a resource. Locational constraints may exist as well as environmental constraints. Suppose that two identical areas of potentially cultivable land, or two identical mineral deposits exist. One is located in a populated area, well served by transport links. The other is in a remote and inaccessible part of the world. Clearly the former is more likely to be utilized than the latter, since transport costs, and probably also the costs of labour and management, will be lower. At the scale of individual countries, the same general point can be illustrated by sand and gravel deposits. These are often widely distributed (i.e. they are commonalities). Since unit value is low and transport costs are high, those deposits located near markets such as large cities are used in preference to more remote deposits. In short, what is an actual, workable resource depends on location as well as on environmental characteristics.

This same general theme is exemplified, in a different way, by potential arable land in a mountainous area. The key to using the mountain grazings may be the production of winter feed for the animals in the small areas of cultivable land in the valleys. In any absolute sense, such land might be of poor quality, and if located elsewhere would probably remain unused. Because of its location, and its key role in the wider agricultural economy of the mountains, it is perceived as a valuable resource. Once again, the value of part of the natural environment as a resource depends on where it is located, and not just on its physical characteristics.

A further aspect of the same general point involves political boundaries and the division of the world into individual states. Many states seek to ensure that their populations are adequately fed and their industries supplied by raw materials. These strategic considerations can mean that certain types of environments, which in other settings would not be regarded as resources, are actively used for agriculture or mining. Political boundaries also assume another form of importance, especially in relation to rarities. If most of the known world reserves of a particular mineral

are concentrated in a few countries, then the potential for geopolitical scarcity (Chapter 1) to be created is obviously much greater than if they were widely scattered among many countries.

If a potential product is sufficiently valuable or attractive, then the resource may be utilized even if formidable barriers of remoteness and inaccessibility have to be overcome. This may involve the costly construction of new roads or railways into previously inaccessible country. These new lines of communications may, in turn, give rise to new resource perceptions and to the use of other previously disregarded resources. For example, a rich mineral deposit in a remote and uninhabited area may stimulate the building of a new road, which traverses potentially cultivable land. Previously this land could not be economically cultivated because the product could not be transported to a market, but now both marketing and cultivation become possible. The perception and use of resource rarities, therefore, can lead in time to changes in the perception and use of other, more widely occurring resources. Sequences of phases of resource development can thus occur, and this has been a pattern in parts of the new lands of the Americas and Australasia 'opened up' in recent centuries.

Mobile resources and mobile humans

A particular group of environmental resources is characterized by mobility or migration. Certain fish species, such as the Atlantic salmon, undertake long-distance migrations and spend different parts of their life cycles in different environments. Certain bird species also migrate over very long distances. These mobile or fugitive resources pose particular problems for management. Private ownership and management cannot be readily applied, and the effects of management or exploitation in one part of the range of the resource may be negated by those in other parts. In the same way, the management of rivers that cross international boundaries is especially difficult. Actions such as water abstraction or the discharge of pollutants in one part of a catchment can compromise the use of the river in other parts downstream. Similarly, the use made of the atmosphere in one area can have significant effects hundreds of miles downwind. The emission of pollutants from power stations in one country, for example, can affect forests in other countries. Again, these trans-national effects can be very difficult to curb because of the difficulty of securing agreement among the various parties with interests in the resource. The problem is often that the costs or externalities are born in a country other than that in which the resource is used (see Chapters 6 and 9).

Some resources are mobile or fugitive. Some resource users are nomadic and some are temporary visitors. Nomadic or shifting patterns of resource use were in the past widely practised. For example, the New England Indians prior to the arrival of the English obtained their food from different areas at different seasons: they had learned to exploit the seasonal diversity of their environment by practising mobility (Cronon, 1983). In much of the New World, such mobility was constrained or halted by the new systems of land tenure introduced by the colonists, but in some areas it has survived to the present. This is especially so in areas of harsh environments such as the margins of the tundra or the desert. In such settings, hunters or pastoralists make use of areas that are more extensive than would be possible from fixed sites. For example, neither grazings nor water supply might be adequate to support permanent settlement at a particular point, but seasonal migration allows the grazings and water of a number of different points to be used in succession. Serious conflicts can sometimes occur if part of the territory used in this way is 'lost', for example by a private regime being imposed (Chapter 2). Even if the 'loss' is small in terms of area, it can mean complete disruption of a traditional resource system, and lead to changes (and perhaps extensions) to migration patterns.

Movement is also typical of many mountainous areas, where the seasonal grazings on the mountain pastures can be used in summer and the animals brought down to the low ground in winter. Another aspect of movement by the resource users is reflected by shifting cultivators. Shifting cultivation has been widely practised in many parts of the world, especially by peoples whose agricultural technology would not permit permanent cultivation at acceptable levels of productivity. Again, these types of movements are usually associated with common-property or other traditional resource regimes, and privatization or enclosure can have serious effects.

A more modern form of mobile resource users is represented by climbers, by those seeking wilderness experiences, and indeed also by tourists. In none of these cases can the environmental resource be transported to the user: instead the user has to be transported to the resource. As traditional forms of nomadism have dwindled and shrunk, the modern form of recreation-related mobility has burgeoned.

This growth has been made possible by the great improvements in transport in recent times, and by great decreases in the real costs of travel.

The same general trend has led to tremendous expansion of the 'area of search' or spatial scale of resource use over recent centuries. In the past, humans were restricted to local areas in their search for environmental resources; today, we can truly talk of world systems of resource use. The spatial restrictions on resource use have been progressively relaxed by a combination of change in transport and the emergence of a world-economy, in which resource-using and other economic activities in widely differing parts of the world are closely inter-related.

The spatial expansion of resource use

The use of environmental resources has become more widespread and extensive with the passage of time. Individual resource processes such as agriculture have expanded, probably from a vary small number of highly localized areas, to the extent that the greater part of the Earth's land surface is now used for some form of cultivation or grazing. Accompanying this expansion, the human transformation of the natural environment has proceeded apace.

Human use of the environment: the great transitions

The relationship between humans and their environment has undergone a series of major transitions down through human history. In the words of Bennett and Dahlberg (1990: 69), 'Each transition has witnessed a progressive incorporation of natural substances into culture, thus extending the dominion of humankind over the earth and its creatures.' In other words, more and more 'neutral stuff' has become an environmental resource, and more and more human influence has been exerted on the environment.

Five major transitions have been identified (Bennett, 1976). These are to some extent associated with changing resource regimes (Chapter 2), but they are of major importance in terms of changing types and levels of the demand for environmental resources. The first involves the change from subsistence foraging by family units to more organized forms of hunting carried out by larger bands or groups.

Although the change may have been accompanied by improvements in tools and weapons, its main feature was cooperation and superior organization, which could provide a better and more reliable food supply than that obtainable from the uncoordinated efforts of individuals.

The second transition is from organized, nomadic hunting to more sedentary lifestyles and the more intensive use of natural (i.e. unfarmed) foods and other products in particular areas. Territories become more clearly defined, and longer-term encampments, for example at forest margins or lakeshores, become possible as subsistence systems become more specialized. These transitions occurred both in the Old World and in the New, but at different times and perhaps in different ways.

Although both of these transitions involved important changes in the relationship between humans and their environment, the significance of the third transition was much greater and more far-reaching. This was the so-called agricultural revolution. Crops were now raised and animals bred: hunting and gathering were no longer the only means of human support. Clearly this transition involved a major technological change, and it made possible a more sedentary form of life based in villages. The rise of industry and trade was in turn accompanied by the emergence of cities: the fourth transition reflected a change in social organization and what has been termed the urban revolution. This in turn had profound effects on the management and use of environmental resources in the surrounding areas (Box 3.1). The fifth transition is that usually known as the Industrial Revolution. Characterized by vast increases in the use of energy (and especially of fossil fuels) compared with previous times, the Industrial Revolution has transformed the human relationship to the natural environment in general, and in particular has led to completely new perceptions of what constitutes environmental resources. Whether and when a sixth transition – to a post-industrial age or an age of sustainability – will occur remains to be seen: perhaps changing environmental attitudes (Chapter 1) are a portent that it will.

The significance of these transitions in relation to the human use of the environment is variable, but all of them to a greater or lesser extent involve changing technology, changing forms of social organization, expanding spatial range or scale of resource use, and changing range and complexity of what constitutes environmental resources. The general trend is epitomized by the increasing level of energy use by each individual, and by the widening range of energy sources called into use (Table 3.2; see also Chapter 7).

The tremendous contrast between energy use in industrial and hunter-gatherer societies typifies the great increase in human use of the environment – and impact on it – in recent times. It does not necessarily

---

Box 3.1 Environmental resources and the growth of cities

A strong interaction between the birth and expansion of cities and the use of land resources in their environs has occurred throughout history. The origin of a city may in the first place depend on the ability of the surrounding area to produce sufficient food to support a population larger than that involved in its production – that is a non-agrarian, urban population. The existence of an urban market in turn can mean that agriculture is commercialized. In exchange for agricultural commodities, the farmers receive manufactured goods such as cloth or ceramics, or more usually the money with which they can be purchased. A money economy was usually slow to expand far beyond this zone of urban influence, and mainly subsistence systems lasted in many areas until the advent of improved transport links in the eighteenth or nineteenth centuries. Within the zone of urban influence, capital accumulated from urban-based activities such as manufacturing or trade might be invested in agriculture, further intensifying it and heightening the contrast with more outlying areas. In the same way, the production of fuelwood might be commoditized around the city, and more recently much of the demand of urban populations for recreation has been satisfied in the 'peri-urban' zone.

The growth of a city can therefore lead to changes in the ways in which environmental resources in its vicinity are used. The location of an area can thus change (in relative terms) through time, as a city develops and exerts its influence as a centre of demand for a variety of resource products. Another aspect of the interaction is that urban expansion is often at the expense of agricultural land of above-average fertility. Such land helped to make the growth of the city possible in the first place, and its loss may be serious if trends in food production are barely matching those of population increase.

Table 3.2 Energy sources and use

Type of society	Energy sources
Earliest hunter-gatherer	Solar energy via plants and animals
Advanced hunter-gatherer	As above, plus stored energy in biomass – use of fire
Early agriculturalist	As above, plus domestic animals
Advanced agriculturalist (pre-industrial)	As above, plus wind and water power
Early industrialist	As above, plus power from coal
Advanced industrialist	As above, plus oil and natural gas
Contemporary	As above plus nuclear power

Type of society	Energy consumption per person per day	
	(kilocalories)	(Joules $\times 10^6$)
Hunter-gatherers	2 000	8.4
Agriculturalists	10 000–12 000	42–50.4
Nineteenth-century industrial society	70 000	294
Twentieth-century industrial society	120 000	504

Source: Simmons (1987).

follow, however, that humans did not change their environment through the use of resources in the past (Chapter 1). In some areas, such as New Zealand and North America, the extinction of several mammal species seems to have coincided in time with the arrival of humans. Nevertheless, the extent of transformation wrought by humans on the environment in pre-agricultural times was slight compared with what was to follow later.

Hunter-gatherers have been displaced from most of the land which they once occupied, but still exist in some areas. Such areas offer glimpses of lifestyles and patterns of resource use that were once much more extensive. In south-west Africa, for example, the !Kung Bushmen live in an area of savannah with an annual rainfall of around 250 mm. A wide variety of plant species is used for food, and meat, obtained by hunting and snaring, is also a regular part of the diet. This pattern of resource use appears to be stable and sustainable, but it involves each person travelling over a distance of 2000–3500 km annually in the quest for food and water. The environmental carrying capacity for hunter-gatherers is usually very low compared with that under agriculture. It is suggested that as a world average, 26 km^2 is needed to support each hunter, and that overall population densities are of the order of 0.1 person/km^2 or less (Simmons, 1989). Actual population levels in such cultures are often only half or even less of the theoretical carrying

capacity estimated from the potential food supply. Whether this is a deliberate response to environmental fluctuations (and hence fluctuations in food supply from year to year) or a result of deliberate attempts at population control is debatable.

The agricultural revolution

How and why agriculture developed is uncertain in detail, but it is known that the domestication of plant and animal species was under way by around 7000 BC in south-west Asia, 6000 BC in south-east Asia and 5000 BC in Meso-America (Simmons 1989). It has been suggested that agriculture developed in response to population pressure or stress on resources under the prevailing systems of hunting and gathering (Cohen, 1977). At first populations relied on exploiting large mammals. As these became scarcer they shifted towards more broadly based economies with a wider range of food sources, and in particular to the production of storable food from domesticated plant species.

From these centres agriculture spread into adjacent areas in parts of Europe, Africa, south Asia and the Americas (Fig.3.2). The spread was not as comprehensive as Fig.3.2 might seem to imply. The whole land area of Europe, for example, was not used for agriculture by AD 1 (nor will it be by AD 2000): mountains and wetlands, for example, would not have been farmed. Nor was the spread necessarily steady or progressive: at times the agricultural

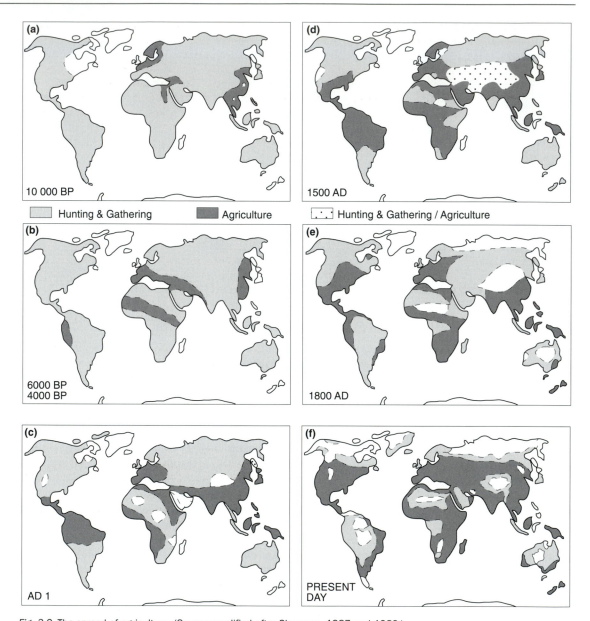

(a) 10 000 BP

☐ Hunting & Gathering   ■ Agriculture

(b) 6000 BP / 4000 BP

(c) AD 1

(d) 1500 AD

☐·· Hunting & Gathering / Agriculture

(e) 1800 AD

(f) PRESENT DAY

Fig. 3.2 The spread of agriculture. (*Source:* modified after Simmons, 1987 and 1989.)

frontier probably contracted. And hunting and gathering probably continued both within and beyond the frontier. Nevertheless, the overall story is one of inexorable expansion, and with it came a new scale of human transformation of nature – a process that has continued at an accelerating pace right to the present (Fig.3.2 (e), (f)). Whereas at 10 000 BC probably the entire human population were hunter-gatherers, by AD 1500 only 1 per cent lived by that mode, and by the end of the nineteenth century the proportion had fallen to 0.001 per cent of the world population (Simmons 1987).

The urban revolution

The advent of agriculture was followed by the growth of towns and cities, and in them the production of crafts and manufactured goods. This in turn had its own consequences for natural resources: the production of ceramics, metal goods and textiles all required the supply of resource products as raw materials or energy sources. Until the modern period, however, both the scale of these industries and of the towns and cities in which they were usually located were small. Even in a country such as France, 80 per cent of the population were engaged in agriculture as late as the latter part of the eighteenth century (Grigg, 1987). Compared with today, towns were few in number and small in size. In 1500, for example, only four cities in Europe had populations in excess of 100 000 (de Vries, 1981). Low yields per hectare and low output per person meant that most of the population had to be engaged in agriculture. Furthermore, transport of food (and indeed of other materials) was difficult and costly. Unless situated on a navigable river allowing less costly transport, towns had in practice to draw their food supply from within a radius of a very few kilometres (see also Box 3.1).

The Industrial Revolution

By the eighteenth century, the major transition of the Industrial Revolution and the associated (second) agricultural revolution were under way, bringing with them a complex set of trends and changes that were to transform the human use of the environment and the perception of environmental resources. Severe environmental impacts occurred around some of the main centres of industrial development, especially in the forms of air and water pollution. Population growth accelerated rapidly, and populations became more highly urbanized. By 1800, the number of European cities with populations in excess of 100 000 had grown to 17 (de Vries, 1981). In industry, raw materials changed from the organic and commonalities to the inorganic and rarities, with profound implications for the spatial patterns of resource-using activities. Farming became less self-contained. Whereas previously it had provided most of its own inputs, such as seed from the previous harvest and manure from livestock, by the mid-nineteenth century it was purchasing more and more inputs. The overall result was that production became more specialized. It became more separated, both occupationally and spatially, from consumption. City dwellers became divorced from the production of food that sustained them. Many of the producers themselves were not the owners of the land or the means of production, and might take little responsibility for conservation of the resource that they exploited. The corollary of the separation was that more and more resource products became commodities, and that the market system for their exchange expanded and developed. At the same time, much land previously farmed or grazed as common property now became private property. In short, the management and production of environmental resources were transformed in the course of this transition. The transition did not, of course, occur everywhere at the same time, any more than did earlier transitions. Beginning in parts of Western Europe in the eighteenth century, it had by the end of the nineteenth century spread to much of Eastern Europe, the United States and Japan.

The world economy

As this transition spread out, it cast a shadow ahead of it, because the development and use of environmental resources in many parts of the world still unaffected in a direct sense became geared to the demands of the industrialized world. In the second half of the nineteenth century, for example, wool was produced in Australia and New Zealand for manufacture into textiles and clothing in Britain. It now became possible to talk of a veritable world economy (e.g. Wallerstein, 1974). By this is meant more than simply resource production and consumption carried on at a global scale: a global system of the production, processing and consumption of resource products was now in operation, initially focused on (and controlled by) north-west Europe. Bulk cargoes of resource products such as wool and wheat could now be shipped around the world, while previously only high-value products such as silks and spices were traded over long distances. Previously, much of the trade was in luxury goods that could not be produced in the home country. Now, it expanded to include basic resource products (such as wheat or wool) that could be produced at home, but which could perhaps be grown more cheaply overseas.

Some disagreement remains both about the timing of the arrival of the world economy, and about the extent of trading of resource products prior to the eighteenth or nineteenth centuries. For example attention has been drawn to the huge exports of wine from south-west France as early as the fourteenth century, to the export of bulk commodities such as bricks and tiles as ballast from countries such as Holland, and to extensive trading systems in Asia, Africa and Latin America (e.g. Chisholm, 1990). Nevertheless, there is little doubt that trade in resource products greatly expanded both in volume

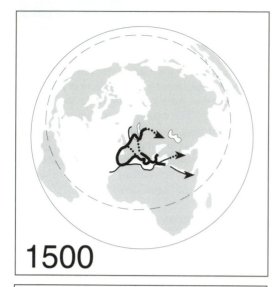

1500

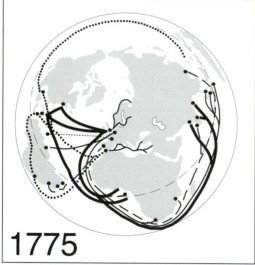

1775

Fig. 3.3 European expansion and the world economy 1500 and 1775. (*Source:* Braudel, 1979.)

and in spatial scale from the eighteenth century, and that most of the trade was focused on European markets. The industrialized parts of Europe had begun to function as a core of the world economy, to which the production of resource products in the extensive periphery was directed. In the centuries leading up to this period, various European groups beginning with the Portuguese had begun to transfer crops and their products around much of the (expanding) known world (Fig.3.3). The process that Crosby (1986) has described as 'ecological

imperialism' practised by Europeans overseas was under way by the fifteenth century with sugar production in Madeira, and culminated with wool and grain production in the Americas and the Antipodes in the nineteenth century and logging in the tropical forest in the twentieth century.

The transformation of environments was one of the results of this 'ecological imperialism'. New resource perceptions, new resource regimes and new systems of production combined to transform many environments, which though by no means always pristine, were previously usually only partly modified by humans. For example, the New England as described by the first English visitors around 1600 had become a very different place by 1800 (Cronon, 1983): the forests were smaller and different in composition, and large areas were now devoid of animals such as beaver, bear, wolf and turkey. On the other hand, large numbers of European grazing animals were now present and other alien plants and animals had been introduced. Another result was the widening separation of production and consumption of resource products. Changes in technology, especially in shipping and later in other forms of transport, made the growth of the world economy possible. Behind these was the underlying driving force of capitalism.

Improvements in transport and the shrinking of space

Maritime freight rates decreased at a rate of 0.5-1.0 per cent per year from 1650 to 1770, and at 3.5 per cent per year from 1814 to 1860 (Grigg, 1987). The introduction of steamships and the construction of the Suez Canal underlay major reductions in freight costs in the nineteenth century, and especially in the second half of the century. With the opening of the Suez Canal, for example, freight rates from what was then Burma (now Myanmar) to Britain dropped by more than half (Richards, 1990). The trans-Atlantic freight rate for wheat from New York to Liverpool in 1900 was one-third of what it had been in the 1870s. Similarly, the overland transport cost from Chicago to New York in 1905 was around one-quarter of what it had been in the early 1870s. In effect, the world was shrinking: terrestrial space was being conquered. At the same time, the world was expanding – at least in terms of the area from which resources could be drawn.

In a succession of new lands such as Argentina, Australia and Canada the expansion of railways was accompanied by that of wheat production (Fig. 3.4).

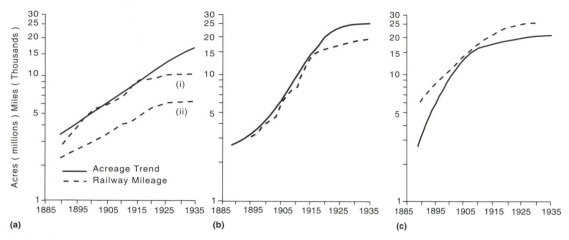

Fig. 3.4 The expansion of the railway network and wheat acreage: (*a*) Australia: (i) Victoria and Western Australia (ii) New South Wales; (*b*) Argentina; (*c*) Canada. (*Source:* compiled from data in Rostow, 1978.)

Much of the capital for the railways and other forms of development in the new lands originated in old-world countries such as Britain, in which the returns to capital in due course accumulated. In the new world economy, a reciprocal relationship of dependency developed, and indeed survives to the present. The nature of this relationship of dependency – and in particular its degree of symmetry and equity – has been the subject of much debate, but its significance from the viewpoint of global environmental resources is not in doubt. Potential supply areas for resource products expanded enormously, with profound consequences for the new and old areas of supply alike. In short, the resource base of the Western world was greatly extended (North, 1958).

Table 3.3 illustrates the increasing average distance of transport of agricultural products imported into Britain during the nineteenth century. This average distance more than trebled between the 1830s and the early part of the twentieth century. At the same time, the proportion of temperate agricultural produce imported into Britain rose from 26 per cent of apparent consumption to 46 per cent between 1870–76 and 1910–14. As semi-natural environments were transformed in the new lands, the older-established agricultural areas struggled to compete. Unless protection against cheap imports was provided, the main alternatives were to switch to more specialized forms of production, where competition was less severe, or to withdraw land from production (Chapter 4).

A similar trend was apparent in timber imports into Britain, and indeed also into neighbouring countries such as Holland. Forest resources were depleted early, and in particular the large timbers required for naval shipbuilding had to be imported. These imports came initially from the southern Baltic around Danzig, and subsequently from further east around Riga, then from the north side of the Baltic. The frontier was expanding: when suitable materials could not be obtained from the existing area of supply, a new or extended area could be found. When Baltic supplies were interrupted during the Napoleonic Wars, Britain became much more dependent on North American sources than it had been previously. Timber from planted forests in Britain could not compete in price with that from the virgin natural forests of Canada, and there was therefore little incentive to plant forests for timber production.

At both national and global scales, the new forms of transport and accompanying spatial organization resulted in increasingly specialized production. As the shackles of distance were relaxed and the imperatives of self-sufficiency weakened, environmental resources could be obtained from the areas that offered them most cheaply. The same process also gave rise to a degree of regularity in spatial patterns of the specialized zones of production. It is apparent from Table 3.3, for example, that perishable products such as fruit and vegetables were typically hauled over shorter distances than more transportable materials such as wheat and wool. Roughly concentric patterns of production had been recognized around some European cities since early modern times, and is another aspect of the influence of the growth of towns on the use of environmental resources (Box 3.1).

Table 3.3 Average distance between London and regions of origin of agricultural imports into Britain (km)

	1831–35	1856–60	1871–75	1891–95	1909–13
Fruit & vegetables	n.s.i.	521	861	1 851	3 025
Live animals	n.s.i.	1 014	1 400	5 681	7 242
Butter, cheese, eggs	422	853	2 156	2 591	5 021
Feed grains	1 384	3 267	3 911	5 214	7 773
Meat & tallow	3 219	4 667	6 019	8 127	10 068
Wheat & flour	3 911	3 492	6 759	8 288	9 575
Wool & hides	3 750	14 210	16 093	17 718	17 541
Weighted average	2 929	5 874	6 920	8 127	9 463

Source: Peet (1969).
n.s.i., no significant imports.

Intensive market gardening was a distinctive feature of a core area of rural land adjacent to the city, while grain and livestock production (in that order) often occupied more distant zones.

With the full development of the world economy, a hierarchy of scales of core and periphery ranged from the local to the global. North-west Europe formed a global core, but within it there was a smaller-scale pattern of cores and peripheries around individual cities. Similarly, in the new lands and other parts of the global periphery, core areas developed around the main cities and ports, exerting their own influence on their peripheries as the global core exerted on the global periphery of which they were part.

The period from around 1870 to the outbreak of the First World War saw the culmination of a process of expansion of the capitalist world economy that had been under way for some centuries. The grand finale to the long period of expansion was a surge of empire-building and colonialism. In addition to the longer established imperial and colonial powers, newly unified countries such as Germany and Italy (as well as the United States) now sought overseas empires and spheres of influence from which resource products could be derived and in which manufactured goods sold. New territorial boundaries resulted from this partitioning, and many of them have survived in the form of the boundaries of the post-independence states. Within individual territories, land was also partitioned in a new way. The land-holding systems and resource regimes of native peoples were widely disregarded, being replaced by new Western-style legal and tenurial systems. This did not always happen immediately after conquest or annexation. In both Mexico and New England, for example, several decades elapsed before European ownership of land became extensive. During these decades populations of indigenous peoples were devastated by diseases brought by the Europeans, and their grip on their traditional resources – already under threat from a superior power and alien concepts of property and resources – was further weakened (Cronon, 1983; Premm, 1992). In some other areas, the process operated more rapidly: the British Crown in New Zealand, for example, assumed ownership of huge tracts of territory and proceeded to hand over, by sale or lease, blocks to private resource developers. In Australia, aboriginal people could only claim parcels of land that had been defined by European law, which completely failed to reflect an understanding of aboriginal land tenure and concepts of land units (Baker, 1992).

An expanding world

The long-term consequences of this process and of the period culminating just before the First World War have been enormous. Firstly, the world – and its endowment of environmental resources – was in effect expanding. Over a period of several centuries, but especially in the late nineteenth century, the area from which countries such as Britain could draw resources was growing larger almost by the year. This process was mirrored by what was happening in individual New World countries such as the United States, where the frontier was progressively extending until the end of the nineteenth century. An essentially similar process had been occurring in the Old World for hundreds of years, under subsistence as well as capitalist systems. German migration into previously thinly-populated land in Eastern Europe had begun to push the frontier eastward by AD 1000, and a similar process was continued by the Russians.

Indeed scores or even hundreds (depending on definitions) of frontiers opened in the course of the last five centuries (Richards, 1990). New lands, new forests and new resources were continually becoming available. In moving to the new lands of the frontier, European settlers all too often abandoned the conservationist agriculture that they had practised in their homelands. Why these practices were abandoned is a fundamental question (Butzer, 1992). Perhaps it is simply that apparent abundance and wastefulness go together, as do scarcity and conservation. Certainly the gloomy conclusion reached by Cronon (1983: 170) on New England was that 'Ecological abundance and economic prodigality went hand in hand: the people of plenty were a people of waste'.

With continually expanding frontiers, the notion of limits was unreal. If a country lacked a particular resource product, it could simply look overseas for it. Furthermore, great increases in yields and productivity could be achieved in resource sectors such as agriculture through new technologies and by the use of imported fertilizers (such as Chilean guano/nitrate) and, later, chemical fertilizers. Since environmental limits were not encountered, it was assumed that they did not exist. The combination of expanding frontiers of land and of advancing technology was indeed a potent one. It is not surprising that the mind-set of unlimited growth became established.

Core and periphery

The second long-term consequence of the set of processes encapsulated in the growth of the world economy is the emergence of a heartland–hinterland or core–periphery spatial relationship. Heartland or core areas, initially in north-west Europe and later including neighbouring parts of Europe as well as the United States and Japan, import food, raw materials and energy resources from hinterland or peripheral countries in other parts of world. In the case of the Russian/Soviet empire, the European area around St Petersburg and Moscow functioned as the core, and Siberia (and central Asia and Trans-Caucasia) the periphery. It is usually assumed that this dependency relationship operates to the disadvantage – both economically and environmentally – of the hinterland or periphery. This is because resource management is geared, wholly or partly and directly or indirectly, to the needs and wants of the core. A few areas have successfully made the transition from peripheral to core status. For example, in Scandinavia the export to

Britain of resource products such as timber and iron provided a springboard for rapid economic development in the nineteenth century. The United States has also passed from 'peripheral' status – providing exports of resource products to 'core' countries – to that of 'core'. Most parts of the world, however, have not succeeded in making this transition. This is especially true in Latin American countries, many of which achieved political independence in the nineteenth century but which have not yet made the transition in status from periphery to core that the United States successfully achieved. Core–periphery trade usually involves resource products moving in one direction, and manufactured goods in the other. Trends in the terms of trade have often operated to the disadvantage of the periphery, and growth in trade of manufactured goods has certainly been far greater than that of resource products. The relationship is therefore not one of equal interdependence, but instead is often one of dependency of a weaker periphery on a stronger core. Indeed the ironic symbol of this unequal relationship is the food exports from the core that are now required to sustain some peripheral populations (Chapter 5).

The environmental side-effects of resource production in the periphery are not always in the forefront of the minds of the core-orientated managers. Furthermore, the use of land for crops for export to the core, rather than for food production for local consumption, has often meant that the local population has been displaced towards environmentally marginal areas. Population growth has been accommodated on, and fed from, such areas, many of which are fragile and easily degraded. There may therefore be a double occurrence of environmental degradation – both on the areas used for production geared to the needs of the core, and on those on which the local population, and their resource-management activities, have been concentrated.

The ultimate expression of core–periphery relationship is in colonialism. In their study of the relationship between colonialism and land degradation, Blaikie and Brookfield (1987) hypothesized that land managers skimped on management practices and sought to use resources for short-term gain. They concluded, in general terms, that their hypotheses were sustained by the evidence they gathered. On the other hand, not all areas were degraded, nor of course is it the case that environmental degradation is found only under capitalist or colonial regimes. As in many questions about resource management, generalizations need to be viewed with caution.

Plate 3.2 Clearance of forest and conversion to grazing land around a homestead in North Island, New Zealand. During the nineteenth century, huge areas of land were transformed in this way. New areas were incorporated into the world economy, in order to supply resource products to Europe. In this instance the products – wool and lamb – were destined for the British market. (*Source:* A. Mather)

Plate 3.3 Hill slopes in Central Otago, New Zealand. Sheep farming was developed on the tussock grasslands of these slopes in the 1850s. Within a few years, the combination of grazing pressures and burning (as a form of land management believed to improve the pastures) had depleted the tussock grass to such an extent that some areas were described as 'man-made deserts'. Major ecological change occurred in numerous similar settings as cheap land was rapidly brought into production geared to the needs of distant countries. (*Source:* A. Mather)

Plate 3.4 Grain elevators and railroad, Canadian prairies. The opening up of the North American prairies in the late nineteenth century was accompanied not only by the local transformation of nature, as grassland passed under the plough, but also by agricultural depression in Britain as British farmers struggled to compete with the cheap imports from overseas. (*Source:* K. Chapman)

The direct political power and control implied by colonialism have, of course, waned during the second half of the twentieth century as many former colonial areas have achieved political independence. The colonial period, however, may have an enduring legacy, and economic independence may be harder to secure. In the words of Carolyn Merchant (1992: 25) 'Instead of enslavement by force or theft of resources, neo-colonialism uses economic investments and foreign aid programs to maintain economic hegemony.' Corporations based in the imperial and colonial powers were actively exploiting resource products such as minerals and tropical crops by the end of the nineteenth century. During the present century many of them have further expanded their activities, and indeed have

become trans-national corporations (TNCs). Many TNCs engaged in resource activities are vertically integrated, in the sense that they are involved in the processing, distribution and marketing of the product, as well as in extracting the original resource. Control is usually concentrated in 'core' countries, where much of the processing value is also added to the product. The country of extraction may therefore lose out on some of the economic and social benefits that might otherwise accrue (for example in employment). The pricing policies used in intra-corporation transfers of resource products may also give rise to resentment: for example the raw material represented by the unprocessed resource product may be assigned low values in the accounting systems of the TNCs, so that the benefits to the host country in terms of royalties or other payments are small (Chapter 8). A further issue is the economic strength and political influence of TNCs. In some cases their annual turnovers greatly exceed the budgets of many independent states. It is not surprising, therefore, that the ability of the latter to enforce environmental regulations is as weak as their overall bargaining strength over resource use.

Over the last few centuries, a complex system has developed whereby matter and energy flow from peripheries to cores. The environmental implications of this system are as great as their social, political and economic consequences are profound. In the core areas themselves, the waste products emitted on the use of these material and energy resources may accumulate. In the peripheral areas, the natural environments from which the resources are drawn are likely to be affected by processes of resource extraction, and may be degraded. In effect an ecological shadow is cast by the core on its peripheral supply areas (see McNeill *et al.*, 1991, for further discussion of this concept, in the context of Japanese imports of timber and other resource products). The social climates of peripheral areas may also be affected. Indigenous societies had usually exploited a wide range of resource products at relatively low intensities and often under conditions where regeneration was possible. Under the new regimes interest tended to focus on a very few products, perhaps at intensities far beyond their capacities for regeneration. At the same time, the social institutions within which the traditional resource use took place decayed or were destroyed. Successive phases of resource exploitation may compound the damage: for example the course – and effects – of these phases in Amazonia over the last three centuries is discussed by Bunker (1984).

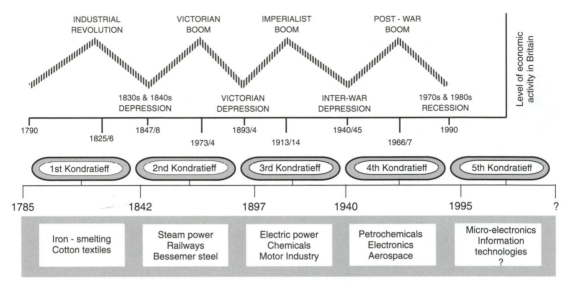

Fig. 3.5 Innovation cycles and long waves of economic activity. (*Source:* Healey and Ilbery, 1990.)

Temporal cycles of resource use

The world economy and Kondratieff cycles

The continuing growth of the economies of the core requires a continuing supply of resource products (including food, raw materials and energy). This supply can be derived from within the core area itself, from the periphery, or from some combination of both.

Economic growth in the core areas has not been constant or continuous, but has proceeded in a series of fluctuations. Some commentators see these fluctuations as cycles with some degree of regularity. A Russian economist, Nikolai Kondratieff (1935), identified three waves of economic expansion and contraction on the basis of fluctuations in commodity prices between the late eighteenth century and the time of his work in the 1920s. The complete cycle of each of these waves occupied approximately 50 years (Fig. 3.5).

The first wave coincided with the Industrial Revolution in Britain, with a crest in the early part of the nineteenth century followed by a trough around the 1840s. A second boom in the 1870s, coinciding with industrial expansion in Europe, was followed by a trough in the 1890s. The third cycle consisted of a boom phase in the period leading up to the First World War, followed by the depression of

the 1930s. The phase of post-Second World War expansion lasted until the beginning of the 1970s, and was followed by international recession. While the existence of regular cycles of the kind illustrated in Fig. 3.5 is disputed by some commentators, there is little disagreement that successive waves have occurred, and that each wave has involved different industrial products and resource demands.

The fluctuations in commodity prices originally identified by Kondratieff in effect reflect fluctuations in the abundance and scarcity of resource products. During phases of scarcity (high prices), new areas are brought into production. For example at the beginning of the nineteenth century in Britain, the combination of wars and high prices for resource products such as grain dampened economic growth, but encouraged the expansion of the cropping area. Thereafter, commodity prices decreased and economic growth rates increased, with the cycle repeating itself in the second half of the nineteenth century.

Kondratieff cycles are the subject of a huge literature, and controversy surrounds their existence, their nature, and especially their determinants. In particular, there has been much debate on the role of technological innovations in relation to the cycles. New technologies induce new types and levels of resource demands, as well as triggering accelerated economic growth. Then the new demands result in increasing scarcity of resources and high resource prices, which in turn reduce growth rates and complete the cycle. A new cycle is then initiated by innovations during the period of recession.

The economic fortunes of resource-producing areas have fluctuated through time, the areas have expanded and contracted, and individual resource products have fluctuated between scarcity and abundance. It follows that the relative fortunes of core and periphery can also fluctuate, but it is important to emphasize that these medium-term waves or cycles (such as 50–55 year Kondratieff cycles) are superimposed on longer-term trends. It has been shown that the world economy developed over a longer term of around 400 years, between 1500 and 1900. During that period, environmental resources (at least from the European/Western viewpoint) became more abundant as potential supply areas expanded. Zobler (1962) sees this as one phase in a long-term cycle – a phase of relative abundance following the long period of relative scarcity that had occupied most of human history. It was in turn followed during the twentieth century by another, modern phase of relative resource scarcity, characterized in Zobler's scheme by increasing government intervention in resource issues. The precise timing of the beginning and end of these phases varied to some degree between Britain, mainland Europe and the United States, but the same order applied throughout these areas.

Resource cycles and conservation movements

Cyclical patterns in resource use are further complicated by cycles in levels of concern about resource adequacy, and in environmental attitudes in general. The factors contributing to these cycles can sometimes be identified with hindsight, but their causes and their relationship to economic fluctuations are poorly understood as is their relationship to the economic fluctuations just discussed. Carville Earle (1988) has suggested that cycles of environmental impact are related to economic fluctuations such as Kondratieff waves. The reasoning is as follows. During periods of depression, experimentation takes place and agricultural innovations occur. Those seen as successful then diffuse widely during the ensuing economic upswing. The innovations, however, may trigger a new phase of environmental impact or resource degradation. This, in turn, may be a contributory factor leading to economic downturn and depression, and hence to a new phase of innovation and the repetition of the cycle.

Earle's theory of the linkage of economic cycles and cycles of environmental impact was devised in relation to the use of the land resource in the American South. Whether it is more widely applicable is debatable. What is also unclear is the relationship between objective, measurable environmental impact and resource problems on the one hand, and the subjective perception of these impacts and problems by humans on the other. Over the last 100 years, the level of concern about environmental resources has fluctuated widely, especially in the United States. These fluctuations are neither regular nor symmetrical, but there is no doubt that they have had a significant effect on the use of environmental resources both in the United States and in the wider world.

Environmental movements

The emergence of modern environmentalism and of concern about environment and resources in modern times is traced by some to experience in small tropical islands such as Mauritius, St Helena and Barbados (e.g. Grove, 1990, 1992). Within a short time of having been incorporated by the imperial powers into the world economy, some of these islands had suffered environmental devastation as a result of over-exploitation of resources such as forests and of the introduction of alien species of animals. The obvious signs of environmental degradation and devastation in these restricted land areas led during the eighteenth century to various government measures such as the declaration of forest reserves. From their original island settings, the concept of the forest reserve spread to India in the nineteenth century, and to South-East Asia and parts of Africa (Chapter 6).

Environmental concern in such limited settings is not surprising. What is more surprising is the key role that the United States has played in providing the stage for environmental movements, several of which have had world-wide ramifications. The large, relatively lightly peopled land of America is a less likely stage than St Helena or other oceanic islands, but it is no less real. Indeed the United States has been the source of the best-documented and probably most influential environmental movements over the last century, and the effects of American movements have, to varying degrees, diffused world-wide. Why that land should have been such a fertile ground for these movements is debatable, and is probably due to a number of coincident factors: conspicuous consumption of resource products; obvious signs of degradation of forests and agricultural land; and key personalities to give philosophical foundations and practical leadership.

Three or four major movements have developed in the United States since the end of the nineteenth century. The first occurred from about 1890 to 1915, peaking during the first decade of the present century. The second was in the 1930s. The third developed during the late 1960s and early 1970s. Although it waned to some extent in the later 1970s, it did not fade away as the two earlier movements had done, and it waxed again during the 1980s. (It is therefore debatable whether a single movement or two separate movements have occurred since the late 1960s. Here a single movement, i.e. the third, is assumed.)

Although the nature, duration and intensity of the three movements have differed, some similarities can be noted between them. O'Riordan (1971) notes three elements that were common to the movements, and two factors that were associated with the waning of the first two. The first of the three common elements was the recognition of impending scarcity of environmental resources, accompanied by reconsideration of traditional means of allocating and using these resources. As will be shown, specific events helped to trigger these concerns. The second was a belief that science and technology could improve resource management. Thirdly, public concern was accompanied by political support – at the highest levels – for conservation policies. Factors common to the waning phases of the first two movements were increasing fuzziness and vagueness of the meaning of conservation (and problems in translating general sentiments into specific management policies and guidelines), and also the onset of major wars, which quite apart from any other effects tended to divert public and government attention to other issues and priorities.

The first conservation movement

The first movement waxed at the end of the nineteenth century. Three 'reigning' assumptions are identified by Clayton Koppes (1988) as having dominated the use of natural resources in the United States from the early seventeenth century: there was an abundance of resources in an unclaimed state (i.e. unclaimed by Europeans); the resources were inexhaustible; and it was best to use the resources immediately. Until the late nineteenth

century, the frontier was extending, and resources must indeed have seemed limitless. By the latter part of the century, however, the frontier was no longer expanding, and so extensive had been the clearing of the forests that the spectre of a timber famine loomed, at least in the eyes of some influential observers. One of these, at the beginning of the twentieth century, was President Theodore Roosevelt himself. Another, and probably the main driving force at the height of the movement, was Gifford Pinchot, head of the state forest service, and influential associate of the president.

Fears of resource scarcity were not the sole driving force of the movement. Up until that time, environmental resources had been allocated by the market system with little regulation by government. Some extremely wasteful forms of exploitation had occurred, especially in forest logging areas, and at the same time huge fortunes had been amassed from exploiting resources such as coal and oil. Control of some natural resources was concentrated in the hands of a few rich and powerful capitalists. The conservation movement had a political flavour: it was concerned with how resources should be allocated and how the benefits should be distributed, as well as with technical aspects of resource management. Human domination of the environment, or the concept of nature as 'neutral stuff' that could be exploited to serve human needs, was not questioned by the mainstream movement. Efficiency in resource use and the avoidance of waste were the main goals (Hays, 1959). Conservation was not seen in terms of preservation or hoarding for future use: instead it was seen in terms of development and active use of resources for the benefit of the people now alive – 'for the greatest good of the greatest number'. By implication, the American environment was seen as a public heritage, but one for practical use rather than of intrinsic value.

Some practical achievements resulted. National forest reserves were established, and an embryonic national park system was developed (the first formal designations had occurred some decades previously). Federal government agencies such as the Forest Service, the Bureau of Reclamation and the National Parks Service were established. The professional resource managers staffing these new agencies, it was implied, would be the interpreters of the 'greatest good'. There was some international spin-off in the form of the diffusion of the national-park idea overseas to countries such as New Zealand and South Africa. The momentum of the movement, however, faded during the second decade of this century, and the advent of the First World War and America's entry to it in 1917 marked its definite waning.

The focus of the movement was on development and efficiency of resource use, rather than on preservation, but there were at the same time strong voices such as those of John Muir who urged a different form of conservation. These voices were essentially of those whose attitudes to the environment, and in particular to that of the American West, were quite different. Non-use, rather than efficiency of use, was their primary concern. Their concept of nature was not one of 'neutral stuff', but rather one of great reverence. The voices of these advocates were not well heard above the clamour of development and exploitation at the time, but since then they have not been forgotten nor have they been without influence. This is especially true in relation to national parks and wilderness areas (Chapter 6).

The second conservation movement

The first conservation movement had largely run its course by 1920, but a second was to build up in the 1930s. The Great Depression of the 1930s highlighted the prevailing reliance on the market place as a means of allocating resource products and of arbitrating economic activities. In some parts of the country, the economic and social distress of the period was accompanied by evident environmental problems, such as the soil erosion and flooding. The disastrous episode of soil erosion in the Dust Bowl was an unmistakable sign that all was not well in the relationship between humans and the environment in the United States. Dust storms blotted out the sun: dust was carried from the centre of the country to the great cities of the east. As in many environmental disasters, different levels of causes could be identified. Blame could readily be attached to wind and drought and 'natural' causes. Some would consider that over-expansion of cropping had occurred in a fragile environment. Donald Worster (1979) argues that there was a close link between the Dust Bowl and the Depression – that 'the same society produced them both, and for similar reasons' (p. 5), and that the ultimate explanation was 'the alienation of man from the land, its commercialization, and its consequent abuse (p. 63).

While by no means all Americans perceived the same connections or identified the culprit simply as capitalism, there is no doubt that the combination of economic, social and environmental problems in the 1930s provided fertile ground for new attitudes and policies towards resource management. Equity and efficiency were both sought: better resource management would lead to better social and economic conditions.

The scene was set for massive government intervention in the management of the economy and of environmental resources. Again, the period was marked by the influence of men well disposed towards planning, active resource management and 'conservation'. This time the President was F. D. Roosevelt, who when governor of New York had set aside large tracts of public land for recreational purposes and who now introduced the 'New Deal'. The role played by Gifford Pinchot in the first movement was to some extent emulated by H. H. Bennett, an active and vigorous promoter of soil conservation and head of the new US Soil Conservation Service.

Several features characterize this second conservation movement. An obvious one was public works and river basin schemes such as that of the Tennessee Valley. New federal agencies with responsibilities for individual sectors of resource management were established. Idle labour was employed in various forms of resource management such as tree planting, terracing of eroded land, and in particular in construction. Dams were built, rivers harnessed, and in general an increased control or domination over an apparently wayward and turbulent nature was sought. Conservation was not generally interpreted as preservation, but rather in terms of use and development; it was anthropocentric rather than biocentric, and pragmatic rather than philosophical.

Another innovation of the 1930s was government intervention in agriculture. In response to the evident distress in farming communities, the federal government intervened in the market for agricultural products – an intervention that has persisted, and which has had profound implications in terms of levels of production, the taking of land out of production, and intertwined economic and soil conservation policies.

A third feature was the diffusion around the world of concerns originating in the United States. While the Depression may have stimulated new perceptions around the developed world, it is almost certainly the case that the publicity focused on processes such as soil erosion in the United States led to increased concern also in a variety of other countries. The threat of widespread land degradation became widely perceived in Australia, for example, during the 1930s (Heathcote and Mabbutt, 1988). It is noticeable that in many countries around the world, legislation for public funding for soil conservation was first enacted during the 1930s and 1940s. Examples include Sri Lanka (1938), New Zealand (1941), Mexico (1942) and Kenya (1943) (Hudson, 1985). In many cases, state soil conservation services were also established during that same period. Even in countries where soil erosion was not a major concern, concern about environmental issues grew. In Britain, data presented by Lowe and Goyder (1979) show that more new environmental groups were established in the decade ending in the mid-1930s than in any prior period. Previously, the period of greatest activity in this respect, around the turn of the century, had coincided with the first conservation movement.

The environmental movement

If ripples from the conservation movement of the 1930s spread around the world, the same is true to an even greater extent of the third conservation movement (or more accurately environmental movement). Although its rise was especially clear in the United States during the 1960s, its scope was truly international in scale. It was also much bigger and wider than the two previous movements in terms of the strands of concerns and environmental attitudes of which it was composed. As is discussed in Chapter 1, a new environmental paradigm in recent decades has been challenging the dominant social paradigm in which continuing growth has featured so strongly.

Although the second movement failed to survive the 1930s and the outbreak of the Second World War, by the 1960s numerous environmental issues were once more coming to the forefront of public attention. Environmental pollution in areas such as the Great Lakes and the Los Angeles basin was becoming all too obvious. Growing prosperity and shortening working weeks were leading to greater demands for pleasant recreational environments in particular, and for better environmental quality in general. Material goods might be plentiful, but non-market environmental resources such as wilderness, open space, and clean air and water were clearly becoming scarcer (e.g. Andrews, 1980). Rachel Carson's The Silent Spring, describing the environmental effects of pesticides, found a massive and receptive audience in the early 1960s (Carson, 1963). In the later 1960s, the images provided by space missions highlighted in a new way the finitude of

Planet Earth. Perhaps there was a parallel with the passing of the American frontier and the first conservation movement: now the whole Earth was seen to be small, finite and limited, in the same way as the passing of the frontier had stimulated a sense of limitation three-quarters of a century earlier. Rapid population growth and fears of shortages of basic resource products further contributed to the growing disquiet about the relationship between humans and their environment. Disillusionment with the materialism of the urban-industrial lifestyles that had typified the period of post-war prosperity was probably also a contributor to the new climate of environmental opinion.

There are some similarities between this movement and the earlier ones. For example renewed government intervention in environmental resource issues took place, new legislation was enacted and new government agencies were established – not only in the United States but also in numerous other countries around the world. There are, however, also significant differences.

On previous occasions wars diverted attention away from environmental concerns. Now the Vietnam War served to strengthen the questioning of national priorities: in particular environmental warfare carried out by an industrial superpower in an agrarian setting seemed to some to symbolize a malaise that pervaded Western society.

The role of war is by no means the only difference between this and previous movements. Compared with earlier periods, the popular base was now far wider. Numerous interest groups and pressure groups burgeoned, not only in the United States but also in many other countries. In Britain, for example, the number of new environmental groups formed during the ten years from 1966 was more than twice that of any previous decade (Lowe and Goyder, 1983). During the 1970s there was much talk of exponential growth – of population and of demand for environmental resources – but one of the most spectacular forms of growth was in the memberships of environmental interest groups (Table 3.4). This growth was sustained down through the 1980s, in some cases after a slacker period during the later 1970s. Growth seems to have accelerated again during the later 1980s. With growing membership went expanding budgets, some of which are now of impressive size. In the United States, for example, the annual budgets of the Sierra Club and the Wilderness Society in 1990 were 32 and 20 million dollars, respectively (Kuzmiak, 1991). In Britain, the Royal

Table 3.4 Membership of some environmental interest groups (in thousands)

	1968	1972	1976	1980	1984	1990
Britain						
Royal Society for the Protection of Birds	41	108	204	300	340	844
National Trust*	170	346	548	1 000	1 460	2 023
Royal Society for Nature Conservation	35	75	109	140	180	250
Worldwide Fund for Nature		12		60		247
Friends of the Earth*		1		18		110
USA						
Sierra Club	68	136	165	182	348	553
National Audubon Society	66	164	269	400	450	580
National Wildlife Foundation	364	524	620	818	820	3 000

Source: McCormick (1989), Kuzmiak (1991), Central Statistical Office (1992).
Note: Data for Friends of the Earth and the Worldwide Fund for Nature are for 1971 and 1981. Asterisk indicates for England and Wales only.

Society for the Protection of Birds had an income of around £13 million in 1990 – a budget of the same order of magnitude as that of the state conservation service (i.e. the Nature Conservancy Council).

Although the memberships of environmental groups comprise only a small percentage of the total population, they are nevertheless now counted in hundreds of thousands or even millions, as Table 3.4 indicates. Environmental groups are very large in relation to most other interest groups. In Britain, for example, the Royal Society for the Protection of Birds has more than six times as many members as the National Farmers' Union, and the National Trust twenty times as many. Furthermore, environmental groups are often characterized by an influence that is disproportionate to size and budget. Many members are articulate, knowledgeable about the machinery of government, and familiar with the corridors of power. In short, the recent environmental movement has been – and is – large, powerful and influential. It is a potent force in the evolution of new policies towards the management of environmental resources. And group membership is perhaps only the tip of the iceberg: various surveys have indicated widespread support for individual 'green' positions. The fact that more than half of the American population expressed belief in the existence of limits to growth (Chapter 1) is a case in point. At the same time, however, the environmental movement is far from monolithic. Perhaps it is even misleading to describe it as a movement, since it embraces a wide range of environmental attitudes and political affiliations or inclinations (from the far Left to the far Right).

This third movement is more diverse than its predecessors. It is perhaps significant that professional resource managers such as Pinchot and Bennett were at the forefront of the two earlier movements, with their emphases on efficiency of resource use. The third movement is a much broader church: professional resource managers have generally not been its leaders, but have had to respond to it. It is significant that it is usually referred to as an environmental (rather than conservation) movement. Samuel Hays (1987) suggests that conservation gave way to environment after the Second World War as aesthetic and amenity values increased in relation to those of materials or commodities. The concerns of the movement include not only resource management for utilitarian ends but also more ecocentric issues. Its spectrum of environmental attitudes is far wider, ranging from efficient use at one end to preservation and deep ecology at the other, and from those who advocate minor reforms to existing practices and systems of resource use to the more radical who call for more revolutionary change and for completely new paradigms. A spatial dimension may exist in this pattern: for example movements inspired by deep ecology and the more radical philosophies have developed mainly in Australia, California and Scandinavia, where significant areas of wilderness are to be found (Young, 1990). At the broader scale, the movement as a whole is essentially one of the developed world, as its American origin indicates. It has, however, begun to spread to developing countries such as Indonesia (Cribb, 1988) and Malaysia (Aiken and Leigh, 1992). Indeed it is now possible to talk of a global environmental movement (McCormick, 1989). And in the same way in which the first two Amercian conservation movements gave rise to national government agencies with responsibilities for resource management, the recent period has seen the beginnings of supra-national organizations as well as national bodies. The United Nations Environment Program was launched after the major environmental conference at Stockholm in 1972, and more recently the 'Earth Summit' at Rio de Janeiro in 1992 has given rise to new international conventions on environmental issues and to a UN Commission on Sustainable Development (Chapter 12).

Growth and development in the environmental movement have not been continuous since the 1960s, but there has been no period of definite waning. The third conservation movement (if it can be called such) has survived the New Right governments of the 1980s, and many of their attempts to dismantle the regulatory frameworks established in the late 1960s and 1970s. It seems set to become a permanent or at least long-lasting factor in the climate of resource management in the years ahead.

Environmental change and environmental resources

Change through time is a common factor uniting economic activity and environmental attitudes as reflected in conservation movements. It may also affect the environment itself, and in particular climate. At the end of the twentieth century, the prospect of climatic change, in the sense of possible global warming, is a familiar one. On the geological time scale, the concept is also readily accepted: climates have fluctuated over the last few tens of thousands of years as glacial and interglacial conditions have alternated.

The notion of possible climatic change in historical times is not a new one, but for several decades around the middle of the present century it was forgotten or abandoned. More recently, the idea has emerged again, particularly in relation to areas such as northern Europe where the success or failure of cereal crops in particular can hinge on slight variations in temperature. For example Parry (1978) concluded that during the historical epoch, around 12 per cent of the British mainland area has been marginal in the sense of being cultivable only at certain times. He found that change in the altitudinal limits of cultivation in upland areas such as southern Scotland could be linked to climatic variations.

The idea of climate change and resource repercussions is also prominent in Scandinavian writings (e.g. Utterström, 1955). In particular, times of famine have been attributed to climatic deterioration, not least in the seventeenth century. Clear evidence exists of how farms established in Iceland in the eleventh century had been overrun by ice by the late seventeenth or early eighteenth century. Indeed Greenland settlements established in the eleventh century were abandoned two or three centuries later. This was after clear signs of impoverishment and of diminution in the physical size of their inhabitants, probably as a result of climatic deterioration and ensuing resource scarcity. In Iceland, economic life began to change markedly during the first half of the fourteenth century. Grain growing became less extensive, and fishing increased: the main focus of resource activity shifted from the interior to the coast. Grain imports increased, as did fish exports (Utterström, 1955). The relative roles of climatic change and political–economic changes are difficult to disentangle in this case and most other similar cases, but there is little doubt that the former was a significant factor in the changing use of environmental resources.

It is not only in the northern margins of Europe that such climatic change may have occurred. Other marginal areas such as the desert fringes are obviously also sensitive to slight climatic change or variation, with dire consequences for those relying on their environmental resources. Climatic change and its consequences have also been suggested as factors contributing to the decline of some cultures and civilizations. Climatic change has also been linked to some of the fluctuations in population that have occurred during the last millennium, in both Europe and China (e.g. Braudel, 1979).

Such hypotheses are difficult to confirm or refute, especially in the absence of reliable data on population and climatic trends. It would, however, be as unwise to rule out the possible contribution of climatic change to resource problems in the past as it would be to discount the possibility of global warming, and its resource implications, in the future. Very slight changes can be critical if they occur around threshold values for successful crop growth. Changes may not be uniform, either in space or in time. Shifting wind belts, for example, can mean greater humidity in one area and less in another. Furthermore, there is evidence that even the same type of change can occur at different times in different areas. For example the 'Little Ice Age' appears to have occurred some centuries later in Scotland and

Iceland than in the Alps (Lamb, 1977). In short, climatic change in the historical past has been a very complex process. Sweeping generalizations and hypotheses about the role of climatic change in famines and in the fortunes of humankind are probably best avoided. Equally, however, it would be imprudent to assume that climate has been constant over the last few centuries and that it will remain so in the foreseeable future.

This chapter has been intentionally wide ranging. Its aim has been to demonstrate that the spatial and historical dimensions of environmental resources are of fundamental importance in the understanding of modern resource issues, and that these two dimensions are closely inter-linked. Current patterns of resource use are no more than a snapshot in time, and we cannot even begin to understand them without an awareness of how they have evolved. An understanding of how we have got here may not fully explain current resource patterns and certainly does not in itself offer a solution to modern resource problems. Without it, however, there is little hope of either explanation or solution.

Further reading

Crosby, A.W. (1986) *Ecological imperialism 900–1900*. Cambridge: CUP.

Grove, R.H. (1992) Origins of Western environmentalism. *Scientific American*, **267**(1), 22–7.

MacCormick, J. (1989) *The global environmental movement: reclaiming paradise*. London: Belhaven.

Simmons, I.G. (1989) *Changing the face of the earth*. Oxford: Blackwell.

Turner, B.L. II *et al.* (1990) *The earth as transformed by human history*. Cambridge: CUP (especially Chapters 1–11).

Wallerstein, I. (1974) *The modern world system: 1 Capitalist agriculture and the origins of the European world-economy in the sixteenth century*. New York: Academic Press.

CHAPTER 4

The land resource: food production

The land surface of the Earth is humankind's principal resource base. The biological productivity that results from the interaction of plants, soil and solar energy provides the supplies of wood and most of the food that humankind demands. Increasingly, land is also valued for its environmental role as well as for material production. This chapter focuses on agricultural land resources, while the next two consider forests and areas protected for reasons of biodiversity and environmental conservation.

In recent centuries human use of the Earth's surface has been characterized by two major trends. These are an intensification of food production on long-occupied areas, and an extension of production into new parts of the world in which the imprint of human activity was previously faint. It is impossible to chart these trends precisely over periods of hundreds of years, and indeed it is difficult to do so even for recent decades. The estimated general trends, however, are impressive, as Table 4.1 indicates. Over the last three centuries, the cropland area has increased approximately five-fold. At the same time, the forest area has declined by nearly 20 per cent. In absolute terms, the expansion of cropland and decline of forest are similar, each amounting to around 1.2 billion ha, while the extent of grassland seems relatively steady. It does not follow, however, that all the growth in cropland has been directly at the expense of the forest, or that the area of grassland has been constant: much cropland has come from grassland (for example in the prairies) and much of the cleared forest area has become grassland. It is also apparent from the table that although the increase in cropland has been spectacular, its extent is still small in relation to the total land area. Even after three centuries of expansion it is still little more than 10 per cent of the total land area. Nevertheless, this area is of crucial importance

for humankind: it is estimated that around 92 per cent of our food, on a dry matter basis, comes from arable land compared with 7 per cent in the form of livestock products from non-arable land and 1 per cent in the the form of fish (Buringh, 1985; Buringh and Dudal, 1987).

Biological production and human impact

Areal trends constitute only one dimension of human influence on the Earth's land surface. Another important dimension is that of biological productivity. Annual net primary productivity (NPP) is the net amount of (mostly solar) energy fixed by biological processes (see Fig.7.1), or the rate of increment of plant material. NPP varies widely according to climate and type of crop or vegetation. The estimated annual global production amounts to around 225 Pg (Pg – petagram – 10^{15} g), of which approximately 132 Pg relate to land areas, 92 Pg to the ocean and the small balance to freshwater ecosystems (Vitousek et al., 1986). Humans directly use only around 7 Pg or about 3 per cent of this total, in the form of food and wood (Table 4.2).

In addition, however, large amounts are used indirectly by humans, or lost through degradation as a consequence of human activities. Direct and indirect human use and losses through human-induced changes in ecosystems may amount to as much as 39 per cent of net primary production. Very approximate though these figures may be, they do suggest very significant overall human impact, even if direct consumption accounts for a very small proportion of net primary production. The figures also provide a context for considering the carrying capacity of the Earth, though it needs to be borne in

Table 4.1 Estimated global land use 1700–1980

	Area (million ha)					Percentage change				
	1700	1850	1920	1950	1980	1700–1850	1850–1920	1920–50	1950–80	1700–1980
Forest and woodlands	6 215	5 965	5 678	5 389	5 053	−4.0	−4.8	−5.1	−6.2	−18.7
Grassland and pasture	6 860	6 837	6 748	6 780	6 788	−0.3	−1.3	+0.5	+0.1	−1.0
Croplands	265	537	913	1 170	1 501	+102.6	+70.0	+28.1	+28.3	+466.4

Source: based on Richards (1990).

Table 4.2 Humans and global net primary productivity (NPP, units of 10^{15} g organic matter per year)

Terrestrial	132.1
Marine	91.6
Fresh water	0.8
Total	224.5
Directly used by humans and domestic animals	
Plants eaten	3.0
Fish eaten	2.0
Wood	2.2
	7.2
Diverted by human activities (e.g. forest destruction during timber harvesting or clearing for cultivation)	35.4
Reduced by human activities (e.g. conversion to deserts, roads)	17.5
	60.1

Source: modified after Vitousek *et al.* (1986) and Diamond (1987).

mind that carrying capacity depends on consumption patterns and levels and on how land is managed. At a very crude level, however, the present human population can be regarded as 'appropriating' around 40 per cent of the potential net primary productivity of the Earth's land surface.

Malthus

The most basic physical requirement of humankind is food, and the adequacy of food supply has been a fundamental concern throughout human history. During the last 200 years, theoretical arguments have raged about the prospects for the adequacy of food supply. Scenes of famine relayed world-wide by the mass media have served to perpetuate the controversy.

The debate focuses on the comparative trends of demand and supply, or of population and food production. It is closely associated with the famous essays of Thomas Robert Malthus, written between 1798 and 1823. In Malthus's view, population and food supply grew in different ways. Population increased in what he called a geometric progression (i.e. 1, 2, 4, 8, 16, 32 . . .), while food production increased only in an arithmetic progression (1, 2, 3, 4, 5, 6 . . .). In other words, the former grows more rapidly than the latter, and tends to outstrip the latter. Unless humans can limit their reproduction voluntarily, through self-restraint ('preventive checks'), their numbers will be reduced by starvation, misery and disease ('positive checks').

Malthusian views have given rise to endless debate at all levels from their ideological basis to their validity in the light of historical trends and future prospects. In recent decades, 'neo-Malthusians' have revived some of these views, and the debate has also widened from its initial focus on food to apply also to other resource sectors. Malthus lived in a particular historical and political setting. He wrote at a time of great social and economic upheaval in England: rapid change was under way in both agriculture and industry. The casualties of this upheaval were numerous: displaced and destitute migrants roamed the land. Debates about fit and proper social policies towards the welfare of the poor provided the background for the essays.

Within a few decades, agricultural yields had been transformed at home, and cheap agricultural products flooded in from new lands overseas. Although population grew rapidly, food supply expanded even more so, and by the twentieth century it seemed that over-production, rather than under-production, of food was to be the insoluble problem. The gloomy prospects of the Malthusian doctrine, it seemed, were not to materialize.

Yet although it was apparently not vindicated by the trends of the nineteenth and twentieth centuries, nor was it entirely invalidated. Famines afflicted the peripheral areas of Britain, in western Ireland and north-west Scotland, in the mid-nineteenth century even as new agricultural areas were being opened up overseas and the technical advances of home-based agriculture were reaching fruition. Since then, the world has rarely been free of regional famines, and few have had the confidence (or complacency?) to reject Malthus outright. Indeed, unprecedented rates of global population growth in the second half of the twentieth century have sustained a continuing interest in Malthusian ideas.

Recent trends in population and food production

The world population reached 5 billion in 1987. In 1950 it was 2.5 billion. This doubling period of 37 years compares with the previous one – when population grew from 1.25 to 2.5 billion – of more than 100 years. Rapid growth is indeed an outstanding characteristic of world population during the second half of the twentieth century. The significance of population growth in relation to food production in particular and to environmental resources in general can scarcely be exaggerated. Phrases such as 'population explosion' and 'population bomb' highlight both the spectacular nature of growth in recent decades and the apparently alarming implications for the adequacy of food and other resources. It is important, however, to view both population trends and their implications for resources with care and caution.

Recent population trends are depicted in Table 4.3. World population probably did not reach 1 billion until around 1800. The second billion was attained by about 1930, the third by around 1960, and the fourth in 1975. Clearly, growth has accelerated spectacularly. During the first half of the twentieth century, annual growth rates averaged around 0.8 per cent. By the late 1960s and early 1970s, they were around 2 per cent. In 1970, world population grew by 2.06 per cent. By that time, growth rates in the developed world had already fallen, but those in the developing world were exceeding 2.5 per cent. Perhaps it is significant that the growth of the environmental movement and of neo-Malthusian views coincided with the peak period of population growth. With population doubling-times

Table 4.3 World population trends

Totals (billions)	1950	1960	1970	1980	1990
World	2.5	3.0	3.7	4.4	5.3
Developed world	0.83	0.94	1.05	1.14	1.21
Developing world	1.68	2.07	2.65	3.31	4.09
Annual rates of increase (%)					
World		1.8	2.0	1.9	1.7
Developed world		1.3	1.0	0.8	0.6
Developing world		2.1	2.5	2.3	2.1

Percentage increases are decennial averages.

falling to less than 35 years, it is perhaps not surprising that concern was widespread and that alarming projections were made, such that SRO (standing room only) day would be reached by about AD 2600.

By the mid-1970s, growth rates were already beginning to decline. The world rate is now around 1.7 per cent per year, while that for the developing world has fallen to 2.0 and in the developed world growth is at a rate of well under 1 per cent. Because of the age structure of the population, however, it will take many years for the population to stabilize. UN medium projections now indicate that the world population will probably rise to around 6 billion by AD 2000 and 8 billion by 2025 before stabilizing around 10 billion by the end of next century.

Differential growth rates have already resulted in large-scale changes in the relative distribution of the global population (Table 4.3). In 1950, the developing world contained approximately two-thirds of the global population, but by the mid-1980s that share had increased to three-quarters and by the end of the century will probably have increased to nearly four-fifths.

Food production

The human population has increased enormously in recent decades. How have trends in food production compared? Food production is even harder to measure than population growth. Much of it is consumed directly in the area of production, and is not easily monitored by census or trade statistics. Estimates of volumes of production therefore need to be viewed with caution.

Nevertheless, most official figures indicate that growth in food production has actually exceeded population growth in recent decades. Population may have grown rapidly, but food production has

Plate 4.1 Small-scale bean production by family labour in Durango, Mexico. (*Source:* J. Fairbairn)

Plate 4.2 Wheat production in the humid highlands of Michoacan, Mexico. The agricultural village in the background is an *ejido* community: a programme of land reform in the early twentieth century resulted in formerly private land being turned over to communes or *ejidos*. (*Source:* J. Fairbairn)

(a)   Food production   **(b)**   Food production per capita

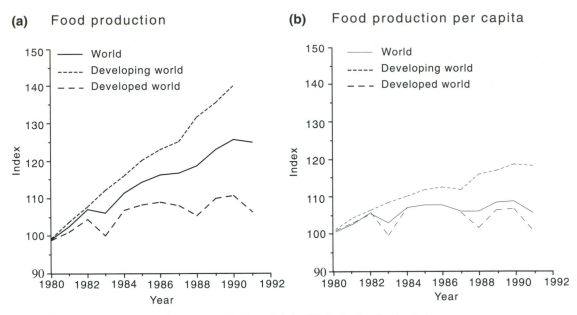

Fig. 4.1 Food production indices. (*Source:* compiled from data in FAO *Production Yearbooks.*)

apparently increased even more rapidly, and has done so in the developed and developing worlds alike (Fig.4.1(a)). Over the last two decades, the expansion of world food output has outpaced that of demand, resulting in declining prices for major food crops in international markets. Price indices for 'total food' and cereals in 1989 stood at 60 and 64 respectively, compared with bases of 100 in the period 1979–81 (WRI, 1992).

This expansion has extended to most of the major food groups, including cereals, root crops, meat, milk and fish, and has continued despite measures to curb production in parts of the developed world. The amount of food produced per capita has also increased in both developed and developing worlds (Fig.4.1(b)). At a time when world population has grown at what is probably an unprecedented rate (at least in modern times), food supply has grown even more rapidly. Does this mean that the Malthusian hypothesis is disproved for once and for all?

A moment's reflection on scenes of African famines is sufficient to deter an unqualified positive answer, and is an effective antidote to over-optimism. If food production is increasing faster than population, why do famines and food shortages still occur? And even if famines did not occur, some of the issues lying behind or below the trends might discourage complacency.

If an adequate basic diet consists of 2350 calories per person per day, then sufficient food is produced

for a population of at least 6 billion (Kates, 1992). At the beginning of the 1990s, almost 2700 calories per person per day were available, compared with 2300 in the early 1960s. In the developing world, the number of calories available per person per day rose from under 2000 in 1961–63 to nearly 2500 by 1988–90 (Table 4.4). World-wide, protein availability also increased. In absolute terms, therefore, food has become more abundant rather than scarcer, and there is not a world food shortage. Even after a period of very rapid population growth, there is, in theory, enough food to sustain everyone, at least at the level of a basic diet. Practical adequacy, however, depends on how the food is distributed: the fact that gross supply matches gross demand is of little consolation to a starving child in Africa. In practice there are major disparities between supply and demand. Mal-distribution is at the heart of the problem of adequacy of food supplies. This mal-distribution occurs both in spatial and in socio-economic dimensions, and at different scales.

Spatial imbalance

Spatial variations in the availability of food have probably existed throughout human history. Wide-spread famine and shortage were common in Europe until the present century, but today they are largely confined to parts of the developing world.

Food production per capita has increased in the

Table 4.4 Food supply in calories per capita per day

	1961–63	1969–71	1979–81	1988–90
World	2 287	2 433	2 579	2 697
Developed world	3 031	3 195	3 287	3 404
Developing world	1 940	2 117	2 324	2 473
Africa	2 155	2 211	2 315	2 348
Asia	1 888	2 086	2 302	2 494

Source: FAO (1992) *Production Yearbook 1991.*

developing world as a whole (as well as in the developed world), but it has not done so in all countries or even in all continents. Africa stands out as the major exception to the overall trend (Fig. 4.2). Over the last 20 years, food production per capita has declined both in many individual African countries and at the level of the continent as a whole. By 1990, the index of food production per capita in several African countries had fallen below 80 compared with a base of 100 ten years earlier (Table 4.5). Clearly, food production had failed to keep up with popula- tion growth in these countries, although it had forged ahead in many other parts of the developing world, especially in Asia. Numerous reasons may have contributed to this dire state of affairs, including

both human factors such as wars and environmental factors such as droughts. Whatever the reasons may be, the result for many has been hunger and starvation.

Africa may be an anomalous case, deviating as it does from the general global trends of recent decades. It serves as a reminder, nevertheless, that the Malthusian hypothesis cannot be lightly rejected, although recent trends at the global level do not support it.

Socio-economic imbalance

Even in continents other than Africa, the fact that food production has grown more rapidly than population does not necessarily mean that everyone is adequately fed. Throughout history, the pain of hunger, whether as a chronic problem or in acute form during times of famine, has been felt mainly by the poor. Indeed it is frequently asserted that the fundamental cause of hunger is the poverty of specific groups of people, rather than a general shortage of food.

Many of today's poor and hungry live in low-income countries. Several African countries, as well as other poor countries such as Bangladesh and Haiti, have per capita calorie consumptions of under 2000 per day (i.e. potential consumptions, on the basis of

(a) Food production Africa and Asia   **(b)** Food poduction per capita Africa and Asia

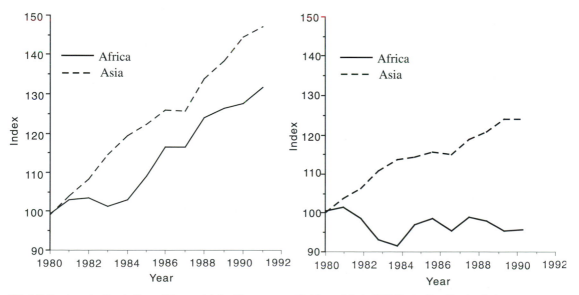

Fig. 4.2 Food production indices, Africa and Asia. (*Source:* compiled from data in FAO *Production Yearbooks.*)

Table 4.5 Indices of food production per capita 1989–91: some national examples (1979–81 = 100)

Angola	80	Brazil	134
Cameroon	81	Cambodia	145
Liberia	75	China	134
Malawi	74	India	120
Mozambique	82	Indonesia	132
Niger	71	Malaysia	160
Sudan	71	Vietnam	123

Source: extracted from FAO (1992) *State of Food and Agriculture*.

estimated food supply and population totals). Countries with per capita incomes of US $400 (1983 figures) or less accounted for about 80 per cent of the people with insufficient calorific intake to support an active working life, and about the same percentage of those obtaining too few calories to prevent stunted growth. Developing countries can be divided into various groups on the basis of per capita income. In the low-income group, 51 per cent of the population is estimated to receive insufficient calories to permit an active working life, compared with 14 per cent in the middle-income group of developing countries (WRI, 1988).

Within both low-income and other countries, the poorest bear the brunt. Grigg (1985) presents data showing that per capita calorie consumption of the lowest income group in India in the 1970s was less than one-third of that of the highest group. The corresponding proportion from a survey in England at the end of last century was one-half. In rural areas, landless households typically have lower calorific intakes than households with land: in Bangladesh, for example, the comparative figures of 1925 and 2375 calories per person per day have been quoted for the two groups (WRI, 1988). In urban areas, hunger and malnutrition may be at least twice as great among the dwellers in the slums and shanty towns as in other parts of the cities. In urban and rural areas alike in the developing world, food consumption increases with income.

The relationship between economic poverty and food poverty is difficult to analyse in quantitative terms, not least because of problems of definition. It is apparent that those who are poor in economic terms are most likely to suffer from food shortage, and that food needs cannot always be effectively expressed in terms of the (economic) demand which might encourage increases in supply. On the other hand, some commentators contend that it is possible to separate food poverty from economic poverty, quoting experiences in China, India and Sri Lanka where hunger has been reduced even in the face of acute poverty (e.g. Kates, 1992). Provision for adequate distribution is probably more significant than absolute level of income.

Hunger: increasing or diminishing?

The scale of the problem of hunger is diminishing, but the picture is complicated by conflicting definitions of hunger, and by contrasts between trends of absolute and relative numbers of people affected.

The shadow of a famous statement by the first Director of the Food and Agriculture Organization, Sir John Boyd Orr, has hung over the debate about world food supply. This statement, made in 1950, was to the effect that a lifetime of malnutrition and actual hunger was the lot of at least two-thirds of mankind. That two-thirds of humankind are hungry has become one of many myths about modern population, food and hunger (Box 4.1). That estimate would be sustainable only with unrealistically high thresholds for defining dietary adequacy. Even with more recent and more realistic definitions, however, wide discrepancies remain. Recent estimates of the numbers affected by hunger or under-nourishment, for example, range from 500 million to 1 billion (WRI, 1990).

Adequacy can be defined in various ways. Different definitions lead to different estimates of the numbers receiving inadequate diets. FAO now defines the prevalence of under-nutrition by the proportion of the population who, on average during the course of a year, do not have enough food to maintain body weight and support light activity. On the basis of this definition, the numbers suffering under-nutrition in the developing world were 940, 840 and 790 millions respectively in 1969–71, 1979–81 and 1988–90. These represent 36, 26 and 20 per cent of the respective (developing-world) populations. Using earlier published FAO surveys and definitions, David Grigg (1985) has attempted to estimate the population receiving less than minimum food energy requirements for 1948–50 and 1978–80. His estimates are, respectively, 550 million (23 per cent of the population) and 534 million (12 per cent). (See also Grigg, 1981 and 1982.)

These short-term trends are located by Grigg in their longer-term context. Over the longer term, it is

Box 4.1 World hunger: six myths

Myth 1 Food-producing resources are stretched to the limits and there is not enough food to go round.
Response: There is enough food to go around, and indeed abundance in some parts of the world.

Myth 2 Nature is to blame (famine has environmental causes such as droughts).
Response: Famines usually have human causes. Natural events may be the final blow.

Myth 3 There are too many mouths to feed (i.e. hunger is caused by rapid population growth).
Response: At the global scale there is no direct correlation between population density and hunger.

Myth 4 Pressure to feed the world's hungry is destroying the very resources needed to grow food.
Response: As in many myths, there may be an element of truth, but similar processes might continue even if the world population were half of its actual level.

Myth 5 The 'green revolution' is the answer.
Response: Technological achievements have been impressive but usually have their price: dependency on suppliers, social stresses as the poor are further marginalized.

Myth 6 Free trade and free markets are the answer.
Response: Why has hunger continued or increased in some developing countries while exports have boomed? Why do many millions in market economies have inadequate food supplies?

Source: adapted from F.M. Lappé and J. Collins (1988).

differences. By the end of that century these food supplies were generally adequate if distributed according to need, but malnutrition was probably still widespread in the developed world, and continued to be found until the 1950s and 1960s. Grigg notes that in some European countries, a time lag of up to a century existed between adequate food supplies becoming available and malnutrition being eliminated. Adequate food supplies are now available at the world scale, but malnutrition and other problems of food adequacy have certainly not been eliminated. Optimists might consider it to be a only matter of time for hunger to be banished from the world, as it has been (to a very large degree) from Europe. Pessimists, on the other hand, might interpret recent trends in population and food supply rather differently. Some may see sufficient signs of stress in the food system to make them question whether the trends of the last few decades can be expected to continue over the next half century.

Why and how has food production increased?

Our view of future prospects for the adequacy of food supply will depend on whether we can reasonably expect recent trends to continue into the future. This in turn requires a basic understanding of the factors that have led to the increases in food production that have been achieved in recent decades. Different levels of factors can be identified, including the immediate or proximate causes and the ultimate, underlying ones. Some uncertainty exists at all levels, not only about how causal factors operate individually, but more especially about how they have interacted in the past and how they will interact in the future. In general terms, however, the increases have been achieved less by expanding the area of cropland than by improving yields (as a result of the application of science and technology).

Expansion of the cropland area

Since most food originates on cropland (whether consumed in the form of cereals, roots or grain-fed livestock products), the most obvious variable affecting food production is the extent of cropland. As Table 4.1 shows, the cropland area has increased significantly, although it still occupies only little more than 10 per cent of the global land surface.

As Chapter 3 indicates, the resource base for cropland is defined by environmental factors of climate, soil and terrain. Most of the Earth's surface

clear that hunger has become increasingly localized. In the past, it was a world-wide phenomenon, whereas it is now largely restricted to some parts of the developing world. As recently as the early nineteenth century, food supplies in several countries in western Europe were below national minimum requirements. Much of the continent had insufficient food supplies for its population, regardless of income

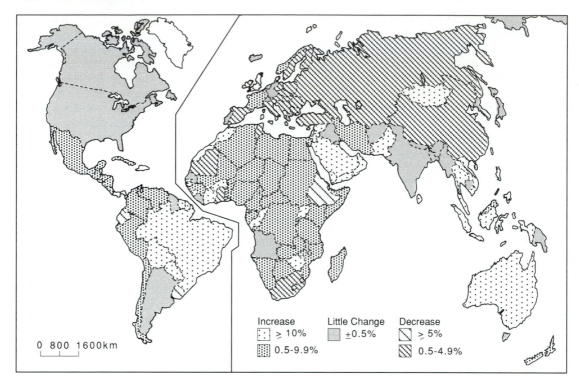

Fig. 4.3 Trends in cropland area 1989–90. (*Source:* compiled from data in FAO *Production Yearbooks*.)

is simply too cold, too dry, or too steep for crop production. Since humans acquired knowledge of agriculture, more and more of the potential cropland resource base has become an actual resource. Throughout most of human history, this process has usually been slow and discontinuous. Over the last 150 years, it has accelerated dramatically. This period has witnessed the opening up of huge areas of cropland, as new areas have been incorporated into the world economy and as the development of railways and steamships allowed the products of cropland to be transported over thousands of miles (Chapter 3). Between 1850 and 1920, for example, the cropland area more than trebled in North America as the Prairies were opened up, and more than doubled in Russia. More recently, increases of over 50 per cent between 1950 and 1980 have been recorded from Latin America, Africa and South and South-East Asia (Richards, 1990).

During the second half of the present century, most of the continuing expansion of the cropland area has been in the developing world. In much of the developed world the extent of cropland has declined (Fig.4.3). In some developing countries substantial

increases have occurred, but two significant points are worth emphasizing.

Firstly, the rate of cropland expansion is much less than that of population growth, both at the global scale and at the level of most individual countries. At the global scale, cropland has expanded at roughly one-tenth of the rate of population. Between 1980 and 1990, for example, the cropland area increased by 1.9 per cent while population grew by 19 per cent. Improvements in food production per capita cannot therefore have come primarily from expansion of the cropland area. This fact is clearly reflected in the second point: the area of cropland per head of population has been shrinking for many years. Whereas it was approximately 0.5 ha per capita at mid-century, it now stands at little more than half of that figure (0.28 ha in 1989). In other words, the average area of cropland per capita measures little more than 50 m by 50 m. Furthermore, in some parts of the world it is much less: the average for Asia is 0.15 ha, and in China as well as in several Middle Eastern countries it falls to less than 0.10 ha.

Cropland will probably continue to expand over the next few decades, but arguably at slower rates

Plate 4.3 Small-scale, unplanned agricultural colonization of the forest, Puerto Rico. (*Source:* K. Chapman)

Table 4.6 Productivity of actual and potential cropland (million ha, rounded)

	Productivity			
	High	Medium	Low	Total
Actual area	400	500	600	1 500
Potential area	450	900	1 950	3 300

Source: based on data in Buringh and Dudal (1987) and Pierce (1990).
Note: No allowance is made for land lost to urbanization.

than over the last 150 years. Few parts of the world today offer the same potential for the expansion of cropping that the prairies, pampas and steppes did a century ago. Estimates of the potential cropland area depend on definitions and assumptions, and hence vary. Most, however, lie between 3000 and 3500 million ha, or approximately twice the current cropland area and under one-quarter of the total land area (e.g. Buringh and Dudal, 1987). Whereas over 80 per cent of the actual cropland area in Europe and Asia is already in use, the proportion is under 25 per cent in Africa, Latin America and Australasia. Potential cropland is available in some parts of the world where population is growing rapidly, for example in Africa and Latin America, but not in others. There is little scope for expansion in Asia, which contains three-quarters of the world's population.

The productivity of cropland varies according to its environmental or biophysical characteristics (as well as with its management). Table 4.6 shows one classification of actual and potential cropland on the basis of its productivity. It suggests that most of the scope for expansion is in the low productivity class: over 75 per cent of the 'potential' area is of low productivity. In other words, most of the more productive land is already in use. Cities have usually grown up in areas of more fertile land, and hence urban expansion is likely to affect above-average agricultural land. The estimate for potential high-productivity cropland does not take account of this serious loss: Buringh and Dudal (1987) suggest that its area may decline from 400 million ha in 1975 to 345 million ha in 2000.

In the late nineteenth and early twentieth centuries, expansion of cropland was the main source of increased food production. Since around 1950, the contribution from expansion of cropland has dwindled. It has been estimated that areal increases have contributed only 15 per cent to growth in world cereal output, for example, compared with 85 per cent from yield increases (Grigg, 1985). In the 20 years to the mid-1980s, the area of cereals harvested rose by 6 per cent, while yields increased by 66 per cent (Tarrant, 1987).

Drainage and irrigation: pressing against the limits?

Over the centuries, humans have sought to farm areas that under natural conditions are too wet or too dry to permit cultivation, as well as to improve yields on existing cropland areas by managing moisture conditions. In The Netherlands and eastern England, for example, large areas of wetland along some of the rivers and parts of the coast have been converted into fertile cropland. This process has now largely ceased in north-west Europe, and indeed some drained areas are now reverting back to wetlands to be used for purposes of nature conservation or recreation. At the global scale, the value of wetlands as an environmental resource has appreciated, and large-scale drainage schemes now encounter formidable opposition from environmental interests.

Irrigation has been practised in some parts of the world for many centuries, but its use has increased greatly during the second half of the twentieth century. In 1900 the irrigated area was less than 50 million ha. It had approximately doubled by 1950, and had grown to 170 million ha by 1970. It is now around 250 million ha. While the cropland area as a whole has grown only slowly, the rate of increase of the irrigated area exceeds that of world population. Perhaps this rapid growth can be interpreted as a symptom of stress in the world food-supply system: certainly the significance of irrigation is not in question. One-sixth of the cultivated area is irrigated, but that area produces one-third of the world's food (Tolba and El-Kholy 1992).

An increasing dependence on irrigation has some worrying aspects. Irrigation is obviously most likely to be necessary or desirable in areas of low rainfall and therefore of potential water shortage. Problems of water supply are aggravated by the fact that water used for irrigation cannot be recycled, unlike much of the water used for domestic and industrial purposes (Chapter 9). A further problem is that political tension and conflict can develop around the management of major trans-boundary river systems. If irrigation is provided from groundwater, there may be a risk that these sources are depleted, and that the systems are not sustainable. Furthermore, serious problems of salinization of soils and declines in crop yields have been encountered in some irrigated areas.

Yields

On both irrigated and rain-fed areas, crop yields have increased dramatically in recent decades as capital inputs into agriculture have risen enormously. These inputs take various forms, including machinery, fertilizers, pesticides and new varieties of seeds.

Ester Boserup (1983) considers that because it was believed that land reserves were small, research in agriculture has focused during the last half century on raising crop yields by means of purchased inputs, such as fertilizers. A factor common to many of these inputs is energy in the form of fossil fuels and their derivatives. Energy consumption is a general summary indicator. Huge variations in energy consumption in agriculture exist between the developed and developing worlds, and within each of these regions. Overall, however, the trend has been steeply upwards.

Commercial energy use in agriculture in the developing world is only about one-quarter of that in the developed world, but is rapidly increasing. Buringh and Dudal (1987) suggest that it may rise from 36 million tonnes oil equivalent (mtoe) in 1980 to 178 mtoe in 2000. In the developed world (OECD countries) the level of energy consumption was already high by 1970. Between then and 1990, it further rose by around 40 per cent. A major factor in this increase was the substitution of machines for manual labour: during that same period, for every square kilometre of agricultural land, one worker was replaced by two machines (OECD, 1991a). In the world as a whole, the number of tractors increased by over 20 per cent between 1980 and 1990.

Fertilizers

The use of chemical fertilizers to supply plant nutrients (mainly nitrogen, phosphorus and potassium) for crop growth is now an integral feature of agriculture. Between 1970 and 1990, world consumption of chemical fertilizers more than doubled, and consumption in the developing world increased more than three-fold. In the North, the level of use per hectare has been levelling off in recent years, but it is rapidly increasing in the South. On average, however, roughly twice as much fertilizer per hectare is still used in the developed world as in the developing world.

At the end of last century, concern was expressed that supplies of nitrogenous fertilizers, which then originated mainly in guano deposits on the west coast of South America, were limited, and hence that agricultural systems based on them were not sustainable. This limit was removed just before the First World War when a chemical process was developed for producing synthetic fertilizer using atmospheric nitrogen. Nevertheless, some concern still exists over the long-term availability of the oil and gas used in that process (Chapter 7), and of phosphates from mineral deposits.

The side-effects of nitrates from fertilizers leached into watercourses can give rise to several problems. Drinking water may be contaminated and fish and other aquatic life harmed. The concern with which this issue is viewed in some European countries is reflected in the curbs that have recently been introduced on the use of fertilizers in some areas in countries such as England.

Pesticides

The rapidly increasing use of chemicals to control pests has also contributed to improved yields in recent decades. The main groups of pesticides are herbicides (weed killers), insecticides and fungicides. The relative use of these groups varies from area to area, but the common feature is rapid expansion. This is more easily charted in terms of value of sales than in terms of volume or tonnage. In real terms, these increased three-fold between the mid-1970s and 1990. As in the case of fertilizers, the rate of increase has been greater in the developing world (7–8 per cent per year) than in the developed world (2–4 per cent), but the latter still accounts for more than three-quarters of the global use of pesticides (Tolba and El-Kholy, 1992).

Again, undesirable side-effects are encountered. A large proportion of the pesticide used does not reach the target pest. Natural predators of potential pests may be unintentionally eliminated, leading to further outbreaks and further use of pesticides. An ever-increasing number of pest species have developed resistance to pesticides, posing serious problems for control. And it is perhaps worrying to find that the relative importance of crop losses to pests has not decreased over the last 50 years (Brader, 1987). Furthermore, pesticides and their residues can give rise to other environmental problems by contaminating surface or groundwater and in some cases directly or indirectly harming the health of farm workers.

High-yield varieties

A third contribution to yield increases has come from plant breeding. This has been a major factor in both the developed and developing worlds. In the former, new varieties of wheat and maize have led to big increases in yields in countries such as Britain and the United States. In the developing world, much attention has been focused on the use of high-yield varieties (HYVs) of cereals developed during the 1960s and 1970s in response to the problem of feeding a rapidly expanding world population. These varieties, associated with the 'Green Revolution', were rapidly adopted in many of the countries where there was stress in the food supply system. By the early 1980s more than half of the wheat and rice areas of the developing world were planted with HYVs. Some apparently spectacular successes have been achieved: for example wheat production in India doubled over a

---

Box 4.2 The expansion of rice production in Indonesia

In the 1970s, Indonesia was the world's largest rice importer. Rice yields were low, population growth rates were high and millions dependent on agriculture lived in absolute poverty. In response to these problems, the government implemented agricultural policies designed to increase production and to improve rural incomes. Rice production grew by 6 per cent per year during the 1980s. By the mid-1980s, the country had become self-sufficient in rice and has remained so except in occasional drought years.

Rice intensification projects provided irrigation, fertilizer, pesticides, HYV seeds, credit, and technical assistance and extension services. Irrigation alone is estimated to have accounted for 50 per cent of the increase in rice production achieved since the mid-1970s. Irrigation water, fertilizer and pesticides were all heavily subsidized, and fertilizer applications, for example, increased by as much as 500 per cent in some areas during the 1980s.

When rice self-sufficiency was achieved, government attention began to turn towards environmental and resource-efficiency issues. The dramatic increase in rice production had been 'bought' at a heavy environmental cost. It was accompanied by high rates of soil erosion and by water pollution from fertilizer and pesticide runoff. An outbreak of a pesticide-resistant strain of insect pest had resulted in large losses in rice production by the second half of the 1980s. As a result of these issues, concern began to swing away from simply maximizing rice production to mitigating the adverse environmental impacts of 'industrialized' production, and to ensuring a greater degree of sustainability. In particular, the mismanagement of subsidized irrigation water, fertilizers and pesticides attracted attention. Heavily subsidized inputs often meant careless use. Water charges levied on farmers, for example, were only 20 per cent or less of the real cost. Pesticide subsidies were phased out in 1989, after having been reduced to 40 per cent of the full retail price in 1986–87 compared with 80 per cent just three years earlier. A three-year programme for eliminating fertilizer subsidies began in 1991. An integrated pest management programme provided farmers with innovative technology which allowed them to maintain yields with smaller inputs of pesticides and fertilizer. The production of other food crops is also being encouraged, in order to reduce dependence on rice and to reduce the risks of monocultures.

Sources: FAO (1992); WRI (1988).

six-year period. These achievements, however, have not been secured without costs. The new varieties have often supplanted native varieties that were resistant to pests and diseases. The narrowing of the genetic base has been accompanied by increasing use of pesticides. HYVs also usually require more fertilizers and often more irrigation and machinery: they are part of a technological package that can become a technological treadmill as more and more inputs are required to maintain yield increases. Furthermore, the technological package is likely to be less affordable on small or marginal holdings and by farmers with poor access to credit. In other words, there are socio-economic problems as well as problems of sustainability of production and of environmental side-effects. Attempts by governments to alleviate some of these problems, for example by subsidizing the use of fertilizers and pesticides, can exacerbate some of the other problems, for example by encouraging heavy or wasteful use of pesticides (Box 4.2).

Overall, yield increases have resulted primarily from the industrialization of agriculture and from increasing inputs of fossil-fuel energy. A growing reliance on purchased inputs in the form of fertilizers, pesticides and new technology has spread from the developed to the developing world. In a sense, the great increase in food production over the last few decades represents the conversion of oil and natural gas, via land, to cereals: fossil-fuel energy, as well as solar energy, has been a key ingredient. In cereals, average yields rose by approximately 66 per cent between the mid-1960s and mid-1980s, and by as much as 80 per cent in Asia. During the 1980s, the decennial increase was around 20 per cent for the world as a whole, and 33 per cent in Asia (WRI, 1992). For root and tuber crops, the increases were smaller but still substantial at approximately 25 per cent for the world as a whole between the mid-1960s and mid-1980s, and over 50 per cent in Asia.

Yields have increased under various forms of resource regimes. An (unweighted) average annual rate of growth in 12 countries with communal tenure is quoted at 2.7 per cent by Griffin (1987), compared with 2.5 and 1.5 per cent respectively for 'other low-income countries' and 'industrial market economies', respectively. They have also increased almost everywhere, and have done so at an accelerating rate since the mid-twentieth century. In England, wheat yield in the 1980s was three times that of the 1930s, five times that of 1800 and ten times that of 1300 (Grigg, 1992).

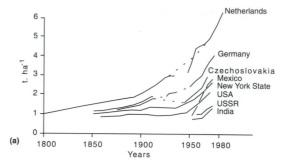

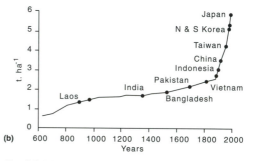

Fig. 4.4 (a) trends in wheat yields; (b) trends in rice yields in Japan. Average (mid-1980s) rice yields in other countries are superimposed on the trend curve for Japan. (*Source:* based on Grigg, 1987.)

In Japan, the trend of rice yield has followed a similar course. Yet in many rice-growing countries, current yields are still at the levels achieved in Japan some centuries ago, and in many developing countries wheat yields remain low compared with those achieved in countries such as England and The Netherlands (Fig.4.4).

The rate of increase in yields, and hence in total food production, is now slowing down in some countries and perhaps also in the world as a whole. For example, annual increases in food production averaged 3.7, 2.5 and 2.1 per cent respectively in the 1960s, 1970s and first half of the 1980s (Crosson and Rosenberg, 1989). Furthermore, there is some evidence that variability in yields is increasing, in both space and time (Tarrant, 1987). The significance of such variability, in the context of steady growth of population, is obvious. The gap between high- and low-yield countries has widened over the last half century, and there is little likelihood that convergence will occur in the foreseeable future. Nevertheless, if the yields currently obtained in countries such as Japan and the Netherlands were achieved worldwide, total production would be increased enormously. At least this would be so if the production systems were sustainable.

A crisis in agriculture?

Impressive as the technical achievements of modern agriculture are, disquiet is felt by many at increasing reliance on 'industrial' agriculture, and at the export of that model of agriculture from the developed to the developing world. In particular, doubts exist about the long-term sustainability of high levels of production in some areas. These doubts exist in a variety of forms, including, for example, concern about the emergence of resistance to pesticides and about contamination of water bodies by fertilizers and pesticide residues. They are also expressed in concern about the possibility that returns to the industrialization are diminishing. For example, in the United States, each additional input of 1 million tonnes of fertilizer yielded an additional 10 million tonnes of grain in the 1950s but only 5.8 million tonnes in the 1970s (Hays, 1987). The transfer of fertile agricultural land to urban uses can also lead to a loss of production. This issue now attracts less attention in countries such as Britain and the United States than it did ten or twenty years ago (as surplus production in these lands has become an increasing problem). On the other hand, the loss of land around rapidly growing cities is an urgent problem in parts of the developing world. Perhaps the most serious issue, however, is land degradation through processes such as soil erosion.

Land degradation in various guises affects much of the world's agricultural land, including rangeland used for grazing as well as both rain-fed and irrigated cropland. It affects both the developed and the developing world, and in some areas (such as the Mediterranean Basin) it has a long history dating back at least as far as classical times. Controversy exists as to the nature of its causes and of possible remedies. Some see it as the inevitable consequence of certain political systems and modes of production: for example Karl Marx blamed capitalism, considering that progress in capitalistic agriculture depended on 'robbing' the soil (as well as the workers). In reality, however, land degradation is associated with a wide variety of political and geographical settings, and simple, single causes can rarely be identified (e.g. Blaikie and Brookfield, 1987).

Part of the problem is that it is often difficult to separate the immediate causes from the underlying ones, and to distinguish between purely technical issues (such as cultivation practices) and the wider economic, political and social aspects of their setting. For example, cultivation of steep and potentially unstable slopes may be the immediate cause of land degradation, but underlying issues of population growth and inequitable distribution of land can mean that the cultivators have no alternative but to make use of these slopes. Another reason for difficulty in understanding the causes of land degradation is that its extents and trends are not easy to quantify, and hence are not easy to evaluate and to explain.

Cultivation and cropping almost invariably result in a major increase (by a factor of 10 or 100) in the rate of soil erosion compared with that under natural vegetation. With increased soil loss and only a slow rate of soil formation, the soil becomes a wasting asset. Complete conservation of soil under cultivation is almost impossible, in the sense of soil losses not exceeding soil gains through natural weathering processes. The rate of soil loss, however, depends on terrain, soil types and climatic conditions as well as on the nature of cropping and cultivation practices. Its significance depends on the initial depth of the soil and on whether the plant nutrients lost through erosion can be replaced by chemical fertilizers. To some extent the cost of the process can be measured by the cost of replacing the lost nutrients, but on thin soils the problem of irreversibility may be encountered. Eventually, a stage may be reached at which continued cropping becomes impossible. World-wide, some 5–7 million ha of arable land are estimated to be lost annually through soil degradation and a further 1–5 million ha because of waterlogging and salinization (FAO, 1991). Land degradation through cropping is a particular problem in many of the world's dryland areas, where multiple resource problems are often encountered.

Degradation on rangelands used for grazing is even more extensive, in both developed and developing worlds. It is at least partly due to the increases in numbers of livestock that have occurred in recent decades. In Africa, for example, cattle numbers rose by 12 per cent between the mid-1970s and mid-1980s, and those of sheep and goats by 16 per cent. Open rangeland was the stage for Hardin's 'tragedy of the commons' (Chapter 2), but the process is also to be found on privately owned land. It is widespread on government land leased to private graziers, especially in Australia and in the United States. In the American

West, the juxtaposition of different tenurial systems allows their effects to be compared. Some studies have suggested that range condition is better on private land than on some federal land, but not by statistically significant amounts. On the other hand, there have been reports of big differences in range condition between areas of federal land managed by different federal agencies (Gardner, 1991).

In the drylands, nearly three-quarters of the rangeland is estimated to have suffered degradation of its vegetation, and on one-quarter of the affected area there is also soil degradation, mainly through soil erosion. Annual losses of rangelands may amount to 5 million ha. Overall, the current direct economic loss (in terms of income foregone) resulting from degradation of drylands is estimated at US $42 billion per year (Tolba and El-Kholy, 1992).

In much of the developed world, the effects of land degradation have been masked by the increasing use of fertilizers: potential losses of output have been avoided, at a cost, by supplying additional inputs. Since the Dust Bowl episode in the 1930s (Chapter 3) attempts have been made almost continuously to curb erosion rates on cropland in America. These attempts have met with only limited physical success, but recent estimates of the effects of soil erosion on crop yields (2–3 per cent reductions over 50 years and 5–10 per cent over a century) suggest that some of the 1930s statements of alarm or concern may have turned out to be excessive (Crosson, 1991). Whether similar levels of success will be achieved in curbing soil erosion and other degradational processes in other parts of the world where 'industrial' agriculture has been introduced remains to be seen, as does the significance of these degradational processes for food production over the long term.

By its very nature, agriculture must have a significant impact on the environment, but the nature of its effects depends on the form of technology and land management. The environmental effects of agriculture were important issues in the conservation movement of the 1930s (in the form of soil erosion), and in the early growth of the environmental movement of the 1960s (in the form of the side-effects of farm chemicals and *Silent Spring*) (Chapter 3). They can affect both the productivity of land in the future, and the quality of the environment at present. As we approach the end of the century, growing concern about the relationship between agriculture and the environment is contributing towards a questioning of what has been the conventional wisdom over the last half century.

Various factors have coincided in the latter part of the twentieth century to cast doubt on the sustainability of the modes of agricultural development that have characterized the developed world (and increasingly, also the developing world) since the middle of the twentieth century. These factors include not only an increasing awareness of energy dependence and of land degradation and other environmental side-effects of 'industrial' agriculture, but also serious political and economic problems. In most of the developed world and in many developing countries, agricultural policies have come under increasingly critical review as their weaknesses and flaws have become all too apparent. Agriculture has been an area in which there has been extensive government failure, as well as market failure. Clear signs are now emerging of major policy crises, in the literal sense of turning points.

Government intervention

Throughout the developed world, governments have intervened in agriculture, in a number of ways and for a variety of reasons. One major reason has been concern with food security and stability. Another has been concern with rural welfare.

Under purely market systems, increasing affluence in the richer countries tends to bypass farmers. As the population becomes richer, additional income is spent on consumer goods other than food. Farmers therefore tend to receive a declining share of expenditure, and at the same time have to purchase inputs from other sectors of the economy. They therefore suffer a squeeze between stagnating demand and prices on the one hand, and rising costs on the other. One solution is for farmers and farm workers to leave the land, so that the agricultural cake can be divided into fewer and larger slices. They have done so in their millions during the present century. Rapid rural depopulation, however, has been seen by many governments as undesirable, and they have therefore intervened in order to bolster farm prices for social reasons as well as to stimulate high production for reasons of food security.

Government intervention has been a feature of agriculture in much of the developed world since the Depression of the 1930s, particularly in the United States and subsequently in Europe and, especially, in Japan. It has taken various forms: direct price support, protection against imports, and incentives to modernize and intensify. Another important aspect has been government funding of research and development, which especially in Europe and Japan was for many years directed at the high-input forms of agriculture deemed necessary to increase food self-sufficiency in densely populated lands.

Having once intervened, governments have found it exceedingly difficult to disengage, and have been beset by two inter-related problems. One is the cost of financial support to agriculture, in essence resulting from prices being set above world levels, and the other is the problem of surpluses of grain, beef, butter and other products. The relationship between real demand, supply and price has, in practice, been suspended. For many years farmers in the European Community in effect produced for a guaranteed (government-financed) market, rather than to satisfy economic demand for food. In Europe and Japan, the traumas of wartime and early post-war food shortages, apparently set in an increasingly hungry world, doubtlessly encouraged drives towards increasing self-sufficiency in addition to concern for rural welfare. By the mid-1980s, however, the surpluses of food produced by intensive 'industrial' agriculture were becoming a political and economic embarrassment. The high-cost food produced by these means could not be cleared on the world market, and lower-cost producers in other parts of the world were excluded (to varying degrees) by tariff barriers and other measures designed to protect domestic agriculture in the developed countries. Since the mid-1980s, these problems have at least been addressed, if not solved. The bizarre situation has developed whereby food surpluses and incentives to take cropland out of production exist in some parts of the world, while in others hunger is widespread and desperate attempts are made to increase food production.

It is not only in the developed world that government action has had profound effects on agriculture. Important indirect effects have resulted from government development policies that have favoured industrial growth rather than the agricultural and rural sectors. In many African countries, for example, policies of providing cheap food for urban populations have meant low prices and poor incentives for rural farmers to increase production (e.g. Schiff and Valdes, 1990; Lemma and Malaska, 1989). Similarly, policies of subsidizing the use of fertilizers and pesticides may encourage wasteful use as well as hastening dependence on 'industrial' forms of farming.

Policies towards agriculture and food are complex and tangled. There are clear signs, however, that what

Plate 4.4 Prairie wheatland and grain elevators in Alberta, Canada. North American wheatlands are the major world source of food exports and food aid. (*Source:* K. Chapman)

has been the conventional wisdom of the last few decades is now being questioned. The disadvantages and dangers of the industrial model of agriculture, in the developed and developing worlds alike, are now being recognized. Signs of shifts towards lower-input systems are appearing, and a convergence and integration of agricultural and environmental policies is now becoming a more real prospect. Another feature of this apparent time of transition is the prospect of trade restrictions being relaxed, and of the opening of currently protected markets for agricultural products.

Trade and aid

One of the most striking features of agriculture has been the rapid increase in the volume of world trade since the Second World War. Compared with 1950, trade volume doubled by 1962, tripled by 1972 and quadrupled by 1980 (WRI, 1986). Whereas world trade represented around 10 per cent of agricultural production in the early 1960s, it had risen to around 14 per cent (of the larger output) by the mid-1980s. The rate of growth, however, slowed down during the 1980s, when many potential importing countries experienced a combination of debt problems, slow economic growth and increased domestic production of food (Alexandratos, 1988).

International trade is a key component in the world food-supply system. Numerous countries in the developing world have become heavily dependent on it, and its growth and structure can be interpreted as a symptom of stress in the system. On the other hand, some aspects of trade reflect growing affluence and demand for better diets, and some also simply reflect the operation of comparative advantage whereby some countries, perhaps for reasons of environment, can produce more cheaply than others. The operation of this principle, however, is complicated by a number of factors arising from government intervention. These include the subsidized production of food in many developed countries and the subsidized export of surplus food from these countries, and trade barriers or restrictions of various kind erected by governments to protect their domestic producers. If trade were fully liberalized and agriculture became more strongly oriented to the world market (compared with domestic markets), then comparative advantage would play a stronger part in determining the pattern of world agriculture.

In other words, regions favoured by environment and by other factors such as ease of transport would emerge as the main areas of production. Full liberalization, however, is unlikely to occur. Many countries are likely to continue to maintain high levels of self-sufficiency in basic foods, even if cheaper food were available to import. The fact that the availability of food (and especially of imports) will continue to depend on policies pursued in a few parts of the developed world, such as the United States and the European Community strengthens this tendency.

Around 200 million tonnes of cereals, or approximately one-tenth of total production, now enter international trade each year. At the continental scale, Africa is by far the major importer (Fig.4.5). Cereal imports into Africa more than trebled between the mid-1970s and mid-1980s. The other main group of fast-growing importing countries is in the Middle East (south-west Asia), where much of the grain is converted into the livestock products demanded by an increasingly affluent population. In the rest of Asia by far the biggest importer is Japan. On the other hand imports have declined in several other Asian countries such as India, Pakistan and Vietnam, although they have rapidly growing populations. Much of the developing world, however, now depends on imports of basic foodstuffs, even if specialized tropical crops are at the same time exported to the developed world. By far the largest exporter of cereals is the United States, which accounts for more than half of the volume of grain entering international trade. The European Community is the second major exporter. Many of its constituent countries were for long major importers, but as has been indicated, production has increased rapidly in reponse to price support and other government incentives. Of individual countries, France, Canada and Australia are the next biggest exporters, usually in that order.

Some countries can neither grow nor afford to import enough food to feed their populations. They then have to depend on food given by other countries in the form of aid. By the mid-1980s, volumes of food aid were exceeding 10 million tonnes of grain annually. Again, the United States was by then by far the biggest donor, and most of the receiving countries were in Africa (Fig.4.6).

While emergency food aid during periods of crop failure and famine is almost universally supported, long-term aid may be less desirable. Free or low-cost food can drive down food prices in the receiving countries. Farmers therefore have no price incentive

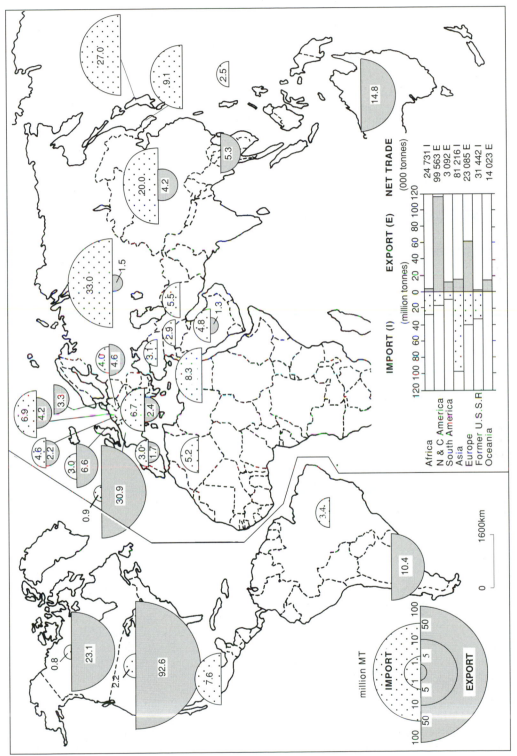

Fig. 4.5 Patterns of world trade in cereals (1990). (Source: compiled from data in FAO *Production Yearbooks*.)

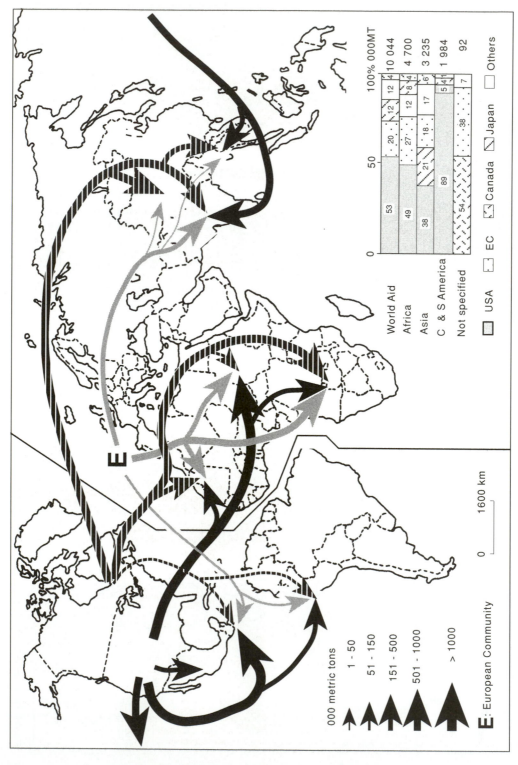

	0			50				100% 000MT
World Aid	53		20		12	12	4	10 044
Africa	49		27		12	8	4	4 700
Asia	38	21	18	17		6		3 235
C & S America	89				5	41	5	1 984
Not specified	54		38			7		92

USA   EC   Canada   Japan   Others

000 metric tons

1 - 50
51 - 150
151 - 500
501 - 1000
> 1000

0   1600 km

E: European Community

Fig. 4.6 Patterns of food aid (cereals) (1987–89). (Source: compiled from data in FAO Production Yearbooks and WRI, 1992.)

to increase production, and food aid can quickly become an essential part of the national food-supply system. There are also problems in ensuring that the aid is distributed to those in greatest need, who often live in remote rural areas rather than in the urban gateways through which the food arrives. Furthermore, there is the danger that food-aid donations are seen simply as a means of getting rid of surpluses that cannot otherwise be sold, and that they are offered in return for political support. Finally, the fact that most food aid originates on the American prairies is disquieting, especially at a time where major climatic change, with its threat to crop yields, is seen as a distinct possibility.

Trends in food production

Several general trends have already been identified, such as increasing intensities of food production, increasing environmental side-effects, and expanding international trade in food. There are also two other important trends.

One is the shift away from producing for home consumption and towards the production of cash crops, for urban markets or export. Commercial agriculture in much of the developing world has tended to develop on the Western, industrial model, rather than in more labour-intensive ways. The bigger and more prosperous farmers have benefited most: often their smaller counterparts have been displaced completely as the bigger units have expanded. Scope for employment on the part of the growing rural population is limited, and there is often little alternative to migration to the city, or to attempts at subsistence farming on marginal and potentially fragile areas where commercial production has not become established. The result may be land degradation (both from intensified production in the commercial areas, and from subsistence farming on marginal land), as well as the loss of forests (Chapter 5).

A second general trend is the increasing use of crops for animal feed. By 1990, 38 per cent of the world's grain was fed to livestock, compared with 33 per cent (of a much smaller output) in 1970 (WRI, 1992). These overall figures conceal huge variations, from 70 per cent in the United States to under 5 per cent in much of Africa and parts of Asia. Nevertheless, the trend operates almost throughout the world, except in some of the richest developed countries where the percentage was already very

high by 1970. It reflects the increasing demand for livestock products that characterizes populations whose incomes are rising, and the fact that the trend has been possible without large increases in hunger and starvation may suggest that severe stress has not been felt in the world food-supply system. On the other hand, consumption of livestock products is ecologically less efficient than direct consumption of cereals. Thus other things such as yield levels being equal, larger areas of land are required to provide a diet containing livestock products than one consisting of cereals and vegetables. As has been indicated, the area of cropland per person has been declining at the same time as demands on it have been increasing.

Recent trends and the question of carrying capacity

By the 1980s, sufficient food was produced to provide a basic diet for a population of at least 6 billion. If, however, the diet were to be somewhat improved beyond that basic level, to a level similar to that consumed in South America, sufficient food would be produced for only 4 billion. The food shortage would be much greater at the dietary level enjoyed in many developed countries: there would be sufficient food for only 2.5 billion or less than half of the current population (Kates, 1992). Food adequacy and the Earth's human carrying capacity therefore depend on the criteria and assumptions employed. Wildly differing estimates of carrying capacities have been made over the last 20–30 years: at one extreme, capacities were estimated at less than present population levels, and if these estimates had been correct, famine would have been widespread by the mid-1970s. At the other extreme are estimates of 40 billion or more.

The definition of the ultimate human carrying capacity of the Earth is a tantalizing goal. It is probably unattainable in any meaningful form, even if the focus is solely on food requirements and other human needs are ignored. Much depends on the assumptions about dietary adequacy, about the availability of inputs such as water, fertilizers and chemicals, about future advances in agricultural technology, and about the role of trade. One of the more sober analyses was carried out under the auspices of FAO (1984). It focuses on the developing world, and is based on the idea of the 'agro-ecological' zones, whose definition rests on information on soil and climate together with crop require-

Table 4.7 Potential population-supporting capacities in developing world

Input levels	Africa	SW Asia	S America	C America	SE Asia	Overall
×1975 populations						
Low	3.0	0.8	5.9	1.6	1.1	2.0
Intermediate	11.6	1.3	23.9	4.2	3.0	6.9
High	33.9	2.0	57.2	11.5	5.1	16.6
×2000 populations						
Low	1.6	0.7	3.5	1.4	1.3	1.6
Intermediate	5.8	0.9	13.3	2.6	2.3	4.2
High	16.5	1.2	31.5	6.0	3.3	9.3

Source: FAO (1984) *Land, food and people.*
Numbers of times populations could be increased. For example, in Africa the estimated 2000 population could be increased by a factor of 1.6 assuming low inputs into food production, and by 16.5 assuming high inputs.

ments. This estimate of the resource base is used in conjunction with different assumptions about the availability of inputs, and the ensuing estimates are then compared with population trends to indicate potential capacities. The work is summarized in Table 4.7: assuming low inputs the potentially supportable population is twice that of 1975 and 1.6 times that projected for 2000. With assumptions of high inputs, the factors are, respectively, 16.6 and 9.3.

The study included 117 developing countries. Of these, 52 could feed their projected year 2000 populations from potentially cultivable land within their borders. With intermediate and high inputs respectively, 28 and 19 others could also do so. The remaining countries could not feed their projected populations even under assumptions of high inputs (Buringh and Dudal, 1987).

Table 4.7 indicates that the potentially supportable population (relative to current populations) in South America, for example, is much greater than in South-East Asia. It also indicates that with low inputs, the supportable population is already less than the actual population in South-West Asia. It is clear, however, that the most acute problems of food supply are not currently suffered in that part of the world, as Middle East countries can generally afford to import food. In other words, it is almost impossible to incorporate the role of trade into the modelling exercise.

Further complications arise from the possibility of climatic change. Global warming resulting from increased CO_2 levels could have a major effect on North American agriculture in particular. It has been estimated that an increase in temperature of 1°C could lead to a reduction of 2 per cent in corn yields in the United States. Another estimate suggests that an increase of 1.5°C combined with a 10 per cent reduction in precipitation could reduce Midwest corn yield by 10 per cent (Pierce and Furuseth, 1986). A replication of the weather conditions of the period from 1933 to 1936, which was characterized by warm, dry summers, could lead to a 27 per cent reduction in US output of corn, wheat, sorghum and soybeans, and a 47 per cent reduction in Canadian wheat (compared with mid-1970s levels) (Pierce and Furuseth, 1986). In the light of the importance of the United States and Canada as grain exporters, the implications for global carrying capacity are obvious. On the other hand, Rosenzweig and Parry (1994) conclude that a doubling of atmospheric CO_2 will lead to only a small decrease in global crop production, but that developing countries are likely to bear the brunt of the problem. At best, the outlook is uncertain.

The world is probably less short of food today than at any time in history. The occurrence of famine has been greatly reduced, not just because of technical advances in food production but also because of better systems for transporting and distributing food. Famine is now almost unknown in Europe, where it was widespread half a millennium ago and where population may have pressed against the capacity for food production until the seventeenth and eighteenth centuries (e.g. Abel, 1980; Rotberg and Rabb, 1985). In many parts of the developing world, the trends have been in the same direction. Of north India, for example, Richards *et al.* (1985: 547) write 'The result [since 1879] ... has been a vast improvement in the productive capacity of society and the land', and that 'Every district now carries a human population far in excess of that it supported eighty to one hundred years earlier'. Development, they stress, has occurred not just in the short term since the Green Revolution, but over a much longer period. The development, however, was at the price of disappearing natural or unmanaged vegetation, which previously yielded a range of useful resources, and of other ecological costs.

Food production has occupied a focal position in the debate about resource adequacy over the last 200 years. In general terms, food production has more than kept up with population, despite rapid population growth. At the same time, millions have faced hunger or starvation, and parts of the physical resource base have been degraded. The interaction between food production and trends in other resource sectors has become more and more complex, and prospects for future resource adequacy may depend as much on what happens in these sectors as it does on trends in population and agricultural land use.

Note: where not otherwise indicated, statistical data are based on FAO annual publications *Production Yearbook* and *State of Food and Agriculture*.

Further reading

Buringh. P. and Dudal, R. (1987) Agricultural land use in space and time. In Wolman, M.G. and Fournier, F.G.A. (eds) *Land transformation in agriculture*. Chichester: Wiley, pp. 9–44.

Dando, W.A. (1980) *The geography of famine*. London: Arnold.

Grigg, D. (1985) *The world food problem*. Oxford: Blackwell.

Grigg, D. (1992) *The transformation of agriculture in the west*. Oxford: Blackwell.

Pierce, J.T. (1990) *The food resource*. Harlow: Longman.

World Resources Institute (biennial) *World resources*. New York: Basic Books; Oxford: OUP.

CHAPTER 5

The land resource: forests

Introduction

The fate of the global forest resource at the end of the twentieth century is a major environmental issue in its own right, and it encapsulates many of the wider modern problems of environmental resources. The way in which the forest resource is perceived has changed dramatically over the centuries. In much of Europe and eastern North America, the forest was at one time dreaded as a place full of wild animals and other dangers. To be useful, it had to be tamed and converted into farmland. Now it is cherished, as a valued environment as much as a source of wood and other products. But big differences in perception still persist. The developing-world forest-dweller perceives the forest quite differently from the developed-world city-dweller. The former obtains from the forest most of the material necessities of life, while the latter looks to it for recreation and spiritual refreshment as well as for material products.

Much of the world's cropland was once forest. As cropland and other human activities have expanded, the forest has contracted, and it has done so at an accelerating rate during the last century. If the forest itself is a symbol of environmental purity, images of unhealthy or devastated forests have become symbols of environmental malaise.

Like cropland, forest land is widely distributed around the Earth's surface but it is not ubiquitous. Large tracts of the Earth's surface are too dry or too cold to support tree growth, and humans have been able to survive in such treeless areas and to exist without the benefit of the forest products that have been available elsewhere. Yet where it exists, the forest resource base has yielded a wide range of useful products, and it is increasingly valued also for the environmental services that it offers. The forest resource has a diverse ecology; it is managed under a wide range of resource regimes; and it offers many different goods and services. Numerous different combinations of forest type, resource regime and resource use exist in different parts of the world.

The extent of the resource base

The extent of forests and woodlands on the Earth's surface is uncertain. Even during recent decades, estimates have ranged from under 3000 to over 6000 million hectares, or from 20 to 45 per cent of the land area. A major reason for this uncertainty is the definition of forest and woodland: boundaries between forest and grassland or woodland and scrub are not always clear and sharp. Table 4.1 in the previous chapter gives the present (1980) area as just over 5 billion ha, but FAO statistics indicate the extent to be 4 billion ha, or 31 per cent of the land surface of the Earth. This is a salutary reminder of the confusion that may arise when data are assembled from different sources and are based on different definitions. These FAO figures are for forest and woodland. Some FAO statistics distinguish between closed forest, where tree crowns cover more than 20 per cent of the land area, and open woodland, where the tree-crown cover is 5–20 per cent. Of the 4000 million ha total, around 2800 million ha are closed forest and the remainder is open woodland. The difference between the forest and woodland total of 4000 million ha and the figure of 5 million ha in Table 4.1 is accounted for by areas under other forms of woody vegetation, including open woodland and scrub.

The distribution of closed forests is shown in Table 5.1. Around 60 per cent of the world's forests are in

Table 5.1 Distribution of forest area by type and wood volume

Type	Area (million ha)	Mean volume (m³/ha)	Total volume (billion m³)
Tropical			
Wet evergreen	560 (20)	350	196 (49)
Moist deciduous	308 (11)	160	49 (12)
Dry*	784 (28)	50	39 (10)
Other*	28 (1)	80–200	5 (2)
Sub-total	1 680 (60)		289 (73)
Temperate and boreal			
Temperate	448 (16)	150	67 (17)
Boreal	672 (24)	60	40 (10)
Sub-total	1 120 (40)		107 (27)
Total	2 800		396

Source: Compiled from data in Persson (1974)
*Including sub-tropical
Note: Data relate to closed forest area only. Figures in brackets are column percentages.

the tropical zone, and the bigger biomass volumes that are found in the humid tropics mean that the tropical share of the wood volume is disproportionately high. Approximately three-quarters of the tropical forests consist of broadleaved species. Most of the world's coniferous forests are in higher latitudes, and especially in the boreal coniferous forest which lies to the north of the temperate deciduous forest. Most of this area lies in two major belts running across North America and in Eurasia from Scandinavia to eastern Siberia.

Plantation forests now extend to around 150 million ha, or about 4 per cent of the total area of forests and woodland. Most of the planted area is in the developed world and in the temperate zone, but in recent years planting rates have increased in many parts of the developing world, and especially in countries such as Brazil and Chile. Plantation forests typically are composed of a narrower range of species than natural forests, and these species are often not native to the locality. They are often characterized by higher productivities, or mean annual increments of wood, than natural forests. This is especially true of plantations in the tropics and sub-tropics (Table 5.2).

Estimating trends in the extent of the resource base

The forest area has been shrinking ever since humans developed agriculture, and probably since they began to use fire as a hunting technique. Since there are

different estimates of the present forest (or forest and woodland) area, it is not surprising that there should be uncertainty about the amount and rate of forest loss resulting from human activities. One study has suggested that 15 per cent of the forest existing in pre-agricultural times had disappeared by 1970 (Matthews, 1983). Another, published in the same year but using a different approach, concluded that 15 per cent of the forest cover existing in 1850 had been lost by 1980 (Houghton *et al.*, 1983). Perhaps only one-quarter or one-fifth of the pre-agricultural forest area has lost its forest cover, during conversion to cropland or as a result of other human activities. This fraction, however, varies greatly between different forest types and areas. Much of the temperate broadleaved forest was some centuries ago converted to cultivation (especially in Western Europe). Conversely, much of the boreal coniferous forest and the tropical forest have survived. This does not necessarily mean that all the forests in these areas have remained in pristine form. Much of the tropical forest in Africa, for example, has been affected by shifting cultivation and other human activities.

During the last few decades the tropical forest has receded just as parts of the temperate forest shrank at earlier points in time. Deforestation in the Brazilian Amazon, for example, has proceeded in the second half of the twentieth century at rates similar to those in the United States during the nineteenth century (Williams, 1989a). Recently quoted rates of loss of tropical forest have sometimes been couched in terms

Table 5.2 Forestry productivities (annual wood increment)

Type of forest		Mean annual increment (m³/ha/year)
Tropics and sub-tropics		
Natural forests		
Tropical high forest		0.5–2
Plantations		
Brazil (Amazonia)	*Gmelina arborea*	35
	Pinus caribaea	27
(Central)	*Eucalyptus*	25–50
Costa Rica	*Pinus caribaea*	40
Temperate regions		
Natural forests		
Canada	(conifers – average)	1
United States	(conifers – average)	2.3
Sweden	(conifers – average)	3.3
Plantations		
Portugal	*Eucalyptus globulus* (average)	13
Chile	*Pinus radiata*	22
New Zealand	*Pinus radiata*	18–30
Britain	(conifers – average)	11

Sources: Compiled from data in Evans (1982), Sedjo (1984) and Kanowski and Savill (1992)

which maximize their dramatic effect – for example 20, 30 or even 40 ha or more per minute, or an area the size of Wales per month. Such figures, however, need to be viewed in the light of their origins and bases, and in the wider context of how much of the forest has already disappeared, and of how much remains. A distinction also needs to be drawn between deforestation and disturbance. The former implies the removal of the forest, while the latter may relate to resource practices such as selective logging or shifting cultivation. These practices may modify the composition of the forest, but need not necessarily cause its permanent removal. Rates of disturbance are usually much higher than rates of deforestation. (Whereas logging in the temperate forest has often involved clear felling of stands of individual species and hence deforestation, it does so less frequently in the tropical forest. Here only a few of the many tree species are commercially valuable.) A further problem is that in one standard and convenient source of data – FAO *Yearbooks* – the area of forest and woodland includes land from which trees have been cleared but which will be reforested in the foreseeable future. In addition, censuses of forest areas are difficult and costly to carry out. The use of remote-sensing techniques should now help in this respect, but effective monitoring systems have been slow to develop. In short, problems of data on trends in forest

areas are even more serious than those on the extent of the forest area.

FAO *Yearbooks* suggest that the net area of forest and woodland (approximately 4 billion ha) declined by around 70 million ha or under 2 per cent between 1980 and 1990 (Table 5.3). (Cropland increased by around 30 million ha during the same period). This figure, however, understates the gross rate of deforestation. In Europe and some other parts of the developed world, the forest area has been expanding as land is released from agriculture. In much of the developing world, deforestation has been proceeding more rapidly than the net global figure would suggest (Fig. 5.1 and Table 5.3). A clear contrast exists: forests are shrinking in the developing world and expanding in the developed world. This pattern is the converse of trends in cropland (compare Fig.4.1).

Forest trends through time

A good guide through the jungle of estimates of tropical deforestation rates is provided by Grainger (1993). His conclusion is that all the estimates are unsatisfactory, but that some are more unsatisfactory than others. Most estimates of (moist) tropical deforestation rates over the last 20 years have been between 6 and 17 million ha per year, with the most

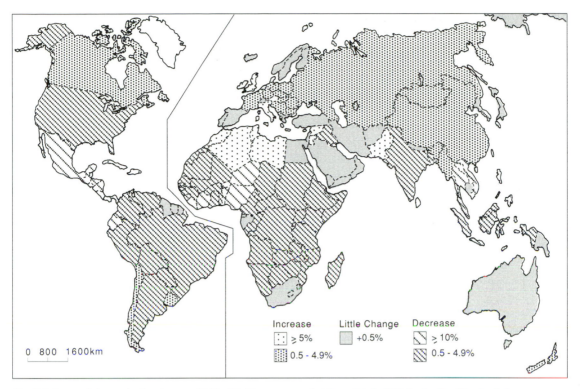

Fig. 5.1 Rates of change in forest areas 1980–90. (*Source:* compiled from data in FAO *Production Yearbooks*.)

likely figures probably lying between 10 and 12 million ha. (Care is required in distinguishing between estimates that relate to tropical rain forests, to tropical moist forests, and to tropical forests and woodland in total.) FAO estimates supplied for the UN Conference on Environment and Development in 1992 suggest an annual deforestation rate for the 1980s of 12.2 million ha for the humid tropics, and 16.9 million ha for all tropical forests. In round figures, the former estimate represents the loss of about 1 per cent of the forest area each year.

It is easy to succumb to the temptation to extrapolate such rates of loss and to conclude that the forest will completely disappear, perhaps by the mid-twenty-first century. Such projections, however, should not be confused with predictions. As predictions, they are valid only if it can be safely assumed that similar rates will apply over several decades. For this assumption to be valid, a good understanding of the driving forces of deforestation is required. Few would be confident that we have achieved such an understanding.

What is clear is that rates of deforestation vary greatly, at all spatial scales. Deforestation is primarily a feature of the developing world, but its annual rates in developing countries range from under 0.1 to over 5 per cent. A few 'hotspots' account for a high proportion of the deforestation, and on the other hand some areas are little affected (Fig.5.2). In general terms, small countries have high rates, especially if they are located on islands or peninsulas, and large, continental countries have lower rates. Within individual countries, deforestation tends to progress from the respective national cores towards the more inaccessible peripheral and mountainous areas.

The reverse process operates in the developed world. In much of the North, and especially in Europe, the forest area is expanding. In many developed countries, much of this expansion has been on marginal farmland, often located in the remoter or more inaccessible areas. If uncertainty exists over the nature of the driving forces of deforestation, our understanding of the factors that bring about a change from deforestation to reforestation is even poorer.

It is usually assumed that tropical deforestation rates have accelerated over the last two or three decades. The differences between estimates are so

Table 5.3 Trends in forest and woodland areas 1980-90

	Area (million ha)		Change (million ha)	Change (%)
	1980	1990		
World	4 100.3	4 027.8	−72.5	−1.8
Developed world	1 868.4	1 899.9	+31.5	+1.7
Developing world	2 231.8	2 127.7	−104.1	−4.7
Europe*	155.7	157.1	+1.4	+0.9
Latin America	946.2	892.8	−53.4	−5.6
Africa	659.4	635.1	−24.3	−3.7
Asia	558.9	535.6	−23.3	−4.2

Source: compiled from FAO *Production Yearbooks*.
*Excluding former Soviet Union.

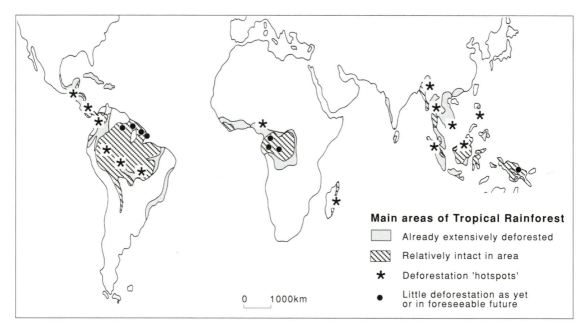

Fig. 5.2 The tropical forest and deforestation hotspots. (*Source:* modified after Myers, 1993.)

great that it is difficult to establish categorically whether deforestation rates are accelerating or decelerating over the short term of a few years. Nevertheless, even if some uncertainty exists over the last 20 years, it is clear that radical changes have occurred over the last century or two.

As recently as the late nineteenth century, rapid deforestation was occurring not only in the new lands such as North America, Australia and New Zealand, but also in parts of long-settled European countries such as France and Italy. During the present century, this process has given way to one of reforestation or afforestation. Much deforested land in the eastern part of the United States, for example, was abandoned as better cropland was taken into production elsewhere, and has now been reforested. In countries such as France and Italy also, agricultural land has been abandoned and has reverted to forest, either through natural regeneration or by afforestation.

Two major transitions are involved in these processes of deforestation related to agricultural expansion, and of the subsequent reforestation. The 'land-use transition' is used by Grainger (1993) to describe the major phase of agricultural expansion and forest contraction that most countries have experienced or are experiencing. The 'forest transition' is the change from shrinking to expanding forests that has been experienced in most of the developed world and which may now be in progress in parts of the developing world (Mather, 1990). The forest transition marks the turning point from the destruction to the re-creation of the resource base, along the lines of the model outlined by Whitaker (1940) (Chapter 1; Fig.1.1). A key question is whether a similar transition will occur in the the developing world, and if so when.

Why is the forest area changing?

Human activity has had only a modest effect on the areal extent of forests, and perhaps as much as 80 per cent of the pre-agricultural forest area still survives. Yet the extent of the forest resource base has changed over time and is continuing to change at present, probably at an accelerating rate. Why is the forest area shrinking in some parts of the world, and expanding in others?

As in many resource issues, different factors probably operate at different levels, and distinctions need to be drawn between immediate or proximate factors and underlying or ultimate causes. It is also possible, and indeed quite probable, that the same end result (in terms of forest expansion or contraction) can be reached by different routes and from different causes.

Immediate causes

A comparison of Figs 4.1 and 5.1 suggests that a definite relationship exists between trends in cropland and trends in forest area. Figure 5.3 indicates that expanding cropland and shrinking forests are associated with countries and continents in the developing world, and that reverse trends are common in the developed world. This association in turn implies that population trends may be an important factor. Countries with rapidly growing populations require more cropland, and deforestation occurs. Conversely, countries with slowly growing populations have been able to release land from agriculture and to expand their forests. The population hypothesis becomes even more credible when the historical dimension is considered. As recently as the late nineteenth

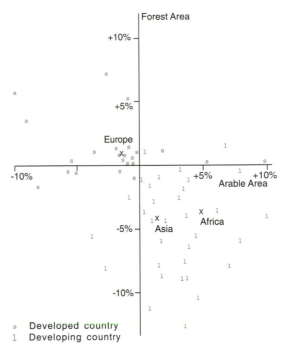

Fig. 5.3 Rates of change in forest and cropland areas 1980–90. (*Source:* compiled from data in FAO *Production Yearbooks.*)

century, rapid population growth in parts of countries such as France and Italy was accompanied by an expansion of cropland, especially in the remoter mountainous areas, and by deforestation.

While at the global scale a relationship clearly exists between trends in population and forest area, it is a mistake to conclude that deforestation is caused simply by population growth. Malthusian interpretations of tropical deforestation are widely held, but to be fully tenable they would require to be judged on at least two criteria. One is that a process or mechanism linking population growth and deforestation is identified, and the other is whether the 'population' explanation is consistent and comprehensive.

In theory, a need to feed more mouths might be translated into a need to bring more cropland into production, and hence to convert forest into cropland. As Chapter 4 shows, however, food production has increased much more because of increases in yields than because of extensions of cropland. In addition, it has already been indicated in this chapter that the forest area declined by around 70 million ha during the 1980s, while cropland increased by only (approximately) 30 million ha. In other words, deforestation is unlikely to result solely or simply

Plate 5.1 Forest clearance for agriculture in the Catlins district, South Island, New Zealand. Clearance for agriculture has probably been the main reason for deforestation throughout history. It was widely practised in the New World and in Australia and New Zealand in the nineteenth century, and continued as recently as the 1980s in this area. In recent times, however, such forest clearance is more usually associated with the tropics. (*Source:* A. Mather)

from population growth – which is not to say that it is quite unrelated to population growth.

Deforestation is also associated in some parts of the world with agricultural expansion geared not to feeding local populations, but rather to providing exports, especially of beef, to developed countries. This has been an important causal factor in Central America and Brazil, but less so in African and Asian forests. Commercial logging has played a bigger role in these latter areas, and like export-oriented agricultural expansion in Latin America, can be seen as representing the exploitation of peripheral resources by external (national or international) interests. On the other hand, attrition on the forest through the collecting of fuelwood results from the activities of local populations, and may be a significant cause of deforestation on the drier margins of the tropical forest. In short, deforestation results from various processes, operating singly or in combination:

- Agricultural expansion, geared to local needs and pressures
- Agricultural expansion, geared to export
- Logging
- Fuelwood gathering.

Some attempts have been made to estimate the relative contributions from these different causes. It is widely agreed that much of the deforestation is related to the activities of small-scale cultivators. Different terms are used to describe these activites. One is subsistence cultivation, but in some situations at least commercial crops (including narcotics) may be produced as well as food for subsistence consumption. 'Encroaching cultivation' is suggested by Grainger (1993), and the term successfully distinguishes the unplanned and uncontrolled activity from that of government settlement schemes.

Small-scale cultivators are considered by most commentators to be the biggest single cause of tropical deforestation. For example it is estimated by Rowe *et al.* (1992) that they account for 60 per cent of the annual loss of tropical forests. Marked differences exist from area to area. The same authors suggest that smallholder agriculture may have been responsible for only 35 per cent of the deforestation in Latin America in recent years, with cattle ranching accounting for the remainder. In Africa and Asia, on the other hand, cattle ranching is not a significant factor, but some areas of forest in these continents have been replaced by commercial plantations of tree crops. Another estimate, by Harrison (1993: 109), suggests that population growth was responsible for 79 per cent of tropical deforestation between 1973 and 1988. The remainder was attributed to ranching in Latin America (13 per cent) and 'to increased consumption of agricultural products'(8 per cent). At the more local scale, peasant agriculture is identified as being responsible for about half of the deforestation in Jamaica during the mid-1980s, with expansion of pasture land and commercial (export-oriented) agriculture accounting for much of the remainder (Eyre, 1987).

Quantitative estimates of the relative importance of causal factors (expressed in terms of the land uses to which the forest is converted) need to be viewed and interpreted with care. Some commentators take the view that quantification is not possible (e.g. Mahar, 1989). In the first place, combinations of activities are sometimes responsible, rather than single causes. For example, previously inaccessible forests have been opened up for commercial logging, and the new roads and trails have allowed small settlers to penetrate. Selective logging may have caused little deforestation directly, but without it, deforestation would have been less likely. Secondly, the interpretation of such estimates is difficult. Smallholders or subsistence cultivators may be the immediate or proximate cause of forest removal, but to blame them is perhaps to blame the victims of circumstances, and especially of the political economies, of their countries. If small cultivators are displaced by commercial agriculture elsewhere, or if they are otherwise landless, they may have little alternative but to occupy and clear forest land. Indeed in some countries forest has to be cleared before secure title to the land can be obtained.

Underlying causes

Five main sets of explanations have been suggested as the underlying causes of shrinkage of tropical forests. Each may be valid to some degree, but each also has limitations. None stands up to rigorous testing at the global scale. In other words, no single cause may comprehensively explain deforestation: different causal factors, and different combinations of factors, may apply in different areas.

1. One is simply the Malthusian view that population pressure is responsible. If 60 per cent of deforestation is caused directly by smallholders, this explanation would seem plausible. As has

already been indicated, however, it is an imperfect explanation. In Brazil, for example, Amazonia comprises 60 per cent of the national territory but only 10 per cent of the population. The population density of Brazil, excluding the Amazon forest, is similar to that of the United States, the world's greatest food exporter (Anderson, 1990). Furthermore, some tropical forest areas (such as southern Mexico) once supported far greater populations than exist at present, without known detrimental effects (Gomez-Pompa and Kaus, 1990).

2. A second is that deforestation results from the penetration of capitalism, and of the modern debt crisis suffered by many developing countries. There is no doubt that many areas of forest were cleared for the production of cash crops: for example tea in Assam, rice in Burma and cotton around Bombay. The processing of some of these products also consumed fuelwood from neighbouring forests, while both the construction and operation of the railways which linked these area with the ports, and hence with the world economy, also brought new pressures to bear on forests (e.g. Tucker and Richards, 1983; Richards and Tucker, 1988). More recently, the need to service debts has been blamed for the growth of export-oriented production which has directly or indirectly caused deforestation. In a study of 103 developing countries, Inman (1993) concluded that deforestation rates are correlated with both increasing rural population and increasing debt. Huge areas of forest, however, were cleared before debt became a major issue, and in the case of Brazil, for example, most of the driving force for forest clearance has come from national rather than international capitalists (e.g. Hecht and Cockburn, 1989). On the other hand, Rich (1990) concludes that the World Bank and the other development banks have played key roles in deforestation, by providing funding for agricultural and other projects that have impinged on tropical forests.

3. A third group of explanations focuses on inappropriate technology. Encroaching cultivators are not always skilled farmers. The agriculture that they practise in some areas sometimes turns out to be unsustainable beyond a very few years. They then have to move on and repeat the process. Better agricultural skills could, it is argued, reduce the impact. On the other hand where appropriate technologies are available, for

example in agroforestry, they have not always been used.

4. Probably more significant have been inappropriate government policies. There is little doubt that many government policies have hastened deforestation. Some, for example in Brazil, have subsidized the creation of cattle ranches and hence have encouraged deforestation. Government resettlement schemes, for example again in Brazil and in Indonesia, have also directly resulted in deforestation. In addition, inadequacy, malfeasance or outright corruption on the part of governments has also played a part. For example, Kummer (1992) concludes that deforestation in the Philippines has resulted more from illegal logging and exporting and similar manifestations of corruption than it has from population pressures. More generally, rich élites have been blamed for much of the deforestation in Asia (Thapa and Weber, 1990). They have used their influence within government to ensure that logging continued unhindered and that land reform was not carried out. In short, the rich left the poor with little option but to encroach into forest land. Failure to implement policies that should help to protect the forest and to enforce existing legislation has also been widespread.

5. 'Tragedy of the commons'. Clearly there is overlap between these last two sets of explanations. Overlap also exists with the fifth type of explanation. This is related to the 'tragedy of the commons' (Chapter 2). Deforestation is seen as the result of inadequate management and control, and privatization is sometimes suggested as the answer. The awkward fact that most deforestation in the Brazilian Amazon has been on private land weakens this view, and indeed privatization is sometimes more likely to be a cause than a cure. In many developing countries (especially in Latin America), small cultivators have been able to acquire private property rights in 'idle' forest land by converting it to agricultural use. In addition to encouraging deforestation, this discourages them from conserving existing farmland (Southgate, 1990).

There may be disagreement about the significance of privatization and of the commons, but there is little doubt that the nature of the resource regime has a significant effect on the manner in which the resource base has been used.

Forest resource regimes

Forest resource regimes vary in time and space. Over most of human history, forests have been open-access or common-property resources. More recently, many have passed into state ownership, or into the hands of private owners or users.

Common-property ownership by indigenous peoples accounted for most of the global forest area until recently. It survived in parts of Europe until modern times, although extensive privatization had occurred by the eighteenth or early nineteenth centuries. In much of what is now the developing world, it lasted until the late nineteenth century, and indeed in some areas has survived until the present. This type of ownership did not guarantee the survival of the forest: for example some forest land was cleared by native Americans so that agriculture could be practised. Nevertheless, the use of the resource was usually regulated by rules or guidelines accepted by the group and handed down from generation to generation. While some trees might be owned by individuals or families, the concept of ownership, and hence of transferability, of forest land was largely unknown. Nature was not yet a commodity.

The arrival of colonial powers in Africa and elsewhere brought an abrupt change. In general terms, they regarded common-property land as ownerless or unoccupied, and in effect appropriated it (e.g. de Saussay, 1987). Post-independence governments have usually retained at least nominal ownership. The exact extent of state ownership is unknown but is probably between 70 and 80 per cent of the world forest area (Mather, 1990). A major contrast therefore exists between cropland, most of which is privately owned, and forests, most of which are state owned.

State ownership, however, does not necessarily mean effective state management. In many cases, a managerial vacuum developed after appropriation by the state. The breakdown in traditional management and regulation was not offset by the advent of effective management by the state. Local management was replaced by remote, state control, and often a sense of alienation set in. In practice, what had previously been a common-property resource became an open-access resource as a free-for-all developed. The tragedy was in the breakdown of the commons, rather than in its existence. As an area of commons became smaller, the pressures on it became larger. Sustainable management became more difficult. The advent of state ownership was sometimes followed by the granting of logging concessions to private companies, or indeed by the transfer of land directly into the private sector.

Within private regimes, wide differences in resource perception, use and management can exist. For example, the objectives of a large industrial forest owner are likely to be very different from those of small farmers or owners of residential properties. Major differences therefore exist within resource sectors as well as between these sectors. As a general principle, however, the management, use and conservation of a forest depend on the nature of the resource regime. Deficiencies in these regimes contribute in significant ways to the impoverishment of the forest resource base.

Forest concessions

In much of the developing world as well as in some parts of the developed world, concessions provide the framework within which the forest resource is used. As much as 90 per cent of industrial wood in the tropics is harvested under concession agreements by which the state, as forest owner, grants rights to a logging company. Between 1960 and 1980, for example, concessions were granted to around 120 million ha of tropical forest, often in areas of 50 000–100 000 ha in extent. Up until the mid-twentieth century, concession lengths frequently extended to 75–100 years, but since then they have usually been for much shorter durations of 20–25 years. In most tropical rain forests, selective logging of the commercially valuable species can be carried out in cycles of perhaps 35 years. If the concession holder can only carry out one cutting cycle, there is little incentive to do so in a careful manner which ensures long-term sustainability. On the contrary, there is an incentive to exploit the concession as fully as possible. It is therefore sometimes argued that concession lengths should be at least as long as cutting cycles.

Concessions are often associated with large foreign companies based in the core region of the world economy, but in recent years these have become less prominent in most countries. Whether the concession holder is foreign or domestic, in theory it has to abide by prescribed forms of management and operation. In practice enforcement is often difficult, especially in large concessions. Payment for the logging concession usually involves royalties based on the amount and type of timber extracted. Terms have not always been

appropriate. Governments in tropical logging countries have often captured an inadequate share of the value of the product: in this respect clear parallels exist with oil and minerals concessions (Chapters 7 and 8). With serious under-pricing of the forest resource, wasteful forms of exploitation have occurred (e.g. Gillis, 1992). The Philippines government managed to capture only 17 per cent of the potential rent from concessions during the period from 1979 to 1982, while in Ghana and Indonesia the corresponding percentage was 38 (Repetto and Gillis, 1988).

This combination of state ownership and private exploitation does not necessarily ensure efficient and sustainable use of the resource. In addition, conflicts have often developed between the concession holder and the indigenous forest dwellers and forest users on whose traditional resource regimes the modern state–private system has been imposed (e.g. Lamb, 1991). These conflicts have stemmed from the different perceptions of resources as well as from the different regimes within which resources have been utilized. In parts of the Philippines and Malaysia, for example, logging operations have seriously disrupted the activities of hunter-gatherers and of shifting cultivators.

Forest concessions sometimes manifest and epitomize the worst aspects of both private and state regimes of forest management. There is widespread evidence that private and state regimes are both flawed in respect of forest management: instances of both market failure and government failure are common, and both can lead to depleted forest areas and to degraded forest resources. It is therefore not surprising that failures, reflected in the squandering of the resource, are common when private and state regimes are combined together in forest concesssions

The use of the forest resource

The forest is a resource base from which a variety of useful goods and services may be obtained (Table 5.4). The range and mix of goods and services vary from place to place, depending on the type of forest and its socio-economic setting. In some instances, the forest is managed primarily for timber production, while in other cases services such as watershed protection are afforded priority. Sometimes the value of the forest is perceived principally or exclusively in terms of the value of timber produc-

Table 5.4 Uses of the forest resource

Traditional use and 'minor products'	Industrial use	Services
Fodder, grazing, shifting cultivation	Sawlogs	Soil conservation
Food – fruit, seeds, nuts, honey, game	Pulpwood and other industrial wood	Water conservation and watershed protection
Medicines	Fuelwood and charcoal	Nature conservation and biodiversity
Fibres	Cork and turpentine	Amenity
Gums, dyes, oils, waxes and resins		Recreation and tourism
Building materials		
Wood for domestic utensils and furnishings		
Fuelwood		

tion, while at other times less tangible values are recognized in terms of recreation and nature conservation. Some of the elements in the *total economic value* of the forest are difficult to estimate in precise terms (Chapter 2), and their values may accrue not just to the forest owners but to a much wider social group. Scope therefore exists for widespread disagreement about the inclusion and quantification of non-material values in cost–benefit analyses of proposed forestry projects and policies, and conflict between different user groups is not uncommon.

The range and mix of goods and services have also varied greatly through time in individual areas, to the extent that a sequential model can be constructed (Fig.5.4). At the pre-industrial stage, a variety of products is obtained, including various foodstuffs and fodder for livestock, medicinal products, wood for domestic utensils and timber for construction, and fuelwood. This pattern of use is usually associated with common-property regimes. It is usually associated with subsistence activities, but commercial elements may also be present. The main characteristic is the diversity of resource products. This stage is today mainly represented by parts of the tropical forest.

In contrast, at the industrial stage, the main emphasis is on the production of timber for industrial purposes. Wood production becomes the primary, if not the sole, aim of management. At this stage, the resource regime is likely to involve state or private control, perhaps involving the use of logging concessions. Much of the forest area of Scandinavia and Eastern Europe, as well as that of part of North America, had passed to this stage by the end of the nineteenth century. During the second half of the twentieth century, this stage has arrived in parts of the tropical forest.

Recent trends in much of Europe and in some other parts of the developed world have suggested that there is also a third phase. At the post-industrial stage, the primacy of wood production weakens. The provision of environmental services, such as recreation and wildlife conservation, becomes increasingly important relative to timber production, as the forest is increasingly valued as an amenity resource, and not just as a source of industrial raw materials. Overall, the post-industrial model of forest management resembles one of the Forestry Principles agreed at the Rio Conference in 1992, in that it embraces the social. ecological. cultural and spiritual needs of the human population, as well as economic needs (Box 5.1).

The practical significance of a transition to the post-industrial forest is that a trade-off has to occur between wood production and other goals. For example, if environments are to be attractive for recreation or useful for wildlife conservation, then open spaces may have to be left in the forests and species other than those that maximize wood production may be selected. If the forests are under private resource regimes, the shift towards the post-industrial stage may be driven by a combination of government regulation of management practices and government incentives such as management grants.

STAGE	RESOURCE SYSTEM	CHARACTERISTICS OF RESOURCE USE
Pre-industrial	Subsistence	Multiple products
Industrial	Commercial	Timber sole or primary product
Post-industrial	Commercial/State	Timber primacy fades Environmental services

Fig. 5.4 Pre-industrial, industrial and post-industrial forests.

Box 5.1 Forestry Principles UNCED Conference, Rio, 1992

At the Rio Conference, a set of general principles for the management, conservation and sustainable development of all types of forests was proposed. These 'Principles' were expressed in a 'non-legally binding authoritative statement', which, it was claimed, represented a 'global consensus'. The statement of principles fell short of a more binding convention, and all countries did not endorse it. Sustainable development is the goal to which the principles are directed, and sound management and conservation are said to be 'of concern to the Governments of the countries to which they belong and are of value to local communities and to the environment as a whole'.

Among the principles are:

- Forest resources and forest lands should be sustainably managed to meet the social, economic, ecological, cultural and spiritual needs of present and future generations.
- Forest policies should be integrated with economic and trade policies.
- The right of states to exploit their own resources is recognized, along with their responsibility to ensure that activities within their jurisdiction do not cause damage to the environment of other states.
- The costs of achieving the benefits of conservation and sustainable development require international cooperation and should be equitably shared by the international community.
- The rights of indigenous peoples are recognized.
- New financial resources, and access to technology, should be made available for developing countries, to help with sustainable management, conservation and development of forest resources.
- Tariff barriers should be reduced or removed to allow easier access to markets and better prices for value-added (i.e. processed) forest products.

Note: These are some of the main aspects of the Forestry Principles, presented in summarized and paraphrased form.

Traditional uses and 'minor' products

Under traditional regimes, the forest was useful in a great variety of different ways. It was the setting for shifting cultivation, and provided, directly or indirectly, most of the materials for hearth and home. It could even yield medicinal products and alcoholic beverages, as well as berries, fruits, nuts, honey and game. In the Gogol Valley of Papua New Guinea, for example, 68 of the 250 species of wildlife are used by local people (Lamb, 1990).

This pattern of use has largely died out in the industrial age in the developed world, but it continues in many other parts of the world. Forest-dwelling shifting cultivators may still number as many as 500 million and are believed to use around one-fifth of the tropical forests. It has recently been suggested that the most significant contributions of moist forests to African national economies are non-industrial goods and benefits, other than timber (Sayer *et al.*, 1992). In both temperate and tropical forests, numerous plants and animals are still utilized, or were utilized until recently. Some of these products are of commercial value: for example nearly half of the 68 wildlife species used in the Gogol Valley had entered the cash economy by the end of the 1980s. Nevertheless, the conventional wisdom has been, at least until very recently, that these 'minor' forest products are relatively insignificant compared with timber. The incorporation of new areas into the world-economy has often brought about a radical change in perception of the forest resource. The perceptions (and needs) of the indigenous peoples have usually been swamped by those of the newcomers, which usually focus on wood.

It is not difficult to see how the transition from the pre-industrial to the industrial stages is likely to be one of conflict. The objectives of the indigenous peoples and of the commercial exploiters are quite different, and mutual incomprehension reigns. Similar conflicts in mineral concessions are discussed in Chapter 8. The power relationship is unequal: indigenous peoples have usually been no match for the logging concessionaire backed up by the state. The main goal now becomes the commercial production of timber: 'minor' products are now largely if not completely irrelevant.

Wood production

The growing stock of the world's forests is estimated at around between 300 and 400 billion m^3. It produces around 11 billion m^3 of new growth each year (Sharma *et al.*, 1992). FAO *Yearbooks* indicate that wood removals by humans in recent years have been at an annual level of approximately 3.5 billion m^3.

This figure may be an understatement, although FAO statistics do include estimated unrecorded volumes in addition to recorded volumes. Other estimates are rather higher: for example one is that annual demand is around 4.4 billion m^3 (Sharma *et al.*, 1992). Even so, it is clear that less than half of the annual growth of wood is harvested. In an absolute sense, therefore, there is no shortage of wood.

Many forests, however, are in remote and inaccessible areas. Some are protected for purposes of conservation. Some contain wood that is unsuitable for commercial purposes. The gap between annual growth and annual removals therefore needs to be viewed with caution. Some parts of the world have always suffered from shortages of wood, and some have begun to suffer as supplies of wood have dwindled. Some areas, such as the Sahel, are suffering from severe wood shortages at present. Britain and some other parts of the world suffered severe shortages of wood in the past. In parts of Europe, the long history of human use of the forest and growing awareness of wood shortages led to increasingly intensive forest management by the nineteenth century, including the creation of planted forests. Since then, management has spread to many other parts of the world, bringing with it increasing yields of timber per unit. In short, forestry is now experiencing a transition that occurred in agriculture much earlier – a transition from the hunter-gatherer to the farmer (Sedjo, 1987). (This management transition should be distinguished from the areal transition discussed on page 97). Nevertheless, if some forests are now 'farmed' for wood, many others are still at the 'hunter/gatherer' stage (Box 5.2).

Trends in wood removals

Wood removals in the developing world amount to just over half (56 per cent) of the world total (Fig.5.5). The developing world's share has been gradually increasing in recent decades, as removals there grow faster than those in the developed world. Between 1980 and 1990, world removals increased by 18 per cent, while those in the developing and developed worlds grew respectively by 23 and 11 per cent. Many other major contrasts exist between the two worlds. One of the most striking is in wood use. In the world as a whole removals are divided almost equally between fuelwood and industrial wood. In the developing world, however, 90 per cent is for fuelwood, while industrial wood accounts for nearly 80 per cent in the developed world.

Box 5.2 Tropical forest management

In the temperate zone, the use of silvicultural techniques has greatly increased the amount of wood that can be produced per hectare, and can apparently do so on a sustainable basis in semi-natural forests as well as in plantations.

Successful forest management in the tropics has proved more elusive. The high diversity of tree species, only some of which are commercially attractive, has been a major problem. After selective logging of these species, they are likely to be replaced by species that are less valuable. There have been some successes, notably in some parts of south Asia, but it has proved very difficult to maintain the species composition and even more difficult to increase wood yields from these species. Attempts have been made at management since the nineteenth century, but it has recently been observed that one of the paradoxes of tropical forestry is that the rise in public interest over the last 30 years has been paralleled by a decline in the application of systematic management (Palmer and Synnott, 1992). Various systems of management of natural or semi-natural forests have largely been abandoned in favour of plantations. Some success was enjoyed as early as the nineteenth century in the teak plantations of south Asia.

In recent years economic, social and environmental cases have been made for basing tropical forest management on non-wood products. Claims have been made that the economic returns from products other than wood could, on a sustainable basis, be much greater than those from wood (Peters *et al.*, 1989). While such claims may be valid for some forest areas, especially near cities, they do not necessarily hold true for the tropical forest as a whole. Markets simply do not exist for such products in many areas.

Industrial wood

Industrial wood, including that used for constructional purposes as well as for wood panels, pulp and paper, is produced mainly from temperate forests in the developed world, notably in Scandinavia, the former USSR, Canada and the United States. These last three countries account for more than half the production. Nearly 70 per cent is from coniferous forests and plantations.

The nature of demand for industrial wood has changed greatly during the present century. Demand for large constructional timber and for pitprops for

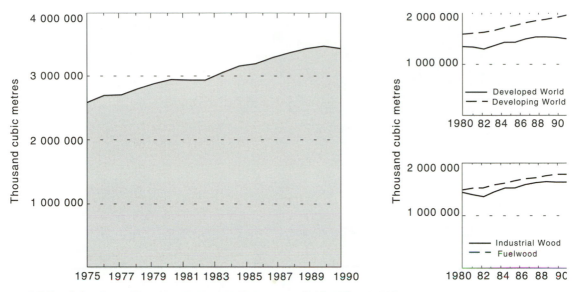

Fig. 5.5 Trends in volume of roundwood removals. (*Source:* compiled from data in FAO *Production Yearbooks*.)

the coal-mining industry has fallen, while that for processing into pulp and manufactured wood panels has increased. This is important in relation to the future adequacy of supplies of industrial wood. Potential supplies of wood for the more highly processed products are relatively plentiful. These uses can draw on an increasing range of tree species and of forest resources as wood technology develops (Chapter 1).

This trend is illustrated in particular by pulping. In the early days of the wood-pulp industry, spruce was the most suitable raw material. Then processes that could use pine were developed, and new areas were opened up, notably in the American South where extensive plantations had been established on degraded farmland (Clark, 1984). Then it became possible to use hardwoods, and especially *Eucalyptus* wood from Australia, which by the 1970s was being imported in the form of chips for pulping in Japan.

There has been a clear trend towards the widening of the range of species perceived as useful (and hence of the geographical area from which supplies can be drawn). There has also been a clear improvement in the efficiency of utilization, at all levels from the forest to the mill. In the United States, for example, the Weyerhauser Corporation reported that the level of timber utilization per unit area increased from 21 per cent in 1950 to 79 per cent in 1975 (Cox *et al.*, 1985). During that period, wood previously regarded as waste was being increasingly utilized. World-wide,

mill residues are used increasingly efficiently, and their contribution to total wood consumption is now over 10 per cent. Similarly, waste paper can be substituted for wood in the manufacturing process, thereby saving on the primary raw material and on energy used in processing.

In short, recycling and increased use of residues are likely to mean that demands on the forest will increase less rapidly than demand for wood products such as paper and wood panels. In this respect, trends in the use of the forest resource resemble those in other sectors such as minerals. With changing technology (superimposed in this case on changing patterns of demand), perceptions of useful resources are revised, and increasing use is made of non-primary inputs. Both of these trends help to reduce pressures on the resource.

Prospects for industrial wood

There may be little prospect of a serious shortage of industrial wood in the foreseeable future, but fears of wood famines have often been expressed. These fears have led some governments to embark on programmes of forest expansion. At the beginning of the twentieth century, it was thought that American forests were being depleted at such a rate that timber would soon become very scarce. As Chapter 3 indicates, this fear had as its backdrop the destructive exploitation of the American forest resource once regarded as almost limitless, and was an important

Plate 5.2 Logging by clear felling in the boreal forest in northern Ontario, Canada. The boreal forest accounts for much of the world's industrial wood, but many of its more accessible areas have been worked over and not all have yet been successfully reforested. (*Source:* A. Mather)

part of the first conservation movement. A wood famine did not materialise, because of major changes in both demand and supply. Demand for construction timber and railway sleepers dwindled as substitutes in the form of steel and concrete were introduced. Demand for paper and processed wood products increased, but part of it could be met by smaller wood, wood from species previously not perceived as useful, and from waste. At the same time more intensive forest management allowed more wood to be produced per unit area.

Sources of supply of industrial wood

Nevertheless, the history of the timber industry is one of ever-widening search for new resources. Britain, having exhausted its accessible home supplies (especially for shipbuilding) turned successively to the remoter parts of Scotland, to the Baltic lands, and to North America. Later, it looked to the tropics and to the Malabar teak from India. Within the United States, successive areas were used: New England, the Lake States of Michigan and Wisconsin, the Pacific Northwest and the South. The same process has operated at the global scale, with the growth of the world economy. Perhaps the growth of tropical logging in the second half of the twentieth century is

the penultimate or even the last chapter in the story (the forests of Siberia and the (former) Soviet Far East may be the setting for the final chapter). As in many previous instances, the incorporation of new areas into the world economy has brought with it new resource practices and new environmental pressures. Just as successive forest areas were ransacked in Europe and North America, so also have successive tropical lands been destructively exploited for timber. For example Japanese firms have turned their attention successively from the Philippines (the main exporter in the 1950s) to Indonesia, Sabah and Sarawak, and now to Amazonia (Repetto, 1990) The Japanese log market has been supplied by logging contractors who have moved from country to country as resources were exhausted or log export bans imposed (Fig.5.6, Dargavel, 1992). In Indonesia, log production amounted to less than 2 million m^3 in 1967, but had increased to 26 million m^3 by 1973 (Walker and Hoesada, 1986). In Africa, a similar process has been in operation, and attention is now turning to countries such as Gabon and Zaire. Production in Ivory Coast, a focus of logging in the late 1970s, has now fallen to half of its peak in the logging boom in the late 1970s. Some countries that once experienced booms in log exports now have to

Plate 5.3 Timber yard at rail head, northern Ontario, Canada. (*Source:* A. Mather)

Plate 5.4 Pulp mill in British Columbia, Canada. (*Source:* K. Chapman)

import timber. Thailand and Nigeria are two examples. The Philippines is becoming a timber importer: the first log imports from New Zealand have already arrived and Filipino teams have been scouring new areas such as Vietnam for potential sources of imports (Mercado, 1990). Some 33 countries exported tropical timber in 1990, but perhaps as few as ten will have supplies available

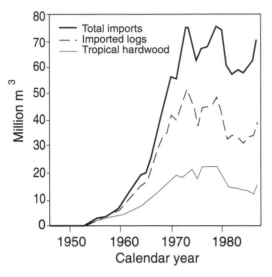
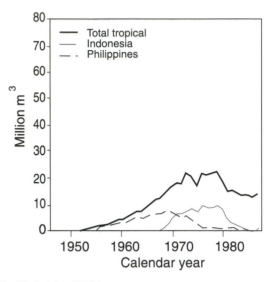

Fig. 5.6 Trends in Japanese timber imports. (*Source:* modified after Kitabatake, 1992.)

for export by the end of the century (Palmer and Synnott, 1992).

As long as new lands were available, to 'cut and run' was economically rational. Wood from planted, intensively managed forests could not normally compete with wood taken from natural forests at no financial cost other than that of felling and transport. There are clear signs, however, that conditions are changing. Few accessible areas of natural forest remain, and those that do are increasingly likely to be protected for purposes of conservation (Chapter 6).

The age of plantations

Plantations are usually characterized by much higher levels of productivity than natural forests (Table 5.2). This productivity is achieved by careful selection of species and by intensive management. Species such as *Pinus radiata* (Monterey pine, originally from California) can, when grown under suitable conditions, produce large amounts of wood in short periods. Short rotations (and hence short pay-back periods) are an important attraction. Until the last few decades, most forest plantations were in Europe or on abandoned farmland in areas such as the American South. Especially since the 1970s, however, huge new areas have been established, particularly in countries such as Brazil, Chile and New Zealand

(Table 5.5). In some cases native forest has been cleared to allow plantations to be established, but more often they have been on degraded pasture land which has not carried trees for many decades.

These plantations are not without their silvicultural problems (particularly in terms of damage from pests, and of long-term sustainability), and in some cases there have also been problems arising from their socio-economic impacts. Nevertheless, they do offer large amounts of wood from small areas. By the 1980s, 30 per cent of Latin America's industrial wood was coming from plantations, which occupied less than 1 per cent of the forest area (Evans, 1987). By 1990, plantations were probably contributing around 10 per cent of industrial roundwood production, and the proportion may rise to up to half of the total world wood supply by the year 2000 (Sedjo, 1987). Wood volumes equivalent to the projected industrial needs for the year 2000 could in theory be met by 100–200 million ha of plantations (around 5 per cent of the total world area of closed forest) (Sedjo and Clawson, 1983).

Trade in forest products: globalization and internationalization

In the long term, the increasing importance of plantations may lead to significant geographical shifts in the world pattern of the forest-products

Table 5.5 Examples of plantation areas and annual rates of establishment (1000 ha)

	Rate (1980s)	Area (*c.* 1985)
China	4 800	13 000/17 000
USA	1 800	12 000
USSR	1 300	21 000
India	138/370	2 000
Brazil	250/450	3 800
Indonesia	131/200	1 900
Japan	200/240	9 600
Rep. of Korea	152	400
Chile	75	1 200
New Zealand	44	1 100
Australia	32	800
Nigeria	26	160
Mexico	22	160
Britain	20	2 000
Kenya	10	200
Tanzania	9	200

Sources: compiled from data in Postel and Heise (1988), Mather (1990), Sharma *et al.* (1992) and World Resource Institute (1992).
Notes: (i) In Britain and some other countries, most of the forest area is in the form of plantations. (ii) Wide variations exist in planting rates reported from some countries, and planting rates and successful establishment rates are not always identical. For a discussion of problems of accurate statistics and for a review of planting policies and progress, see Mather (1993).

industry. Most of the industry has been in the North, and especially in Scandinavia, the former Soviet Union, and in North America. Plantations in the tropics and sub-tropics are likely to be more productive than those in higher latitudes (Table 5.2). They are likely to attract an increasing share of the wood-processing industry, especially if production from natural forests in North America and elsewhere is increasingly constrained by conservation measures. This in turn is likely to lead to an increasing international trade in forest products, and indeed to a globalization of the industry.

At present, less than 20 per cent of wood products enter international trade, although the proportion varies by product. As might be expected, a higher proportion of manufactured products is traded than of bulky, low-value logs. Trade flows are largely concentrated into a number of regional spheres. Three account for about half of world trade in forest products: within North America, within western Europe, and between northern and western Europe (Kornai, 1987). One-half of exports are from North America and northern Europe: half of the imports are to western Europe. Most trade in forest products is between developed countries (Table 5.6), although the developed world's share has been slowly declining (from 85 per cent in 1963 to 78 per cent in 1980 (Nagy, 1988)). This pattern could change further (and perhaps more rapidly) in the long term if extensive and productive tropical pine plantations are established.

One of the main features of international trade in forest products has been the enormous growth of exports of tropical hardwoods. This trade increased 24-fold between 1946 and 1980 (Laarman, 1988). This growth is explained in part by the dwindling availability of certain species and grades of temperate hardwoods, in part by improvements in mechanized logging and transport, which allowed previously inaccessible forests to be used, and in part by changing technology that allowed plywood, panels and pulp to be produced from tropical hardwoods.

The growth of the tropical hardwood trade has been accompanied by the emergence of a fourth major trade sphere over the last few decades. It focuses on east Asia, and especially on Japan. Japan stands out as a major importer of forest products, and especially of logs, despite being one of the world's most extensively forested countries. Its forests were devastated during the Second World War, and could not meet the demand for timber during the rapid economic development in the late 1950s and 1960s. By the time the post-war replantings were becoming productive, the high-cost timber produced in plantations in a high-cost economy (and often on difficult terrain) in Japan could not compete with cheaper imports from lower-cost areas, and especially from natural forests and/or from areas with low labour costs. Timber was imported initially from the Pacific coast of North America, and more recently also from countries such as New Zealand and Chile. Hardwood chips, as has been indicated, are also imported from Australia and other Pacific countries. The most controversial trade is in tropical hardwood logs from South-East Asia to Japan (and to 'transit' processors such as Singapore). (By contrast, flows from African tropical forests to Europe are relatively modest, and those from Latin America to North America are also small.)

Several countries, especially in South-East Asia, have attempted to ban log exports in the hope of establishing domestic processing industries that will provide local social and economic benefits (Table 5.7). It is doubtful whether such attempts have always been as successful as hoped, and there are many

Plate 5.5 The Shell Company's Rucamanqui forest estate in the foothills of the Andes, Chile. The area of industrial plantations (mainly of Radiata pine) increased rapidly in Chile during the 1980s (as it did in countries such as Brazil and New Zealand), and an export trade in logs and in pulp developed. (*Source:* Shell)

Plate 5.6 Afforestation with exotic species – mainly Sitka spruce – in Glen Spean, western Scotland. During the present century, the forest area in Scotland has approximately trebled. Extensive plantations were established initially to safeguard against shortage of timber imports during periods of war, and more recently to produce wood for pulping and for other industrial purposes. (*Source:* A. Mather)

Table 5.6 Major international flows of forest products 1990 (as percentage of trade by volume)

1. Coniferous sawlogs and veneer logs (5%)

Exporters Importers	Developed	Developing
Developed	73	< 1
Developing	23	3

Leading exporters USA 52, USSR 18, NZ 8
Leading importers Japan 47, Korea Rep 14, China 13

2. Non-coniferous sawlogs and veneer logs (12%)

Exporters Importers	Developed	Developing
Developed	15	46
Developing	2	38

Leading exporters Malaysia 64, France 5, Papua New Guinea 4
Leading importers Japan 33, Korea Rep 12, China 11

3. Coniferous sawnwood (20%)

Exporters Importers	Developed	Developing
Developed	91	11
Developing	7	1

Leading exporters Canada 51, USA 9, USSR 8, Sweden 8
Leading importers USA 40, UK 11, Japan 10

4. Non-coniferous sawnwood (12%)

Exporters Importers	Developed	Developing
Developed	31	28
Developing	4	37

Leading exporters Malaysia 11, USA 10, France 5
Leading importers USA 11, Japan 10, Italy 9, Thailand 9

5. Wood pulp (16%)

Exporters Importers	Developed	Developing
Developed	78	6
Developing	14	1

Leading exporters Canada 29, USA 13, Sweden 10
Leading importers USA 17, Germany 14, Japan 11

6. Paper (17%)

Exporters Importers	Developed	Developing
Developed	79	5
Developing	14	4

Leading exporters Finland 16, USA 12, Sweden 11
Leading importers Germany 14, UK 11, USA 11

Main figures indicate structure of trade. For example 79 per cent of the trade in paper is between developed countries and 4 per cent is between developing countries. The figure in parentheses after the sub-heading is the volume of that commodity traded as a percentage of production. The figures indicated for leading exporters and importers are the percentages accounted for by the named countries.

Table 5.7 Examples of log export bans

Country/area	Date
Canada – British Columbia	1906
USA – Oregon state forests	1963
Brazil – entire country	1973
Thailand – entire country	1977 and 1989
Philippines – entire country	1977 and 1982
Indonesia	1985
Malaysia – western	1972
Sabah – restriction	1976
Sarawak – ramin embargo	1980

Source: compiled from A. Nagy (ed.) (1988) and other sources.

of the forest-products industry. Nevertheless, there have been clear signs in recent decades of a growing degree of internationalization, which has in particular involved relatively new forest-product countries such as New Zealand and Chile (e.g. Le Heron, 1988). Japanese companies have become involved in forest ownership and forest-products industries in New Zealand and Chile, as well as in South-East Asia, Brazil, the United States and Canada. Corporations based in the developed world, and especially in the United States and Britain, have for long been involved in the developing world. A more recent trend, however, has been for companies based in countries such as Malaysia and Indonesia to expand into neighbouring lands, as well as into New Zealand.

The forest has often been exploited for industrial wood in destructive ways, and the transition to more sustainable forms of use has not yet been made everywhere. There is little definite evidence to suggest, however, that a serious world wood shortage will develop in the foreseeable future, at least in overall terms. Trends such as the widening range of species perceived as useful and the increasing contributions from highly productive plantations make that unlikely. And if wood did become scarce, the use of substitutes would probably increase rapidly and thus reduce demand. For example steel can be used to frame houses if timber is scarce or very expensive.

Greater problems may arise over the availability of particular types of wood. A recent trend that is likely to continue is the growing use of manufactured wood products as substitutes for solid wood. This trend may facilitate and accelerate a trend of globalization. Whereas trade in lumber and in some other wood products has been international in scope, customer preferences for particular types and specifications of

reports of evasion, especially in countries such as the Philippines (Westoby, 1989; Kummer, 1992). They do, however, represent one obstacle to free flows of forest products, and hence to a genuine globalization

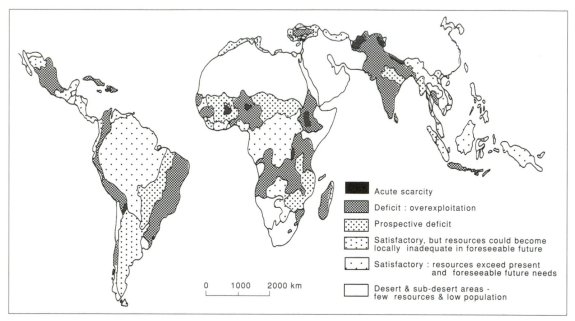

Fig. 5.7 Areas suffering fuelwood scarcity. (*Source:* modified after FAO, 1982.)

timber have inhibited the emergence of a truly global market. As these types and specifications become more difficult to obtain, substitution by manufactured boards and panels is likely to occur. These materials can be readily produced to standard sizes and specifications in many parts of the world, and trade in wood products may thus become more truly globalized.

Fuelwood

The production of fuelwood contrasts with that of industrial wood in numerous ways, including spatial patterns, types of wood and economic systems. Fuelwood is produced mainly in the developing world and from hardwood sources, whereas the production of industrial wood is largely from coniferous sources in the developed world. Whereas most industrial wood is obtained through the market, most fuelwood, in both developed and developing worlds, has traditionally been obtained by self-collection. In the United States, for example, only one-quarter is purchased (Skog and Watterson, 1984), while in India the corresponding proportion in the 1980s was around 13 per cent (Agarwal, 1986).

Contrasts also exist in production trends. Production of industrial wood grew by 14 per cent between 1980 and 1990 – a rate which is less than that of world population. Fuelwood production, however, grew faster than population, and increased by 21 per cent during the same period. It represents only a small percentage of world energy consumption, but over 2000 million persons rely on it and many millions rely on it completely.

Fuelwood shortages exist in some parts of the world today, and have also affected other areas, such as England and Greece, in the past. In Athens more than 2000 years ago, the fuel could cost more than the meat in the pot. In England, the price of firewood rose ten-fold in England between the latter half of the fifteenth century and 1700: as indicated in Chapter 1, the resource crisis which this price rise reflected is regarded by some commentators as having been the trigger for the Industrial Revolution, and for the arrival of the age of fossil fuels.

The 'fuelwood crisis' has attracted widespread concern during the last 20 years. The problem of scarcity of fuelwood is acute in some parts of the developing world (Fig.5.7). Many people are facing severe scarcity, and many others can meet their needs only by depleting the resource beyond the critical zone (Chapter 1). In the early 1980s, FAO estimated that around 100 million people already faced acute scarcity, and 1 billion were facing deficit situations. By 2000, it estimated, nearly 3 billion would face acute scarcity or deficit (FAO, 1982). The underlying causes of resource degradation are complex and interacting:

- Increasing demand, especially from urban areas.
- Breakdown of traditional resource systems and regimes.
- The transition from a free good to a commercial commodity.

The fuelwood problem is difficult to specify and quantify in precise terms, not least because of the enormous differences that exist in energy requirements and consumption levels between different areas. Projections of supply and demand have sometimes been based on dubious assumptions and over-generalizations, and scarcity may have been over-stated in some cases. More recently, it has been more widely acknowledged that great contrasts exist between between rural areas, where fuelwood is usually a non-market good gathered locally, and urban areas, where it is a commodity. Major contrasts also exist between semi-arid areas such as the Sahel and better-watered mountainous regions such as Nepal. What is common to many parts of the developing world is that fuelwood is the main energy source, and that the poor in particular rely heavily on it.

In rural areas, fuelwood was traditionally gathered from trees around the houses and farms and from dead trees and branches. It did not always come from the forest, and deforestation rarely resulted even when it did. Stresses have built up, however, for a number of reasons. One is that privatization of common-property land and forests has sometimes reduced the available areas of search. There is a related gender-based issue: tenure may be vested among the men, while fuel-gathering is more likely to be carried out by the women and children. Another reason is that grazing pressures have increased, hindering natural regeneration of woodland, especially in the drier areas. A third is related to population growth, and hence to growth in demand.

In urban areas, fuelwood is more likely to be a market product. As such it may be produced from forests or woodlands some distance from the city, and transported to the market in the form of wood or charcoal. This pattern of use is more likely to result in deforestation in the producing areas, and to patterns of widening rings of denudation around the main cities (see, for example, Bowonder *et al.*, 1987, for a review of the impact of fuelwood use around some Indian cities).

In theory, the existence of a market ought to stimulate production, perhaps by the creation of commercially managed fuelwood plantations. In practice, the long lead times, which are likely to extend to several years even in the most productive areas and with fast-growing species, are a disincentive. A further problem is the ability of the potential urban customers to purchase the fuel at commercial prices. Nevertheless, some success has been achieved in establishing fuelwood plantations, for example around Addis Ababa and other Ethiopian cities (Mercer and Soussan, 1992). Some successes are also reported from rural areas. Paul Harrison (1993: 112) writes: 'When I visited Kenya in 1980, women in the highlands were spending up to three hours a day scouring the landscape for twigs and brushwood. When I returned only six years later, things had changed dramatically. Wood had to be bought on the market. Husbands had to shell out hard cash if they wanted cooked dinners. Planting trees had become a boom industry.'

Whether the arrival of a market system has had such an impact on fuelwood supply (and gender relations) everywhere is another issue. In many rural areas, continuing fuelwood shortages are simply one dimension of the multiple-resource problems found in regions such as the Sahel, where food and water are often also in short supply. A scarcity of fuelwood not only imposes hardship on those who depend on it (and especially on those responsible for its collection) but it can also exacerbate other resource problems. If wood is scarce, dung may be used instead as a fuel, and hence will not be available to fertilize cropland. And so land degradation and a scarcity of food become more likely.

Services (non-consumptive uses)

In addition to yielding wood and 'minor' forest products, forests also fulfil important service functions. These include environmental protection in general and wildlife conservation in particular: the former includes general protection of slopes or river catchments against soil erosion, while the latter relates to the conservation of particular species and their habitats. They can also include various forms of recreation.

Service functions of forests are difficult to quantify in terms of areas and extents. Recreation, for example, may be a high-intensity primary use in some areas of forest but a low-intensity subsidiary use, requiring no management, in others. Similarly, the provision of protective functions may be the

Table 5.8 Forest and woodland areas by importance of function (1000 ha)

Country/category	Importance of function	Function						Area	
		Wood production	Recreation	Hunting	Protection	Nature conservation	Range	Total	Closed forest
Sweden (closed forest)	High	11 800	5 000	100	500	200	0		
	Medium	10 000	12 400	24 000	1 500	1 000	2 600	24 400	24 400
	Low	2 600	7 000	300	22 400	232 00	232 00		
France (forest & other wooded land)	High	7 140	1 000	11 600	738	95	0		
	Medium	5 200	121 000	2 520	4 500	12 580	620	15 075	13 875
	Low	2 735	1 955	955	9 837	2 400	14 455		
Netherlands (closed forest)	High	120	270	92	5	40	0		
	Medium	130	30	104	15		0	300	300
	Low	50	0	104	180	260	300		
Spain (forest & other wooded land)	High	1 800	100	6 973	5 433	60	750		
	Medium	2 900	1 000	2 693	3 770	40	350	12 511	6 906
	Low	7 811	11 411	2 845	3 308	12 411	11 411		
Former Soviet Union (forest & other wooded land)	High	548 400	18 900	754 600	176 300	–	138 700		
	Medium	442 900	304 900	0	73 600	–	110 600	1 185 900	
	Low	194 600	862 100	431 300	936 600	–	936 500		
United States (forest & other wooded land)	High	141 288	80 898	56 634	87 409	17 300	30 583		
	Medium	56 933	103 134	155 000	44 711	15 030	143 911	298 076	195 256
	Low	99 855	114 044	86 442	165 956	26 555 746	123 582		

Source: Compiled from UNECE/FAO (1985).

primary objective of management in some cases but a secondary one in others. A common feature of forest management is that multiple objectives are sought, including both the production of timber and the provision of various services. Different priorities may be applied to these objectives in different areas of forest.

Table 5.8 illustrates, for a number of countries, the relative extents and importance of different forest functions. In Sweden, for example, wood production is of 'high' importance in approximately one-half of the forest area and of 'low' importance on little more than one-tenth of the area. On the other hand, nature conservation is of 'high' importance in only 200 000 ha but of 'low' importance in over 23 million ha. Even from the small sample of countries illustrated in the table, it is clear that the relative extents of different functions are highly variable. In Europe as a whole, wood production is the major function on 78 per cent of 'total forest and wooded land' while protection and recreation account for 19 and 2 per cent of the area respectively (UNECE/FAO, 1985). In northern Europe, the proportion of forest primarily classed as protective is under 10 per cent, while in southern Europe it is over 40 per cent. In Japan, around one-third of the forest is classed as protective, and the proportion is similar in Malaysia

and Indonesia. In the United States it is around 10 per cent, while in Chile is 60 per cent (Mather, 1990). Since the figures depend so much on definitions, the significance of these wide variations is debatable. Nevertheless protective forests, irrespective of their relative extent, usually share a number of characteristics. Perhaps the most obvious of these is state ownership. In the absence of marketable products, there is little incentive for private ownership.

Recreational use also tends to be associated with forests in the ownership of the state or other public bodies, but it is not confined to that sector. In many countries significant amounts of public recreation take place in private forests, sometimes because it is practically impossible for the owners to prevent access. Types of recreational activities range from low-intensity hunting or hiking to high-intensity walking and picnicking. More intensive recreation, in some cases amounting to the primary use of the forest, is usually restricted to forests near cities. In densely populated countries such as the Netherlands and Britain, some forests are established for this purpose, on public land or on private land with funding from government agencies. Most forms of forest recreation are unlikely to generate revenue in amounts that are attractive to private owners.

The concept of multiple use of forests has attracted

much attention, but it suffers from problems of different interpretations. It is seen by some to mean the use of different sections of a forest for different purposes, while others see it as meaning the use of an area of forest for a variety of purposes. Where several uses are carried on in a particular area, trade-offs usually have to be made between the intensities of management for individual uses. For example in a forest used for both timber production and recreation, a 'loss' of timber production may have to be accepted if the forest design – complete with open spaces and a variety of species – is to produce an attractive recreational environment. Equally, recreational use may have to be discontinued during timber-harvesting operations. Some argue that 'segregated' multiple use is more efficient, allowing the pursuit of individual management goals (such as that of timber production) at maximum intensity. On the other hand, the 'post-industrial' model of forest management increasingly incorporates these service functions alongside the function of wood production for industrial purposes. Equally, increasing areas of forests in many countries are being classed as 'protective', or are having service functions such as conservation upgraded relative to wood production. A classic example of the conflict that can be generated over priorities in forest management is the case of the northern spotted owl controversy in the Pacific Northwest of the United States. In 1990, the federal government added this species to the list of threatened species under the Endangered Species Act of 1973. In order to conserve its habitat, the harvest of timber from the old-growth forests of Oregon had to be reduced. This in turn led to fears of losses of jobs in logging and saw-milling and the social effects of unemployment and community disruption, as well as the direct economic effects. Estimates of the costs of lost jobs and production over the next 20 years could amount to US$12.6 billion (1990 dollars) (Watson and Muraoka, 1992).

Forest resource use: environmental implications

The use of the forest resource can have important environmental implications at all scales from the local to the global. These implications can relate to the future productivity and usefulness of the resource itself, and also to the wider regional or global environment.

Logging has often been associated with strong environmental impact. When carried out in clear-felling mode, the result can be erosion of the exposed ground (erosion which may be aggravated by the disturbance caused by log extraction) and, in turn, impoverishment of the soil within the affected area and silting downstream. In selective-logging mode, as carried out in many parts of the tropical rain forest at present, the removal of a few commercially valuable species can cause damage to other parts of the forest ecosystem. In addition, the places of the individual trees removed may be taken by other, less valuable, species. In either case, the forest resource becomes less valuable.

If the removal of trees can have significant environmental effects, so also can their replacement in plantations. In addition to the ecological changes resulting from the near monoculture of plantations (often consisting of exotic species), ground preparation operations such as ploughing can also trigger episodes of accelerated erosion and silting. These effects associated with logging and planting can be minimized by appropriate management techniques. Some other wider-scale effects, such as the contribution of forest trends to climatic change, are almost impossible to mitigate.

The atmospheric concentration of the most important of the greenhouse gases, carbon dioxide, has increased by about 25 per cent since the mid-nineteenth century (Chapter 7). Deforestation was the dominant source of this increase until the mid-1960s, when it was overtaken as a source by the combustion of fossil fuels. At present, it is estimated that deforestation contributes annually 1–3 billion tonnes of carbon, compared with about 5.6 billion tonnes for fossil fuels. In other words, deforestation has been and is a major source of carbon dioxide.

On the other hand, reforestation has the opposite effect, and some regions have changed from being a CO_2 source to becoming a sink over the last 100 years. The American South is one major example (Delcourt and Harris, 1980). Indeed with the forest transition having been effected in much of the developed world, the contribution of the temperate zone has dwindled (Fig.5.8). In recent years, reforestation has been seen as a partial means of averting the climatic warming that is likely to result from increasing atmospheric concentrations of carbon dioxide. However the contribution it is likely to be able to make is very limited. The reforestation of between 100 and 200 million ha per year would be required to absorb 1 billion tonnes, which is at the lower end of the

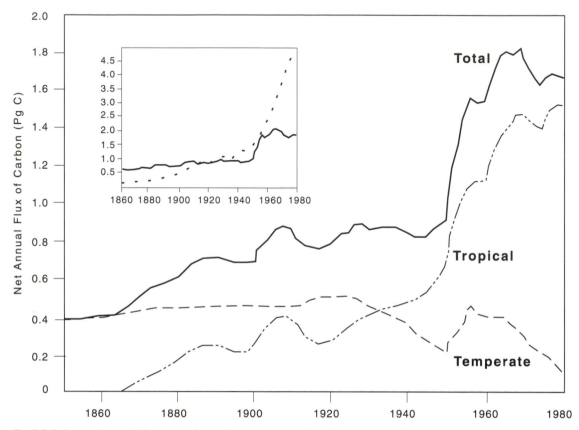

Fig. 5.8 Deforestation and CO_2: carbon fluxes. Main diagram: carbon flux from tropical and temperate forests. Inset: annual carbon flux from biota and soils (solid line) and fossil fuels (dotted line). (*Source:* modified after Woodwell, 1992, and Houghton and Skole, 1990.)

estimates of the output from deforestation. This is aproximately the same as the total area of plantations currently in existence.

If present trends of increasing carbon dioxide continue, an effect is likely to be felt in the forest resource itself. Much uncertainty exists about the probable patterns of climatic change, particularly because of the uncertainty that exists over cloud cover, rainfall and moisture availability as mean temperatures rise. One scenario, however, is that the main Northern Hemisphere temperate forest belt would move northwards, and in its southern margins be replaced by grasslands. Since the Arctic Ocean is a major obstacle to northward migration, the result would be that the forest belt would be narrowed, and the forest area would thus shrink rapidly. In practice, the northward movement of forests is likely to be slow, especially as soils capable of supporting forests will take time to develop (Houghton and Woodwell,

1989). On the other hand forests may respond to warmer temperatures and higher CO_2 concentrations by growing more rapidly.

Perhaps future historians will conclude that a forest revolution occurred in the twentieth century, just as an agricultural revolution occurred in the late eighteenth and nineteenth centuries. Alternatively, they may conclude that the period saw a management failure at the global scale to match that at the local scale in ancient Greece more than 2000 years ago. The legacy of that local failure was a chronic shortage of wood and other forest products, and an environment devastated by soil erosion.

The management of the world forest resource faces major and serious problems as the end of the century approaches. The greater part of the original forest area still survives, but the tropical forest is shrinking, and it is unlikely that this trend can be halted in the short term. Even where the extent of the forest is not dwindling, degradation is often occurring. In parts of the tropical forests, the most valuable species are being removed. In Canada and the former Soviet Union, reforestation is now accelerating in parts of the boreal coniferous forest most exploited for lumber, but large logged-over areas remain to be regenerated. There may be little likelihood of a serious industrial wood shortage in the foreseeable future, but acute fuelwood scarcities already exist in some parts of the world.

Perceptions of the forest resource are changing, and with them attitudes to its management. Active forest management in the past has often focused on wood production, and the other potential goods and services that the forest could provide have received less attention. Perhaps this is because real markets have existed for wood, but not for most other goods and services. Whatever the reason, some would claim that forest managers have sometimes been unable to see the trees for the wood.

In recent years, however, other forest goods and services have begun to attract more attention, and some estimates now indicate that when fully costed, the sustainable production of 'minor' products of tropical rain forests could yield bigger economic returns in the long term than could wood production. Even more striking is the increasing attention falling on services such as recreation, nature conservation and environmental protection. The economic values of these services (from tropical and temperate forests alike) are now far more widely acknowledged than even a decade ago, although they still cannot be easily quantified (Chapter 2).

The total economic value of the forest resource is the sum of use value (including that of 'minor' forest products and of services such as recreation as well as of wood), option value, bequest value and existence value. Few full valuations of this type have been carried out. One example, of Swedish forests, estimated the net value of wood production at 8000 million Swedish crowns, compared with 3325 million for other values. These included berries, mushrooms and game (1150 million), recreation (1600 million) and values of preserving species and virgin forests (750 million) (Jones and Wibe, 1991). In Britain, the recreational 'value' of state forests has been estimated

as lying between £14 million and £45 million per year, compared with timber sales of around £60 million (Willis and Benson, 1989). Debate will continue about the validity and interpretation of such estimates, but they at the very least serve as useful reminders that forests cannot be viewed simply in terms of wood production.

Market failures in the forestry sector have been widespread. Forests have simply disappeared under private enterprise systems in Britain and in many other parts of the world, and have only begun to stabilize when governments intervened. But government failure is also common, not least in many parts of the developing world where the forest resource is continuing to degrade even though under (at least nominal) government ownership. One of the frequently alleged reasons for government failure is that economic factors are not adequately taken into account in decision-making. Conversely, a common reason for market failure is that only economic factors focusing on wood have been acknowledged. The growing interest in environmental economics may permit governments to be more fully informed about the total economic value of the forests about which they must make decisions. This, in turn, may be a step in the direction of sustainable management.

Note: Unless otherwise indicated, data on areas and production are from FAO *Production Yearbooks* and *Forest Products Yearbooks*.

Further reading

Grainger, A. (1993) *Controlling tropical deforestation.* London: Earthscan.

Mather, A.S. (1990) *Global forest resources.* London: Belhaven.

Sharma, N.P. *et al.* (eds) (1992) *Managing the world's forests.* Dubuque, Iowa: Kendall/Hunt.

Westoby, J. (1989) *Introduction to world forestry.* Oxford: Blackwell.

Williams, M. (1989a) *Americans and their forests.* Cambridge and New York: Cambridge University Press.

Williams, M. (1989b) Deforestation: past and present. *Progress in Human Geography,* **13**: 176–208.

CHAPTER 6	# The land resource: biodiversity, protected areas and environmental conservation

Introduction

Human impact on the global environment has increased greatly in recent centuries, and especially during the last 50 years. Forests have been exploited and cleared. Cropland has increased in extent, and its use has intensified. Domestic livestock have increased in number, and their grazing pressures have increased in severity. As human activities exert increasing pressures on the global environment, biological diversity declines as habitats are transformed and the populations of some species dwindle, to the point of extinction in some cases

There are three main aspects of biological diversity (biodiversity): ecosystems; species; and genetic material. Losses of biodiversity are occurring at accelerating rates at all three levels. The immediate causes of this trend include the expansion of agriculture and forestry, and the over-cropping of some species of plants and animals. Like deforestation (Chapter 5) to which it is related, loss of biodiversity results from a variety of causes operating at different levels. One underlying cause is the growth of the human population and its requirements of food, wood and other resource products.

At the more immediate level, the loss of biodiversity is undoubtedly related to intensification of production in existing areas of agriculture and forestry, and the encroachment of these activities into new areas. Biodiversity is under threat from intensification in areas where agriculture is practised. The widespread use of a few varieties of the main cereals, for example, can mean the displacement and eventual loss of local varieties and their genetic material. As Chapter 4 indicates, the increasing use of fertilizers and pesticides has direct ecological effects, as does the increasing use of drainage and

other techniques of land management. These effects can include the loss of habitats and in some cases the eventual loss of species. Similar trends are in operation in areas of intensive forest management.

Biodiversity is also under threat from the encroachment of agriculture and other forms of resource development in forests and other wildlands. Wildlands, or areas that are relatively untouched by human activities, are usually remote and inhospitable to humans, except for the traditional cultures that have evolved ways of living in such environments. Today, the natural barriers that once protected the wildlands are breaking down, and new defences, in the form of legal protection, are increasingly being erected in order to try to safeguard their biodiversity and other natural characteristics.

These characteristics have a wide range of resource attributes and potentials, and thus demands for their conservation come from a wide range of interests. At one end of the spectrum is the view that wildlife should be protected because of the 'rights of nature' and its intrinsic value (Chapter 1). In its preamble, the UN World Charter for Nature (1982) states that 'Every form of life is unique, warranting respect regardless of its worth to man'. A related view is that wildland or wilderness is valuable because of the philosophical or spiritual experiences it may offer. At the other (utilitarian) end of the spectrum, the conservation of species and genetic material is seen as offering practical benefits in terms of economic potential in agriculture and forestry, and in medicine (Box 6.1).

This range of different resource perceptions has an important spatial dimension which complicates the framework for resource management. Those perceiving wilderness or intrinsic values in the tundra or tropical rain forest, for example, are often developed-world city-dwellers living thousands of miles from

Box 6.1 Potential benefits from conserving biodiversity

The conservation of biodiversity is advocated by some for ethical or philosophical reasons. In addition, potential practical and physical benefits may exist, especially in medicine, in plant and animal breeding, and in developing the commercial use of species that are not used commercially at present.

Agriculture
- Only 3000–7000 plant species have been used as food out of the 250 000 or more that have been identified. Only about 150 have been cultivated on a significant scale, and the food supply for over 85 per cent of the human population is based on a mere 20. Yet at least 75 000 plant species have edible parts.
- As plant breeding has proceeded, the gene pools in crops such as rice and maize are becoming smaller. In Indonesia, 1500 local varieties of rice have disappeared since 1975, and nearly three-quarters of the rice planted today is descended from a single maternal plant. In Canada, three-quarters of the wheat crop is derived from four varieties, and the entire soybean production in the United States is derived from six plants from one site in Asia.

  The loss of genetic diversity among domestic plants and animals is seen by some as a greater threat to human welfare than the loss of species: uniformity can mean vulnerability to pest and disease outbreaks. Recently introduced wild relatives of domestic wheat and barley have been estimated to provide disease protection worth US $50–160 million per year for US crops alone. One gene from a single Ethiopian barley plant now protects the Californian barley crop, with an annual worth of US $160 million, from yellow dwarf virus

Forestry and tree crops
- Species may be transferred from one part of the world to another. Examples include Norway spruce (*Picea abies*) from mainland Europe to Britain, Sitka spruce and Lodgepole pine (*Picea sitchensis* and *Pinus contorta*) from western Canada and Alaska to Britain and Scandinavia, and Monterey pine (*Pinus radiata*) from California to numerous other countries, including Chile and New Zealand.

Medicines
- Over 40 per cent of all prescriptions written in the United States contain one or more drugs that originate from wild species, and the annual

sales of drugs are over US $8 billion. Around 3000 plant species are known to have anti-cancer properties: 70 per cent of them are found in tropical forests. It is estimated that the Rosy periwinkle, found in Madagascar, now yields pharmaceutical products worth US $88 million per year (1985 prices).

Tourism and recreation
- The economic value of wildlife and scenery in tourism and recreation is enormous. In some countries, protected areas are major attractions: over three-quarters of tourists in Ecuador and more than half in Mexico and Costa Rica visit national parks or other protected areas. Ecotourism is a major source of foreign exchange in some countries. In Kenya, the viewing value of elephants has been estimated at US $25 million per year. 'Ecotourism' is estimated to have accounted for US $2–12 billion of the US $55 billion that tourism generated for developing countries in the 1988. In effect, nature is 'exported' by countries such as Kenya, and 'imported' by countries such as the United States, Germany and Britain whose citizens make up the majority of 'ecotourists'. In functioning as the basis for the economic activity of tourism, nature is becoming commoditized, albeit in a less direct form than that represented by sale of plants or animals to collectors.

The total economic value of biodiversity or biological material is the sum of direct-use value (e.g. from harvested products), indirect-use value (e.g. as carbon store), option and bequest value, and existence value (Chapter 2). Indirect values and option, bequest and existence values are all difficult to measure, and their measurement often involves major assumptions. Interpretation is often difficult, and quoted figures should be viewed with caution. Examples include an indirect-use value of US $1300/ha/year for tropical forests (for carbon storage) and an option/existence value of US $4.7 million /year for elephants and other species in Thailand, and US $3.2 billion /year for the Amazonia forest. Willingness-to-pay values of US $11 and US $15 respectively for the bald eagle and the grizzly bear have been reported by American adult respondents: the grossed-up values of these species would be enormous.

Source: compiled from Alexandratos (1988), Swanson and Barbier (1992) and other sources.

these areas. Another (and to some extent related) issue is the ownership of genetic material or of plant or animal species, and the issue of who should pay the costs of conservation and who should reap the benefits.

Under many traditional resource systems and common-property regimes, biodiversity was little threatened by activities such as food production and the gathering of forest products. Today, the scale and intensity of these activities have increased far more than the ability of their managers to mitigate the side-effects. (Whether the commercial and other value systems that provide the context for these activities encouraged the managers to mitigate the environmental effects is another matter.) The result is that increasing numbers of species are disappearing. In response, protected areas have been designated. Here, it is hoped, development pressures can be kept at bay and biodiversity and other non-material resources such as scenic beauty can be conserved. At the same time, the last few years have seen determined efforts being made to incorporate non-material values in decision-making about resource management in general (Chapter 2), and to attempt to curb further losses of biodiversity in areas already used for agriculture and forestry.

Biodiversity trends

The estimation of trends

No one knows how many species live on Earth. Around 1.5 million have been identified, but there are probably at least 5 million, and estimates of the total number range from 2 to 30 or even 80 million (Table 6.1). Around 90 per cent of the estimated numbers of mammals, reptiles, amphibians, birds and fish have already been identified. Together, these species number around 40 000. Probably a smaller proportion of the larger number of plant species has as yet been identified, but the greatest uncertainty is over invertebrate species. The danger is that many of these will become extinct before they have even been identified. Equally, there is a danger that many already known plants, birds and animals will also become extinct.

Extinctions related to human activities are not peculiar to the twentieth century. A phase of mammal extinctions occurred in North America around 10 000 BC, when humans first arrived. Similar epi-sodes occurred in Australia rather earlier and in Madagascar and New Zealand more recently. Even early hunting cultures could apparently have serious effects on biodiversity, in driving some species to extinction and in reducing the populations and ranges of others. As cultures developed and civilizations emerged, human capabilities of modifying habitats and of affecting populations of wild species increased. In the Nile Valley, for example, the expansion of agriculture and the draining of marshland, together with organized hunting, led to the elimination of species such as elephants, rhinoceroses and giraffes from the area by the time of the Old Kingdom (2950–2350 BC) (Ponting, 1991). Species such as bears, beavers and wolves were extinct or seriously reduced in population in Scotland by medieval times, as a result of a combination of hunting, loss of habitat and persecution.

A clearer picture emerges over the last few centuries, and it suggests that extinction rates for birds and mammals have increased greatly during the second half of the nineteenth century and in the twentieth century. Around 230 vertebrates (half of them birds) and nearly 600 plants have definitely become extinct since 1600 (Table 6.2). It is estimated that 60 species of birds and mammals became extinct during the first half of the twentieth century, compared with a background extinction rate for these groups of one every 100–1000 years (Reid, 1992). The current rate of loss is probably much higher. At least 14 species of birds became extinct between 1986 and 1990, as did a minimum of 163 plant species between 1990 and 1992 (Smith et al., 1993).

Some commentators suggest that at least 15–20 per cent of the species in existence in the mid-1980s will have become extinct by 2000. A more sober estimate is that 2–7 per cent of species will become extinct in the next quarter century, at a rate of 20–75 species per day. For birds, plants and mammals the correspond-ing rate would be one every 0.5–1.7 days (Reid, 1992). Another view is that half of all present species will become extinct in the next 50–100 years if current rates of tropical deforestation continue. Present rates of extinction are 1 million times faster than those at which new species are emerging (May, 1988). The rate of species loss from tropical deforestation is perhaps 10 000 times greater than the naturally occurring 'background' extinction rate (Wilson, 1989).

These estimates need to be viewed with caution. As has been indicated, there is uncertainty over the total number of species, and Chapter 5 has indicated that

Table 6.1 Estimates of species numbers

	Identified	Per cent identified
Mammals*	13 500	90–95
Birds	9 500	95–100
Fish	24 000	83–100
Plants	240 000	67–100
Invertebrates	1+ million	3–27
Total	1.4+ million	?

Source: compiled from various sources, including Smith et al. (1993) and WRI (1988).
*With reptiles and amphibians.

Table 6.2 Current status of species of plants and animals

	Extinctions*	Extinct*	Threatened
	(species numbers)		
Mammals	59	1	11
Birds	116	1	11
Reptiles	23	0.4	3
Amphibians	2	0.1	2
Fishes	29	0.1	2
Plants	584	0.3	9

Source: compiled from data in Smith et al. (1993).
*Since 1600.

there is also uncertainty over the rate of tropical deforestation. This is a key process, as at least half of all species are to be found in closed tropical forests. But even if the rate of extinction is uncertain, there is widespread agreement that it is at an all-time high in terms of human history, and there is even (disputed) talk of a 'mega-extinction spasm' comparable with those in geological history.

Extinction is the end stage in a process by which the population of a species dwindles. This process is usually accompanied by a shrinking of the range, or the geographical area in which the species is found. For example the African elephant population fell by more than half between 1979 and 1989 (Hosang, 1992). Its range may now be as little as one-third of that four centuries ago (WRI, 1988). The range of the grizzly bear in North America has decreased by a comparable amount (Peters and Lovejoy 1990). Dwindling population and contracting geographical range mean loss of genetic material, even if a species is not at immediate risk of extinction.

Like estimates of species numbers and extinction rates, estimates of numbers of vulnerable, threatened and endangered species need to be viewed with caution, not least because of problems of definition. But it is clear that significant and increasing proportions of species are at risk, in both the developing and developed worlds. The list of threatened animal species increased by around 40 per cent between 1986 and 1990 (Smith et al., 1993). Around 2200 vertebrate species and 22 000 plant species are now threatened, representing nearly 5 per cent of the former and 10 per cent of the latter.

Why are species endangered?

As the introduction to this chapter suggests, different causes operate at different levels. Various factors, operating singly or more usually in combination, can function as immediate causes. They include pressures from hunting, from loss of habitat resulting from the conversion to farmland, or from the effects of pesticides or other hazards (as highlighted in Rachel Carson's Silent Spring (Chapter 3)).

Although some cultures from early times succeeded in devising sustainable hunting strategies, some species of mammals and birds were hunted to extinction even in pre-agricultural times. Over-cropping is still an important cause of depletion towards or beyond the critical zone (Chapter 1). Large mammals such as the elephant and the African rhino are especially at risk, and the populations of some have declined rapidly in recent years (the black rhino population fell by half between 1980 and 1985, and in the Central African Republic went from 3000 to about 150 during that period (WRI, 1988)). Some species of whale have also been badly affected.

Hunting is not only for food. The commercial value of products such as ivory or rhino horn is high, and they make tempting targets for would-be hunters. Other small animals and birds and more especially plants have also been depleted because of their potential value in commercial trade within or between countries. In addition, hunting for sport has been a pastime down through the centuries, especially for élite members of society. Paradoxically, hunting can be both a serious threat to wildlife and a source of attempts at conservation. Over-hunting can deplete animal populations and eventually lead to extinction. On the other hand, hunters sometimes seek to protect their prey species from excessive exploitation or habitat loss, and can thus help to conserve them. This is as true for recreational hunting as it is for hunting for subsistence. While it may seem morally incongruous for blood sports such as hunting to lead to wildlife conservation, the pragmatism of the hunters can give rise to this effect.

While the direct pressures of hunting and gathering are important factors, the modification or loss of habitat is a more pervasive cause of decline in populations. It often occurs in conjunction with increased use of pesticides, and hence wildlife may be under several pressures simultaneously. Habitats are modified in various ways: by deforestation, by drainage of wetlands, and by cultivation of grasslands. The severity of modification varies: strictly speaking habitats are not 'lost' but they are converted from one type to another, which usually accommodates a smaller range of wildlife. The extent of habitat loss varies with country and type of habitat. In some European countries, very little natural habitat remains. In several African and Asian countries, over 80 per cent of the natural wildlife habitat had been lost by the mid-1980s (WRI, 1988). The implications for wildlife depend on what replaces the original habitat and on the species' requirements. In many developed countries, habitat modification has been severe as well as extensive, with fields and buildings having replaced forests, wetlands and other natural habitats. Elsewhere, the severity of modification is often less, and the original habitat may have been replaced by semi-natural grasslands or rangelands. As agriculture expands and urbanization continues, however, these semi-natural habitats come under threats of more severe modification. Certain habitats have been especially vulnerable to severe modification. One example is wetlands, many of which have been drained for agricultural or other purposes. While this process has a long history, extending back for many centuries in countries such as the Netherlands, it has accelerated during the twentieth century as technical capability and demand for resource products have increased. Up to 90 per cent of the wetlands in areas such as California, Italy and New Zealand are reported to have disappeared (Hughes, 1993).

The significance of habitat loss also depends on the setting. Species are not evenly distributed over the Earth's surface. At the global scale, the number of species per unit area tends to increase from either pole towards the equator (Huston, 1993). Within that general pattern, some geographical areas and ecosystems are characterized by low species diversities, and others by high diversities. Deserts are examples of the former, and tropical rain forests of the latter. Within rain forests, for example, there are further variations depending on factors such as climatic history and geographical location. Between 50 and 90 per cent of all species are found in the tropical rain forest, which occupies about 7 per cent of the Earth's land surface. Perhaps around 20 per cent of species are to be found in a few small areas within these forests, collectively extending to about 0.5 per cent of the land area (Tolba and El-Kholy, 1992). In at least some cases these coincide with the 'hotspots' of tropical deforestation (Fig. 5.2).

One of the areas of coincidence is Madagascar, which is one of 12 'megadiversity' countries which account for 70 per cent of the world's biodiversity. Madagascar has 4 per cent of the world's flowering plants, on 0.4 per cent of the land area (Ryan, 1992). Like other oceanic islands, it contains many endemic species (i.e. species not found elsewhere). Around 80 per cent of Madagascar's 10 000–12 000 plant species are endemic (Davis et al., 1986).

Endemic species are especially susceptible to extinction, if the habitat of the restricted area in which they are found is modified. Three-quarters of the vertebrate extinctions since around 1600 have been on oceanic islands (Peters and Lovejoy, 1990). In New Zealand the extinction rate for endemic species has been eight times higher than that for all the species native to the islands (Mark and McSweeney, 1990). In small islands such as St Helena and Mauritius, environmental destruction was wrought both by introduced species and by direct exploitation of forests under colonial rule. In St Helena, for example, 22 of the 33 endemic plant species have become extinct since rabbits were introduced in the nineteenth century. More than a century earlier, it had been considered necessary by the colonial government to introduce measures to preserve forests and to protect soil, such was the ecological devastation on that island after a short period of British exploitation (Grove, 1990).

The incorporation of new lands into the world economy, with the accompanying introduction of new forms of land use and environmental management, has been followed in many instances by extinction or severe loss of populations of native species. In some cases this has been brought about by over-cropping of natural populations, especially if these are now seen as commodities with economic potential. Another is the commoditization of certain environmental resources such as fur-bearing animals. For example, before European settlement became established in North America, there was little incentive to trap more animals than were required for local needs. After markets became established, fur-trappers had an incentive to increase their level of activity, with serious effects on wildlife populations in some areas. More recently, similar effects have resulted from the commoditization of certain plant and animal species for sale for gardens and zoos or as pets. The introduction of alien or exotic species, whether done intentionally or unintentionally, may also have a serious effect, as may the loss of habitat through logging or conversion to cropland or plantations. But it does not follow that pre-European resource management systems were benign in their effect on biodiversity. Debate may continue over the 'Pleistocene overkill' episode in North America, but there is little doubt that Polynesian cultures in New Zealand exterminated the large moa bird and converted large areas of forest to grassland. Whatever their environmental attitudes may have been (Chapter 1), biodiversity was not maintained.

The protection of wildlife and biodiversity

Attempts to protect wildlife and biodiversity have taken two main forms in modern times:

- Through laws that proscribe the taking or killing of named species (species protection), and that control trade in live plants and animals and in products such as ivory.
- Through the protection of habitats (i.e. by focusing on the living spaces or specific areas of land rather than on individual species).

These approaches have been adopted both at national and international levels, and are often employed in combination. The driving force has usually been the voluntary group, operating domestically or internationally.

The roots of both species and habitat protection have long histories, extending back over several centuries. In Britain, the modern nature-conservation effort can be traced back to protection afforded by law to certain species in the mid-nineteenth century. Voluntary bodies such as the Royal Society for the Prevention of Cruelty to Animals and the

Royal Society for the Protection of Birds were instrumental in producing a climate of opinion in which legislation could be enacted in nineteenth-century Britain. Since then the list of protected species has been extended greatly, especially by the Wildlife and Countryside Act of 1981. In recent years environmental groups have become larger, better funded and increasingly effective (Chapter 3). One of the ways this is manifested is in lobbying at both national and international levels, and in extension of species protection at both national and international levels. Species protection on its own may help to reduce threats from activities such as hunting or collecting, but it will not be effective if serious loss of habitat is taking place. It is therefore usually combined with attempts to protect habitats. One approach to habitat protection is through the designation of protected areas.

Protected areas

Many natural areas have been protected for religious or cultural reasons over many centuries. For example, forest groves in Ivory Coast were protected by local people because of their reverence for certain species of monkey found there (Sayer *et al.*, 1992). Whether or not environmental conservation was the main motivating factor, taboos and cultural pressures often helped to ensure protection of both habitat and species. In many parts of the world these taboos and cultural pressures have weakened in recent times, but at the same time the modern concept of the legally protected area has spread from its Western origins to most parts of the world.

The modern concept of the protected area has long and wide-ranging roots. Protected reserves helped to stem peasant encroachment on the forest and its wildlife in medieval Scotland and England, even if the underlying motive was geared to hunting rather than to wildlife conservation. More recently, protected forest and game reserves were established in Africa and India during the colonial era in the nineteenth and early twentieth centuries. Around the same time, the idea of the national park emerged in the New World. Its origins are complex and are probably at least partly related to national histories and identities (see Runte, 1979). Today, conservation of wildlife is usually a major objective. This was not always so. Originally national parks were more concerned with landscape, and it was not until the 1930s that the concept of national parks as nature

Plate 6.1 The idea of the national park originated in the United States during the second half of the nineteenth century, and then spread to countries such as Canada, Australia, New Zealand and South Africa. Other forms of protected area, including provincial parks and nature reserves, soon followed, to the extent that few countries now lack examples of such areas. Some protected areas are intended primarily for purposes of wildlife conservation whereas others, including national and provincial parks, combine objectives of landscape conservation and recreation. (a) Mount Egmont National Park, New Zealand. (*Source:* A. Mather)

Plate 6.1 (b) Rainbow Bridge National Monument, Utah, USA. (*Source:* K. Chapman)

Plate 6.1 (c) Peter Lougheed Provincial Park, Alberta, Canada. (*Source:* K. Chapman)

conservation preserves clearly emerged (Stankey, 1989).

Growth of protected areas

The total area under national or international systems of protection is now over 600 million ha or nearly 5 per cent of the Earth's land surface. The aggregate area of wildlife refuges in the United States is now nearly as large as California (Sax, 1993), and the aggregate protected area in the tropics is as large as Iran or twice the size of Venezuela, Pakistan or Tanzania (McNeely and MacKinnon, 1990). Most of the protected area has been designated since 1950 and much of it since 1970. In 1940, the total extent of the protected area was equal in size to Madagascar: by 1980 it was equal in size to India, Sri Lanka, Pakistan and Bangladesh combined (MacKinnon *et al.*, 1986). The number of protected areas doubled between the early 1970s and early 1980s, but the rate of growth has slowed down since the mid-1980s (Fig. 6.1). This may reflect both a shortage of potential areas for such status, and opposition to designation on the part of local peoples or other interests. Nevertheless, there is little doubt that the growth of protected areas (at both national and international levels) is one of the major features of environmental resource management in the second half of the twentieth century.

Limitations of protected areas

Impressive as the growth in protected areas has been, it needs to be viewed with caution. This is true for a number of reasons.

- There is the problem of 'paper parks', of protected areas that have been legally designated but where designation means little in practical terms.
- There is the question of location. It may be easier to designate a park or protected area in a remote part of a country where there are few pressures from logging, mining or agriculture, than in another area where the pressures, and hence the threats to biodiversity, are greater.
- Third, there is the question of distribution. As has been indicated, species are not evenly distributed. Ideally the distribution of protected areas would cover all the areas of greatest biodiversity, as well as representing all the main types of ecosystems. In Africa, for example, there is more extensive coverage of savanna ecosystems than of rain forests (Fig. 6.1). Nevertheless, protected areas in forested countries in Africa have been found to contain 70–90 per cent of these countries' bird faunas (Whitmore and Sayer, 1992).

In the light of the diversity of its antecedents, it is not surprising that protected areas such as national parks today take different forms (Table 6.3: see also

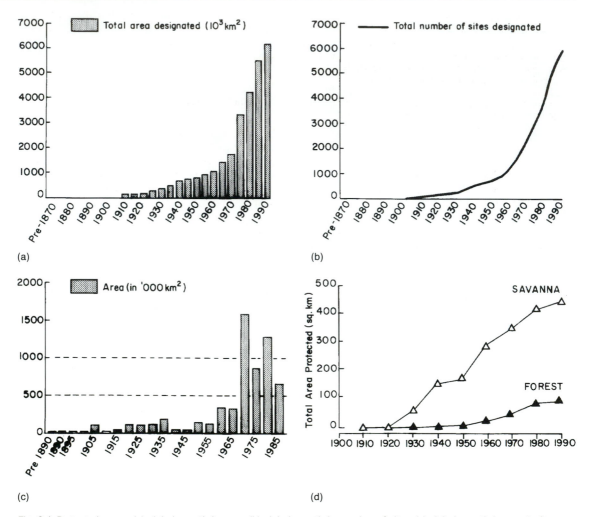

Fig. 6.1 Protected areas: (a) global growth in area; (b) global growth in number of sites; (c) global growth in area in five-year periods; (d) growth in rainforest and savanna protected area in Africa. (*Source:* modified after WRI (1986) and Sayer *et al.*, 1992.)

Box 6.2 for discussion of wilderness as protected area). Protection has different management implications in different types of areas. In some, conservation or preservation is the main management objective, but in others it is pursued alongside other resource activities. The condition of individual species may be influential in the designation of the protected area, but management is usually concerned with conserving entire habitats and whole ecosystems rather than individual species *per se*.

Biosphere reserves

One type of protected area is the biosphere reserve (Category IX in the classification shown in Table 6.3).

Biosphere reserves form an international system of protected areas developed under the auspices of the UNESCO Man and the Biosphere Programme. The goal was to establish a series of protected areas, linked together in an international network, with the aim of demonstrating the benefits of conservation to society in general and to the development process. The concept encompassed three concerns: the protection of genetic resources, species and ecosystems; a 'logistic' role, relating to the interconnected network of research and monitoring; and a 'development' role, relating to human use under sustainable conditions. The distinctive feature of the biosphere reserve is the combined presence of these roles, even if

Table 6.3 Categories of protected areas

Category I Scientific Reserve/Strict Nature Reserve. Natural areas in which natural processes are allowed to operate in the absence of direct human interference. Tourism, recreation and public access are generally proscribed (9%).

Category II National Parks. Relatively large areas where one or several ecosystems are not materially altered by human exploitation and occupation, and in which the highest competent authority of the country has taken steps to prevent exploitation or occupation, Visitors are allowed to enter for inspirational, educative, cultural and recreative purposes (60%).

Category III Natural Monuments/Natural Landmarks. These contain one or more of several specific natural features of outstanding natural significance (1%).

Category IV Nature Conservation Reserve/Managed Nature Reserve/Wildlife Sanctuary Areas. The protection of specific sites or habitats is essential to the continued well being of resident or migratory fauna of national or global significance (25%).

Category V Protected Landscape or Seascape Areas. The landscape possesses special aesthetic qualities which are the result of the interaction of man [*sic*] and land, and those that are primarily natural areas managed intensively for recreational and tourist use (5%).

Source: based on IUCN (1985).
Figure 6.1 (growth of protected areas) is based on these five categories. Among the other categories is IX Biosphere Reserve, where the management objective is to conserve for present and future use the diversity and integrity of biotic communities of plants and animals within natural ecosystems and to safeguard the genetic diversity of species on which their continuing evolution depends.
Figures in parentheses represent the extent of each category as a percentage of the total protected area (*c.* 1985).

the same emphasis is not always placed on each (Batisse, 1990).

The model biosphere reserve consists of a core zone located in a major ecosystem (such as a tropical rain forest), little touched by human activity and large enough to allow its genetic material to be conserved. Around it is a buffer zone where some human activities are carried on, for example in the form of traditional resource use and sustainable development projects (Fig. 6.2). The hope is that this buffer zone will protect the core from pressures of encroachment. (A fourth zone in which restoration or rehabilitation is attempted may also be demarcated.) The extent to which individual reserves conform to this ideal model varies.

The concept was developed in the early 1970s and the first biosphere reserve was designated in 1976. By 1990, there were around 300 reserves in 75 countries, with an aggregate area of 1.5 million km^2 (Batisse, 1990). Around one-quarter of the reserves are in

tropical rain forests. In some countries (especially in the developed world), biosphere reserves are superimposed on pre-existing protected areas such as national parks.

Conflicting perceptions

In national parks and strict reserves, there is likely to be a stronger emphasis on preservation, and the aim may be to keep them completely inviolate. Attempts at preservation have sometimes given rise to conflict between the park managers or conservation officials on the one hand, and local people and development interests on the other. Such conflicts are not surprising in the light of the origins of the reserve concept, and in particular of the social and political context in which reserves were introduced. For example, in southern Africa the British, as colonial power, pioneered reserves in the period between the 1860s and 1890s, and 'in essence, these policies were aimed at excluding all Africans from game reserves and banning African hunting' (Grove, 1990: 39). In the present century, the same pattern has continued, with reserve or park status being imposed by external power (in the form of the post-independence state, rather than by the colonial power). In some cases, indigenous people have been removed from prospective park areas, and in many others their resource-using activities have been constrained. Such conflicts stem in part from the different perceptions of environmental resources held by the local people and by the park managers. The former see the area in question as the source of their needs of food, fuel and other resource products. Since they have successfully conserved the environment, they argue, why should they now be required to give up their land and lifestyles? The park managers, on the other hand, are often from a different culture, their physical resource needs are met from outside the area, and they perceive the rarity value of the ecosystem they are seeking to preserve. In short, there are different attitudes to nature.

The national park or strict reserve is essentially a Western idea, introduced to developing countries by colonial powers and later copied by developing countries under a combination of pressure and encouragement from international organizations and conservationist members of local élite groups (see, for example, Hough, 1988). Frequently, it is imposed on a peripheral area by interests based on the core, at the national or global levels. This process is

Box 6.2 Wilderness areas

Changing attitudes to nature and changing perceptions of resources are nowhere better displayed than in the subject of wilderness. In Old Testament times, wilderness was seen as an inhospitable place of banishment. In medieval Europe and early modern America, it was also feared as a place full of hazards and hardships. Over the last 100 years, however, the wilderness that was once loathed has become cherished, at least by some social groups in some parts of the world.

By the second half of the nineteenth century, attitudes towards wilderness in America were changing. At the philosophical level, writers such as H.D. Thoreau began to highlight the spiritual value of wilderness experience. More pragmatically, the dangers and deprivations of the wilderness were being lessened as travel and provisioning became easier. And wilderness was simply becoming scarcer. By the end of the century, wilderness qualities were being conserved in some of the early national parks such as Yosemite in California. The climax in transformation of wilderness perceptions was marked in the passage of the Wilderness Act by Congress in 1964. Between then and 1990 over 35 million ha have been designated as wilderness in around 450 separate areas in the United States. The aggregate extent is about the same as that of California. In these protected areas, wilderness qualities are conserved for their recreational, spiritual, aesthetic and ecological values. Vehicular access, and activities such as mining and logging are usually prohibited. Recreational use grew rapidly until the 1980s, but more slowly thereafter (Lucas, 1989).

Although originally an American concept, designated wilderness has spread to other comparable parts of the world, such as Canada, Australia, New Zealand and South Africa. In most other parts of the world, separate wilderness designations have not been made, but wilderness areas have been zoned within other protected areas in many other countries. They occupy around half of the aggregate global protected area, or over 200 million ha (Eidsvik, 1989).

In addition to formally designated and zoned wilderness, huge areas of wild land remain. Using small-scale navigation charts of the US Defense Mapping Agency, McCloskey and Spalding (1990) identified land areas of over 400 000 ha with no human artefacts such as buildings, roads or mines. They estimated that around 5 billion ha, or one-third of the Earth's land surface, met these criteria.

Proposed resource developments such as logging or dam construction have been focal points for environmental groups in Australia and North America in particular. Such proposals have been defeated in several major test cases (such as the proposed Franklin River dam in Tasmania), and it has been suggested that some of these cases mark milestones in the transition towards the New Environmental Paradigm (Sewell et al., 1989). On the other hand, wilderness proponents are sometimes seen as élite and influential but small and unrepresentative groups. It is estimated, for example, that only 6–15 per cent of Americans have ever set foot in a wilderness (Manning, 1989). Even so, non-visitors may still perceive option and existence value in protected wilderness.

Wilderness has been described as a finite, non-substitutable, non-renewable, irreversible and common-property resource (Dearden, 1989). There can be little argument that it is finite and non-substitutable. There may be more debate about whether it is renewable or irreversible. Some areas of now-reforested abandoned agricultural land in the eastern half of the United States have now been designated as wilderness areas. 'Common-property' status raises many questions, especially in relation to the need for state action to protect wilderness areas and to the élite nature of wilderness advocates. Whatever view is taken of these questions, there is little doubt that the wilderness epitomizes many of the characteristics of environmental resources, not least in terms of changing perceptions through time and of differences in perceptions and attitudes between different social groups.

---

seen by some as the latest manifestation of a long-standing relationship in which the resource needs and values of the core have been visited on the long-suffering periphery. National parks have even been used in some developing countries as a mechanism for extending central-government influence into the most distant parts of the country (McNeely, 1988).

Furthermore, the military has been used in an attempt to enforce conservation in some countries, and the legitimacy of the use of force in this context (and the direct or indirect encouragement given by Western conservation interests) is questionable (e.g. Peluso, 1993).

The result of the imposition of national parks and

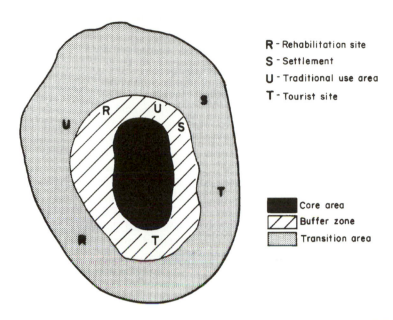

R - Rehabilitation site
S - Settlement
U - Traditional use area
T - Tourist site

Core area
Buffer zone
Transition area

Fig. 6.2 Biosphere reserve.

similar areas can be hostility stemming from the unequal power relationship between park officials and people, from their different cultures and perceptions, and from mutual lack of understanding. Park managers sometimes develop siege mentalities, in the face of pressures of encroachment from all sides. Local people sometimes feel dispossessed, or at least disadvantaged by the constraints placed on their resource activities. The resulting alienation is not conducive to conservation.

Conflicts have been reported from developed and developing worlds alike (e.g. Zube, 1986). One extreme example is the invasion by the Bodo tribe of the Manas National Park in India in 1989. The park, which contained 19 endangered mammal species, was opened to poachers, land was cleared, and 12 park staff were killed (McNeely, 1990). Conflicts between officials and local communities have a long history in Africa in particular. They are not only (or perhaps not even mainly) about conservation, but stem from different ideas about how land and other natural resources should be controlled and used (Beinart, 1989). Nevertheless, they do stem in part from the basic concept of a protected area, and in part from the view that the problem is essentially a biological one which can be solved by scientifically rational means.

Over the last few years the existence and significance of these conflicts has increasingly been recognized, and a shift in thinking has been evident. It reflects a move from a basically ecocentric or biocentric position towards one which is more anthropocentric. This may seem paradoxical in the light of what was said in Chapter 1. (Doubly paradoxically, national parks and similar protected areas would not be necessary if we did not have an exploitative attitude to nature (McNeely, 1988)). It reflects a growing recognition that conservation cannot be achieved simply by preservation, but will have to acknowledge the values, needs and aspirations of local people (Eidsvik, 1980). This shift has been towards the fuller incorporation of local people in the management of parks and similar areas, and of principles of sustainable development rather than non-use. In the view of Jeffrey McNeely, deputy Director of the International Union for the Conservation of Nature, protected areas must offer benefits to humans if they are to survive in a period of increasing pressures: '...protected areas should be designed and managed in ways that contribute to the well-being of the people living around them' (McNeely, 1988: 139). Even if such sentiments might look like compromises from the biocentric or 'deep ecology' viewpoints, there is little doubt that they have gained ground in recent years.

This does not, of course, mean that the concept of the national park or strict reserve has been discarded. It does mean recognition that the 'strict reserve' model of protected area is insufficient in the face of modern pressures. One problem with the conventional national park model is the difficulty of effective policing, especially but not only in the context of conflict that has just been discussed. Threats to biodiversity in national parks and similar areas exist in the developed and developing worlds alike, albeit in different forms. Unlawful entry and the illegal removal of animals (i.e. poaching) are more common in the developing world, while the taking of exotic plants is more widespread in the developed world (Machlis and Tichnell, 1987).

A related issue is that parks are sometimes located in areas that are perceived as having little development potential, rather than where conservation value is greatest. Another problem is that a park is a discrete area, with a boundary, and therefore there is a danger that the impression is given that effective conservation can be achieved solely in such areas, and that it can be ignored elsewhere. It is now increasingly being recognized that conservation cannot successfully focus solely on protected areas (or on protected species), and that it needs to be applied to all aspects of environmental resources. The corollary is that the scientific management of protected areas and species cannot in itself be successful if it is divorced from the wider use of environmental resources. As McNeely (1988: 141) concludes, 'Conservation is far more a social challenge than a biological one', and 'Conservation is too important to leave to scientists' (1989b: 157). The same commentator guesses that there will be no national parks by AD 2100: either they will have merged into more flexible approaches to land use, or we will have been unable to continue being 'biosphere people' and will have been forced back into being 'ecosystem people', who do not and will not need national parks. But, he concludes, 'until people learn to live with nature, or are forced to live with nature, we need national parks more than ever' (McNeely, 1989b: 157).

International conventions and agreements

Origins

The first major international conservation treaty resulted from an international conference on African wildlife, held in London in 1900. All the European powers with African possessions attended, but Britain and Germany assumed leading roles. As has been indicated, the British had been involved in establishing game reserves in southern Africa for several decades. Two large wildlife conservation areas had also been established in German East Africa (now Tanzania), in the 1890s (Lado, 1992). The Convention agreed that the colonial powers should seek to establish protected areas and to protect certain species. Classifications of game and those who might hunt it, and under what conditions, were drawn up in ways that were as revealing of contemporary perceptions of wildlife as they were of the imperial social attitudes of the day. Some species such as crocodiles and venomous snakes were still regarded as harmful and their destruction was encouraged. Reductions in numbers of some other species such as lions and predatory birds were also encouraged.

Following the conference, the British and Germans began to implement the Convention's principles on their territories and to encourage the other African colonial powers to do the same. The Society for the Preservation of the Fauna of the Empire (SPFE), founded in 1903, was an influential group which kept up pressure for game reserves and sanctuaries. Its hunter/conservationist members were sometimes referred to as the 'penitent butchers' (MacKenzie, 1988). (At around the same time, hunter-conservationists had been instrumental in having the first forest reserve established in the United States, and were involved in the first Conservation Movement (Reiger, 1992)). The membership of SPFE grew rapidly after the First World War, and reserves and sanctuaries continued to expand during the 1920s. A second convention was agreed in 1933, and as well as incorporating some additional territories such as Egypt and Sudan, it strengthened the concept of protected zones by means of national parks. It also now referred to plants as well as animals (which the first convention had not done). References to harmful species were dropped.

In the African colonial territories, various differences emerged as national parks were created. The

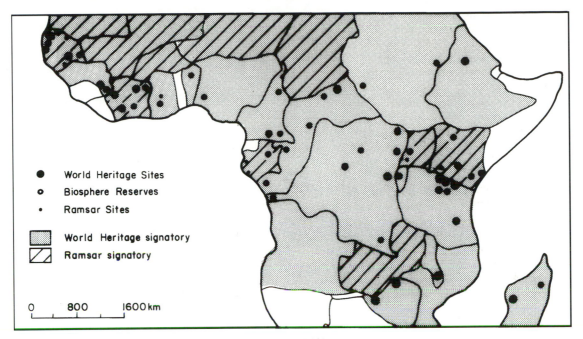

Fig. 6.3 Areas protected under international conventions, tropical Africa.
World Heritage Sites: sites listed in Convention Concerning the Protection of World Cultural and Natural Heritage (Paris, 1972).
(Objective: to protect cultural and natural heritage of outstanding universal value.)
Biosphere Reserves: see text.
Ramsar sites: sites listed under Convention on Wetlands of International Importance especially as Waterfowl Habitat (Ramsar, Iran, 1971) Objective: to stem loss of wetlands.) (*Source:* compiled from data in Sayer *et al.*, 1992 and other sources.)

French, for example, were more active in tropical forests, whereas the British focused more on savannahs and highland areas. These differences have been explained, at least partly, in terms of different attitudes to nature and different approaches to land development (Burnett and Stilwell, 1990). In addition, humid lowland forests were perceived as lacking potential for tourism, unlike savannahs (Pullan, 1988).

Although the concrete achievements of the 1900 and 1933 conventions are debatable (not all the parties ratified them), there is little doubt that they were significant in reflecting changing attitudes to wildlife and its management. Their provisions encompasssed a range of measures that are now in widespread use at both national and international levels. In addition to game reserves and national parks, they included measures for the protection of particular species against hunting and other threats, and also sought to regulate trade in commodities related to hunting (e.g. ivory).

Recent developments in international agreements

At the international level, the conventions agreed during the last 20 years reflect the changing attitudes to nature that have characterized this period. Several international conventions stemmed from the time of the UN Conference on the Human Environment held at Stockholm in 1972, a milestone in the emergence of the environment as an international issue. Ten years later, the UN World Charter for Nature declared that special protection should be given to unique areas, to the habitats of rare or endangered species, and to representative examples of different kinds of ecosystems. A new era in conservation policy had begun, marked by an awareness of the planetary dimension of the issues (Kiss and Shelton, 1991). Among the new measures of the 1970s and 1980s were the Ramsar Convention on wetland conservation (Fig. 6.3), and the Convention on International Trade in Endangered

Species of Wild Fauna and Flora (CITES) concluded in Washington in 1973.

CITES and the Bonn Convention

The goal of the CITES treaty is to control, reduce or eliminate international trade in those species whose numbers or conditions suggest that further removals of individuals would be detrimental to the species' survival. Among the obstacles to this aim, the primary factor as identified by Favre (1989) is human greed – the desire to own the unique or exotic and to make large sums of money. Essentially, CITES controls trade in live specimens of listed species, and also in products deriving from them.

Sustainable production of ivory from African elephants might amount to around 50 tonnes per year, but over 900 tonnes were exported in 1979. In order to curb this trade, a system of ivory export quotas was introduced by the parties to CITES in 1985. At that time, the elephant was listed in Appendix II of the Convention, which permitted limited trade. The quota system proved to be inadequate. In 1989 CITES members (by then numbering over 100 countries) voted to transfer it to Appendix I. This in effect banned commercial trade in ivory. Some African countries report that poaching has declined as a result (Hosang, 1992).

The same kinds of conflicts as those found in protected areas can arise if local peoples are deprived of income from trade in animal products, especially if the species in question is not endangered in their area. Trade bans may be effective short-term or emergency measures for species protection. In the longer term, sustainable utilization, with economic incentives for conservation, may be a more appropriate measure.

Another example of an international measure is the Bonn Convention on Conservation of Migratory Species, dating from 1979. Its objectives are to conserve species of wildlife that migrate across national boundaries, by restricting harvests, conserving habitat, and mitigating other adverse effects. Animals listed in Appendix I of the Convention are in danger of extinction in all or part of their ranges, including migratory routes. They are to receive strict protection from member countries. The Bonn Convention illustrates some of the problems inherent in this type of measure. It is a potentially powerful tool for conserving migratory wildlife, but only about 40 countries have acceded to it because of its require-

ments and enforcement costs. Some major countries such as the United States and Canada have not signed.

The significance of international agreements

Three major weaknesses are widely encountered in international agreements and conventions such as CITES and Bonn.

- One is enforcement. Non-compliance may attract criticism from other members (and from environmental groups), but compliance can rarely be effectively enforced by law.
- A second is in the translation of the agreements in principle that are represented by international conventions into effective national law or action.
- The third weakness is structural: to attract agreement, the provisions of the conventions are sometimes couched in terms which are neither very precise nor very constraining.

Like most international conventions, CITES and Bonn are imperfect, but at least are a significant step in the right direction. They represent adjustments in resource regimes at the international level, and mark the beginnings of a framework whereby the global commons are regulated (Chapter 2). Within groups of countries such as the European Union, the application of conventions such as CITES and Bonn is strengthened by measures such as directives and regulations emanating from the European Commission. For instance, under the Birds Directive, each member state is required to comply with a requirement to protect certain species of wild birds and their habitat, for example by controlling or prohibiting hunting and the taking of eggs, and by establishing special protection areas.

There has also been a rapid development of 'soft law', in the form of international standards, codes and guidelines that are not legally binding but which have been promoted by UNEP in recent years. The Rio Conference marked the culmination of this process of growth of international agreements. An International Convention on Biological Diversity was first proposed in 1974. During 1990 and 1991 it was negotiated under the leadership of UNEP, and was launched at UNCED in 1992 (Table 6.4). The objectives of the Convention are to ensure the conservation of biological diversity and the sustainable use of its components, and to promote a fair and equitable sharing of the benefits arising from genetic resources. Each contracting party is com-

Table 6.4 Biodiversity Convention: main features

- Development of national plans, strategies or programmes for the conservation and sustainable use of biodiversity.
- Inventory and monitoring of biodiversity and of the processes that impact on it.
- Development and strengthening of the current mechanism for conservation of biodiversity both within and outside protected areas, and the development of new mechanisms.
- Restoration of degraded ecosystems and endangered species.
- Preservation and maintenance of indigenous and local systems of biological resource management and equitable sharing of benefits with local communities.
- Assessment of impacts on biodiversity of proposed projects, programmes and policies.
- Recognition of the sovereign right of states over their natural resources.
- Sharing in a fair and equitable way the results of research and development and the benefits arising from commercial and other utilization of genetic resources.
- Regulation of the release of genetically modified organisms.

Source: compiled from Fowler (1993) and other sources.

mitted to developing national strategies for the conservation and sustainable use of biological diversity, and to integrate conservation and sustainable use into their relevant sectoral policies (e.g. for agriculture).

The question of a 'fair and equitable sharing' and the wider question of the ownership and resource regime are important and thorny issues. With the development of market economies, the exclusive possession of plant species could lead to control over production and prices, and hence profits. The colonial period saw the rapid development of these economies, and the active involvement of colonial powers in evaluating and exploiting plant species. For example rubber was smuggled from its natural home in Brazil, and developed through a network of British botanical gardens and commercialized in South-East Asia, to the serious detriment of Brazilian rubber production. In this and many other examples, genetic resources were simply appropriated wherever they were been found, without benefit to the host country.

As has been indicated, some developing countries (for example Brazil) are rich in biodiversity, and may be expected by the international community to meet much of the cost of conservation (for example in forgoing benefits that might stem from agricultural development). There is little incentive for

them to do so if the potential economic benefits are to accrue only to plant-breeding or pharmaceutical corporations in the United States or other parts of the developed world. On the other hand, it is argued that the value of a species or of genetic material in its natural setting is small compared with that added in the course of development of new varieties of seeds or new pharmaceuticals. Hence, it is claimed, 'intellectual property' rights are involved, and that these are held by the 'inventors'.

In essence, the argument is whether the benefits should accrue to the country in which the biological material is found, often in the South, or to the companies or individuals, often based in countries in the North, who 'add value'. (Another view is that the benefits should somehow accrue to humankind in general.) In more general terms, the dispute is between 'common heritage' and 'private property' perceptions of plants.

Although 153 countries signed the Convention at Rio in 1992, the United States initially declined to do so because of the issue of the ownership and sharing of the benefits accruing from commercial developments based on biological resources. In response, Venezuela threatened legal action against American pharmaceutical companies in order to prevent them gaining access to the country's native flora and fauna. Eventually the United States did sign, just before the Biodiversity Convention was closed for signature in June 1993 (by which time 163 countries had signed). Like many other countries including the United Kingdom, it has, however, been slow to ratify the Convention. Despite their government's slow response, some American organizations did acknowledge the rights of host countries to secure benefits from biological resources. For example, the National Cancer Institute of the United States negotiated contracts for access to genetic resources in countries such as Madagascar, Tanzania and the Philippines. Merck, a major American pharmaceutical company, has entered into an agreement with the agency responsible for Costa Rica's biodiversity. Under such contracts, provision is made for the payment of royalties on commercial applications (Simpson and Sedjo, 1992). In essence, national ownership of genetic resources is being recognized, and the resources are being commoditized (Chapter 1). Whether this new economic dimension will contribute to their conservation remains to be seen.

Limitations of species and habitat protection

Measures relating to protection of species and habitats are helpful in conserving biodiversity, but on their own are inadequate and insufficient. The aggregate protected area is only about 5 per cent of the Earth's surface, and all species have not yet been identified, far less protected under national or international law. One limitation is that legislation on species protection does not ensure that ecosystem diversity is maintained. Another is that protected areas reflect the 'key area' approach or 'enclave' strategy that has been followed in many developed countries (see, for example, Sax, 1993). This in turn reflects an assumption or implication that conservation can be achieved by protecting certain areas, rather than on a countrywide basis. Over the last few years, the inadequacy of this approach has been increasingly recognized, and some progress has been made towards integrating environmental principles into production-related policies, for example in the agricultural sector in the European Union. Perhaps the broad objective of sustainable development will provide the context for further progress in this direction. Protection for individual species and particular areas will continue to be necessary, but so also will policies that help to ensure that biodiversity is maintained on cropland as well as in wildlands.

The biodiversity issue is full of ironies and paradoxes. One is that some protection measures have originated from hunter/sportsmen who valued wildlife primarily as prey, rather than for its own sake. Another is that conservation has at the same time become more ethical and more pragmatic in its orientation in recent years. The UN Charter refers to the intrinsic value of wildlife, but there has also been increasing recognition that measures based on such perceptions (for example in national parks and strict reserves) cannot succeed without regard to the needs of local people. In other words, conservation has to be integrated into the wider framework of the management of environmental resources, and not be viewed simply as a separate sector.

With an accelerating rate of extinctions and a continuing loss of habitats and of genetic material, the outlook for biodiversity seems bleak. But there are some grounds for optimism. One is that biodiversity is inversely related to soil fertility and agricultural productivity (Huston, 1993): the most fertile and productive areas are generally not the areas with the highest levels of biodiversity. A fully integrated global economy, with unrestricted trade, could mean that the optimal areas for biodiversity remain relatively intact, while the optimal areas in terms of soil fertility are used for intensive food production. Such a scenario raises numerous questions, not the least of which is the nature of the economic benefits that can be reaped by high-diversity countries in return for conserving biodiversity. In principle, however, the operation of comparative advantage could mean that high levels of food production and conservation could be combined.

In Chapter 5, reference was made to the forest transition, or the change from declining to increasing forest areas that has characterized many developed countries during the present century. Perhaps it is possible that a similar transition may yet exist in relation to biodiversity or wildlife. The decline in wildlife in the United States, in the face of loss of habitat and over-exploitation, has been documented by Harrington (1991). He goes on, however, to discuss the recovery that has occurred during the present century. For example, in the United States populations of elk, whitetail deer, wild turkey have all grown substantially during the present century. To be sure, the recovery is only partial, and there is no way back from extinction. Nevertheless, the American experience would suggest that at least some success can be achieved through a combination of hunting controls, species protection and protected areas. Whether the ecological decline associated with tropical deforestation can be halted without catastrophic loss of biodiversity remains to be seen.

Further reading

Kiss, A. and Shelton, D. (1991) *International environmental law*. Ardsley on Hudson: Transnational.

Ledec, G. and Goodland, R. (1988) *Wildlands: their protection and management in economic development*. Washington: World Bank.

May, R.M. (1988) How many species are there on Earth? *Science*, **241**, 1441–8.

McCormick, J. (1989) *The global environmental movement*. London: Belhaven.

Munasinghe, M. (1992) Biodiversity protection policy: environmental valuation and distribution issues, *Ambio*, **21**(3), 227–36.

Swanson, T.M. and Barbier, E.G. (eds) (1992) *Economics for the wilds: wildlife, wildlands, diversity and development.* London: Earthscan.

Whitmore, T.C. and Sayer, J.A. (eds) (1992) *Tropical* deforestation and species extinction. London: Chapman & Hall.

Wilson, E.O. (1989) Threats to biodiversity, *Scientific American*, **261**, 60–66.

CHAPTER 7

Energy resources

The importance of energy resources in human development is undisputed. H.G. Wells (1914: 1) suggests that 'The history of mankind is the history of the attainment of external power'. Many factors contributed to the differentiation of *Homo sapiens* from the other primates, but the ability to supplement the physical strength of the human body by the development of tools and by the exploitation of energy sources additional to the food obtained by hunting animals and collecting edible plants was very important. Indeed the search for new and better sources of 'external power' has been a persistent theme in the endeavour to improve the quality of human life (see Cottrell, 1955; Cook, 1976, 1977). Despite the huge advances that have been made, concerns about the adequacy and availability of energy resources remain while the benefits of these advances appear to be more unevenly distributed among the world's population than ever before. Wells's vision of 'A world set free' by universal access to unlimited energy seems a distant prospect and, in these circumstances, it is important to appreciate the place of energy resources in the modern world.

Sources of energy

Humankind recognizes numerous energy resources and various classification schemes have been devised to represent this diversity. For example, a distinction may be made between 'primary' and 'secondary' sources of energy. Much of the energy we use is 'secondary' in the sense that it has been 'processed' before reaching us; most consumers rarely see crude oil, but are familiar with the products of its refining such as gasoline and paraffin. Another important distinction is that between 'commercial' and 'non-commercial' sources of energy. Commercial energy is bought and sold; non-commercial energy never enters a market in the accepted economic sense. Firewood which is collected by and used within the same family group is non-commercial; firewood bought from a trader is commercial. The distinction is more than semantic. Most published statistics refer to commercial energy, but the daily energy requirements of the majority of the world's population are met from non-commercial sources. In these circumstances, published statistics may give a misleading impression of the relative importance of the various energy sources, especially in developing countries. Most statistics used in this chapter will, however, reflect the availability of information and refer to commercial energy unless stated otherwise.

Figure 7.1 uses another classification system based on the dichotomy between renewable and non-renewable sources of energy. Some of the ambiguities of this distinction as a general principle applying to the classification of all environmental resources were noted in Chapter 1. Nevertheless, it provides a convenient framework for describing the characteristics of energy resources.

Renewable energy resources

Solar energy

Despite the 'spaceship Earth' concept, more than 99 per cent of the energy entering the global ecosystem is derived from an extra-terrestrial source – the sun – with the remainder coming from terrestrial sources in the form of heat from geothermal activity and manifestations of gravitational energy such as falling water (Hubbert, 1971). The sun is, therefore, the ultimate renewable energy source (although it is a

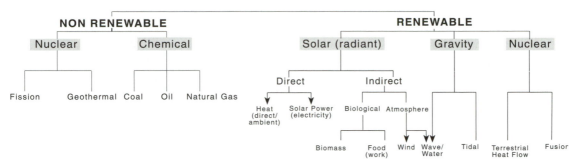

Fig. 7.1 A classification of energy resources.

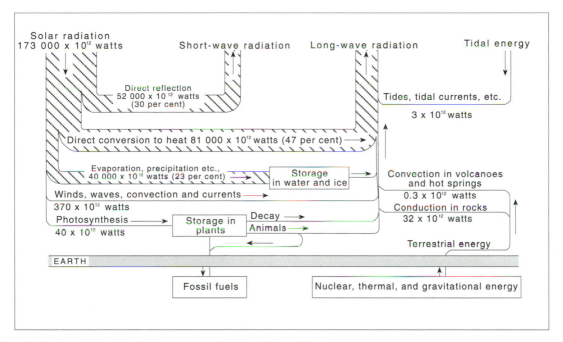

Fig. 7.2 Flow of energy to and from the Earth. (*Source:* Hubbert, 1971.)

wasting asset on an astronomical time scale). Approximately 30 per cent of solar radiation intercepted by the Earth is directly reflected back into space; most of the remaining 70 per cent is absorbed by the atmosphere, land and water surfaces (two-thirds of this is directly converted to heat and one-third drives the hydrologic cycle essential to human life). Two remaining destinations of incoming solar radiation are identified in Fig. 7.2 which represents the Earth's energy balance. A small fraction is responsible for the operation of the atmospheric and oceanic convections and circula-

tions and an even smaller proportion (approximately 0.022 5 per cent) is absorbed in the leaves of plants as a result of photosynthesis. The apparently residual nature of this latter use belies its critical importance in meeting the basic biological requirements of the Earth's plants and animals. Its importance in energy terms is also considerable because part of the solar radiation contributing to photosynthesis is transformed by processes of organic decay into fossil fuels such as coal and oil. The minute proportion of incoming solar radiation stored (i.e. fossilized) in this way is, therefore, effectively converted into non-renewable form and it may be argued that most forms of energy used by humans (except nuclear, tidal and geothermal) are of solar origin.

The solar energy absorbed by the atmosphere maintains the ambient temperature of the planet. The other direct use of the energy of the sun is that which is converted to electricity (i.e. solar power). Despite its appeal as an apparently pollution-free renewable source of electricity, solar power makes an infinitesimal contribution to commercial energy consumption at a global scale. Even in the United States, where considerable research and investment has been directed to its promotion in favourable locations such as southern California and Arizona, solar power scarcely registers in official statistics (US Department of Energy, 1993). It is possible that solar power may have a brighter future in developing countries where it may prove cost-effective in supplying electricity to rural communities. Government agencies in India, for example, are exploring this possibility with financial support from the World Bank. Of course, solar energy may also be used directly to provide space- and water-heating. Careful design of new buildings and modification of existing ones are required to make the most effective use of solar heat. Steps have been taken to utilize solar heat in this way, but such initiatives tend to be locally based, and the exploitation of solar heating as a deliberate strategy of energy use is of little commercial significance.

The life-supporting role of the sun may be regarded as an indirect form of solar energy. This is reflected in the food essential to sustain the human capacity for work, and in fuels such as wood and agricultural waste. Wood and waste are forms of biomass which is identified as a renewable energy source in Fig. 7.1. Others include animal manure, which is dried and burnt as domestic fuel in several countries including India, and crops such as sugar cane, sorghum and maize which may be cultivated for conversion into ethanol by fermentation and distillation. In certain countries, notably Brazil, ethanol is used to supplement oil-derived gasoline as a fuel for transport use. It is important to appreciate that biomass is renewable only if levels of usage remain below the *critical zone* (see Chapter 1). In practice, this threshold is often exceeded, resulting in a progressive elimination of the resource base. This is reflected in, for example, a chronic shortage of woodfuel in many parts of Africa (see Chapter 5). The definition of biomass as renewable is, therefore, problematic and it is more appropriately placed at an intermediate position along the resource continuum (Fig. 1.4).

Wind and water

Solar and gravitational energy combine to create several renewable sources dependent upon the movement of wind and water. Wind power, which ultimately reflects the operation of the atmospheric 'engine' driven by the sun, has been and is harnessed for many purposes including the grinding of grain, the raising of water for both drainage and irrigation and, most important, for marine transportation. Its usefulness is limited by the geography and temporal variability of suitable climatic conditions. These problems continue to frustrate attempts to promote wind power as a source of electricity on anything other than a local basis, although numerous studies have emphasized its potential in specific locations as far apart as Scotland and Israel. Water power may be used directly to drive waterwheels of various types and indirectly to generate electricity. Both applications are renewable to the extent that they rest upon the gravitational energy contained within running or falling water; both are, however, constrained by the availability of suitable sites. This may reflect physical conditions resulting from the intensive development of water-powered factories along specific rivers. It may also reflect organized opposition. Hydro-electric schemes are often controversial because they may involve massive engineering works and the flooding of large areas of land (see Chapter 9).

The same physical principles which permit the generation of electricity from the movement of water in rivers may be applied to the marine environment. Various experimental devices to convert the kinetic energy of waves into electricity have been tested and may ultimately be installed off the coast of such places as the Outer Hebrides which rarely experience a dead calm. Tidal energy, which rests upon the exploitation of the differential between high and low water, proceeded beyond the experimental stage with the commissioning of a plant at La Rance in France in 1968. This remains the only such facility in the world, however, and tidal power is an engineering curiosity rather than a commercial reality.

Non-renewable energy resources

Nuclear power

Nuclear power is shown in Fig. 7.1 as both renewable and non-renewable. As originally conceived, it depended upon the availability of naturally occurring uranium to supply the isotope (^{235}U) responsible for the release of energy by the breakdown (i.e.

fission) of large nuclei. Supplies of ^{235}U were initially thought to be limited. This perception stimulated attempts to develop a 'closed' cycle in which spent fuel from nuclear power stations is reprocessed to create new fuel rods for return to the original power station. This practice was viewed as a means of delaying considerably the rate of depletion of uranium resources. The prospect of infinitely renewable nuclear energy was subsequently raised by further technological developments epitomized in the concept of 'fast-breeder' reactors based on nuclear fusion. This technology remains a theoretical possibility rather than a practical proposition, however, and prevailing applications of nuclear power are essentially non-renewable (see pp. 146–8).

Geothermal energy

Similar ambiguities arise in classifying geothermal energy as a renewable or non-renewable source. As currently used it is non-renewable in the sense that it depends upon using heat contained in water and steam trapped in pockets near the Earth's surface. These pockets are analogous to oil reservoirs and are finite. Although geothermal energy is used to generate electricity and also to provide direct space-heating in several countries including Italy, New Zealand and Iceland, its contribution to global energy production is negligible. Theoretically it could be regarded as a renewable energy source if a commercially acceptable means of exploiting the rise in temperature evident in boreholes penetrating the Earth's crust could be found (i.e. terrestrial heat flow in Fig. 7.1).

'Fossil' fuels

The fossil fuels of coal, oil (including oil shales and tar sands) and natural gas represent forms of chemical energy, ultimately created by solar radiation (Fig. 7.2), and locked by geological processes into energy stores. The energy contained in these stores is released by combustion. They are undeniably non-renewable on a human time scale, although it is difficult to judge the size of their respective resource bases (see p. 143 et seq.). It is estimated that more than 80 per cent of world energy use is derived from non-renewable sources, principally the fossil fuels (World Energy Council, 1993). Indeed, coal, oil and natural gas have provided the overwhelming proportion of the energy which has fuelled the growth of the developed world since the Industrial Revolution and much of this chapter is concerned with some of the consequences of this dependence.

Energy consumption and economic development

Energy and society

The link between human social evolution and the use of energy has already been noted. Essentially, economic growth has required a progressive increase in levels of per capita energy consumption. This increase has not been gradual, but has involved sharp jumps associated with the great transitions in economic and social organization referred to in Chapter 3. Table 3.2 illustrates this relationship and the huge gulf in levels of energy consumption has prompted a distinction between low- (i.e. hunter-gatherer and agricultural) and high- (i.e. industrial) energy societies. It is not only the level, but also the type of energy consumption which distinguishes these societies. Low-energy societies rely upon plants, animals and human labour (see Cottrell, 1955: 15–38; Rapaport, 1971). By contrast, modern high-energy economies are characterized by a dependence upon fossil fuels, a significant proportion of which are diverted to the production of secondary energy in the form of electricity (see Cook, 1971).

The historical perspective upon the connections between energy consumption and economic development suffers from the difficulty of obtaining reliable data, although an interesting attempt has been made to assemble a range of evidence for trends in the United Kingdom since 1700 (Humphrey and Stanislaw, 1979). It also encourages a deterministic view of human evolution and it is important to remember that most of the world's population still live in low-energy societies. Figure 7.3 is based on a representative sample of countries. It suggests a good correlation between levels of energy consumption and economic development (as measured by per capita gross national product (GNP)). The world's poorest countries lie close to the origin of the graph and the richest in the opposite corner. The clear implication is that higher standards of living are sustained by higher levels of energy consumption. While this conclusion is undisputed, energy consumption is influenced by many factors, including the climate and size of a country. Requirements for heating (and cooling) are higher where extremes of climate are experienced and for transport where population centres are separated by great distances. Such factors contribute to Canada's position well above a notional 'best-fit' line on Fig. 7.3. Conversely, Japan lies well below

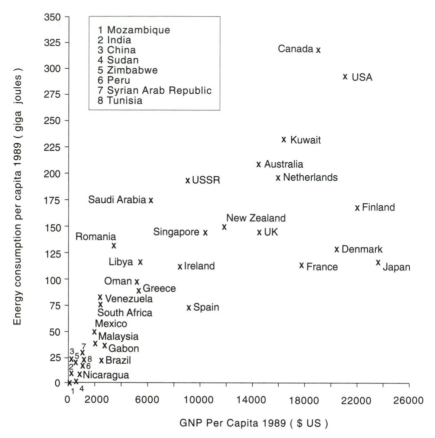

Fig. 7.3 Energy consumption and economic development for selected countries, 1989. (*Source:* compiled from data in World Resources Institute, 1992.)

such a line, partly because of the success of policies designed to promote the more efficient use of energy in a country which is heavily dependent upon imported oil.

It is significant that a large number of countries are positioned in the bottom left corner of Fig. 7.3. The significance of this pattern is reinforced when it is noted that China and India, which rank first and second respectively in population size, are found in this cluster together with other populous countries such as Bangladesh and Nigeria. This suggests that the disparities in wealth between rich and poor countries are matched by very unequal shares of world energy consumption. Figure 7.4 emphasizes this point by plotting the respective shares of the seven most populous countries and the seven largest energy consuming countries in 1991. The fact that the United States accounted for less than 5 per cent of world population, but was responsible for almost 25 per cent of total energy consumption clearly raises moral questions relating to the equitable distribution of resources.

Consumption by energy source

The dependence upon non-renewable energy sources within the developed economies is evident in Fig. 7.5. The United States is of particular interest because it is the country with the largest gross energy consumption (Fig. 7.4). The transformation from a predominantly agricultural and rural society to an urban and industrial one between 1860 and 1914 coincided with a pronounced rise in energy consumption (relative to earlier trends) and with the displacement of wood by coal as the principal fuel. Coal remained the most important single fuel until the late 1940s. Oil was used in kerosene and paraffin lamps as early as the 1860s, but it was the widespread introduction of the motor

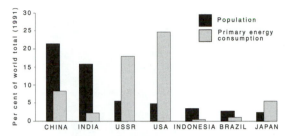

Fig. 7.4 Shares of world population and primary energy consumption, 1991. (*Source:* compiled from data in BP, 1993, and World Resources Institute, 1992.)

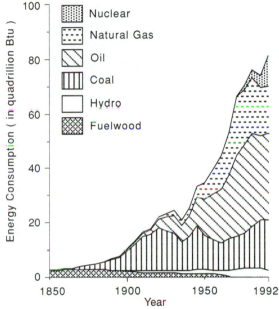

Fig. 7.5 US primary energy consumption by source, 1850–1992. (*Source:* compiled from data in US Bureau of the Census, 1960 and US Department of Energy, 1993.)

car which was mainly responsible for its increasing share of total energy consumption from the late 1920s. Before 1940 natural gas use was restricted to the principal areas of production such as Texas but the construction of long-distance pipelines opened up new markets in the population centres of the North East, leading to a sharp rise in its contribution to aggregate energy consumption. Despite the development of nuclear power, the exploitation of hydro-electric potential in several states and a marked recovery in the contribution of coal since the mid-1970s, the two closely-related fuels of oil and natural gas were responsible for two-thirds of US primary energy consumption in 1992 (Fig. 7.5).

Coal's supremacy as the dominant fuel was challenged by oil and natural gas earlier in the United States than in other developed economies. Nevertheless, the energy 'mix' of the United Kingdom, for example, is now remarkably similar to that of the United States. Other developed economies reveal differences in the relative importance of the three fossil fuels with Japan relying more heavily upon oil whereas coal is much more significant in Eastern Europe (Aitchison and Heal, 1987).

In global terms, the core economies of North America, Western Europe and Japan are accounting for a declining share of world energy consumption, mainly as a result of rapid increases in newly industrializing countries (NICs) such as South Korea and Taiwan. The OECD countries (i.e. the developed countries excluding members of the former communist bloc) accounted for approximately 65 per cent of world energy consumption in 1973; their share is expected to fall to less than 50 per cent during the 1990s (OECD, 1992a). The NICs have followed the path of the 'core' economies in relying heavily upon the fossil fuels and it seems likely that the development aspirations of the poorest countries are equally dependent upon these fuels. The demand for electricity in particular implies a shift to fossil fuels for new and expanding power stations and there is much evidence that urbanization has been accompanied by increasing dependence upon imported oil in countries such as Kenya (Kimuyu, 1993) and Tanzania (Hosier *et al.*, 1993).

Detailed differences in the energy 'mix' of individual countries should not obscure the fact that there is a powerful link between economic growth and the consumption of coal, oil and natural gas. Historical experience suggests that these non-renewable fuels provide the energy essential for the transformation from agricultural to industrial societies. Although renewable sources of energy such as HEP have considerable unexploited potential in many parts of the world including Africa (Lazenby and Jones, 1987), there is good reason to suppose that this experience will be repeated by countries seeking to make the same transformation. Furthermore, while the rate of increase in energy consumption associated with the progression to a post-industrial economy seems to slacken (see below), there is no evidence of any impending reduction in the dependence upon fossil fuels. These fuels will, therefore, continue to be central to the process of economic growth and, in these circumstances, it is appropriate to consider the adequacy of the energy resource base to meet future demands upon it.

Plate 7.1 Natural gas is an important fossil fuel which has been less intensively exploited than oil. Vast underground storage capacity beneath the small town of Mont Belvieu, Texas makes it the unlikely reference point for natural gas pricing in the US. (*Source:* K. Chapman)

Energy resources, reserves and scarcity

The critical importance of energy resources is reflected in periodic fears of scarcity. Such fears have been a particular feature of the US oil industry. Conventional wisdom at the beginning of the 1920s suggested that domestic supplies would be exhausted by the end of the decade and in 1939 the US Department of Interior predicted a similar fate within 13 years. With the benefit of hindsight, it can be seen that US oil production did not peak until 1972 and the country was still the world's second largest producer in 1992 (after Saudi Arabia). This example illustrates the technical difficulties involved in reserve estimation. It also reveals the way that cultural factors may influence the interpretations placed upon such estimates, leading to magnified perceptions of resource scarcity.

Life cycle of a stock energy resource

Figure 7.6 provides a conceptual framework for considering the life cycle of a stock energy resource such as oil. It suggests that production will assume the form of a bell-shaped curve. The form of this curve is a consequence of the relationships between trends in discoveries, reserves and production. The figure suggests that the success rate will accelerate for a period following the initial discovery and then slow down as it becomes more difficult to find further deposits. This logistic curve is paralleled by a similar, lagged curve of production. This lag ensures that the rate of increase in discoveries begins to decline before the same thing happens to production. This in turn affects trends in reserves which continue to increase until the discovery and production curves intersect, approximately half-way between the two peaks. Beyond this point, reserves are gradually depleted because additions resulting from new discoveries are offset by 'losses' to production. Eventually, the reserves are exhausted and production ceases.

Although useful in providing an overview of the relationships involved, this model is difficult to test in reality because of the limitations of available data. Problems also arise with the definition of 'reserves'. Applying the model at the global scale, the reserves in Fig. 7.6 correspond to the ultimate resource base. For practical purposes, however, there is a distinction between resources and reserves.

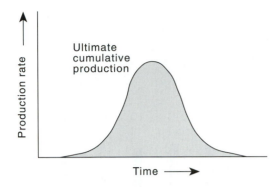

(a) Standard production curve
of an exhaustible energy
source

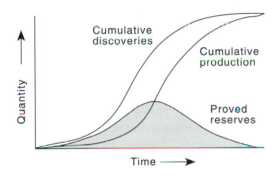

(b) Cumulative discoveries,
production and proved
reserves

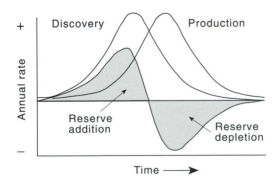

(c) Rates of discovery,
production and change of
proved reserves

Resources and reserves

This distinction was drawn in Chapter 1 which emphasized that, at any point in time, only part of the *resource base* is defined as a *reserve*. The definition of a reserve is itself problematic and Fig. 1.7 represents the gradation from proven through probable, possible to speculative reserves. This is not the only scheme used by geologists seeking to define reserves, but all classification systems involve a progression reflecting increasing degrees of uncertainty incorporated in assumptions about future technical and economic conditions. For example, there have been great advances in the ability to find and produce oil in environments, such as the Arctic Ocean, that would have seemed inconceivable as potential sources 40 or 50 years ago.

Reserve estimates are meaningless in human terms unless some attempt is made to link them to future levels of demand. This is also difficult. The impact of the invention of the internal combustion engine upon the market for oil, for example, could not have been foreseen in 1870. In these circumstances, it is customary to calculate reserves/production ratios based on proven (i.e. most certain) reserves and assuming a continuation of current rates of consumption. Bearing in mind that proven reserves usually represent a small proportion of the resource base, such estimates must be regarded as providing very conservative answers to the question of how long the resource will last. Coal is by far the most abundant fossil fuel and current reserves/production ratios exceed 230 years. The corresponding ratios for oil and natural gas are much lower (Fig. 7.7). This is not surprising since the formation of oil and gas reservoirs requires a much more specific set of favourable geological circumstances than does the creation of coal which is a sedimentary rock. The fact that oil is less abundant, together with its crucial role in modern economies (especially in the United States where fears of oil exhaustion have sometimes seemed to indicate a national paranoia) has meant that many attempts have been made to estimate its ultimately recoverable resource base. It is worth reviewing these efforts to demonstrate the 'scientific' limits of such predictions as well as the opportunities for the intrusion of value-judgements.

Fig. 7.6 Life cycle of a stock energy resource. (*Source:* Hubbert, 1962.)

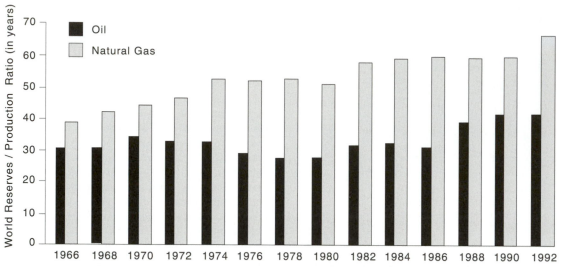

Fig. 7.7 World reserves/production ratios for oil and natural gas, 1966–92. (*Source:* BP, 1993.)

Oil: physical scarcity or abundance?

Numerous estimates of ultimately recoverable reserves of conventional oil (excluding oil shale and tar sands) have been made. These estimates have increased from a consensus figure of 500×10^9 bbls (barrels) in the 1940s to a corresponding value in the range 2000–3000×10^9 bbls in the 1990s. The upward trend does not, however, obscure the wide variation in these estimates. Such variations are not surprising given the diversity of methods and underlying assumptions. An exchange of views between Warman (1972), then chief geologist of BP, and Odell (1973), a frequent but respected critic of the multinational oil companies, drew attention to some of the sources of disagreement between the pessimists and the optimists. The former believe that, since oil is a finite resource, rates of discovery must decline as the industry is forced to move on from the more to the less promising prospects. There is no doubt that this kind of progression is characteristic of commercial judgements relating to the exploitation of any finite resource with the easiest (and often largest) deposits developed first (see Chapter 8). Odell and other optimists do not dispute this progression, but argue that the cycle of development is far less advanced. Thus whereas Warman (1972) believed that the limitations of the resource base would ensure that world oil production could not be maintained beyond a peak around 1990, Odell and Rosing (1983) suggested that this peak will not be reached until

2017 at the earliest and possibly not until 2080. Despite reference to 'volumetric analyses of sedimentary rocks' and 'probabilistic simulation models' we should not be deceived by the apparently scientific basis of the debate. The oil industry has a history of extreme caution in estimating reserves. This is at odds with its record of technical ingenuity in finding and producing oil in unlikely places. There is more than a suspicion that it is frequently in the industry's commercial interest, especially in its dealings with governments, to promote a conservative view of the resource base. This attitude is incorporated into the assumptions which underpin many 'scientific' forecasts of the ultimately recoverable resource base.

Only time will resolve this debate, but if Fig. 7.6 is accepted as a valid model of the cycle of development of a non-renewable resource, clues may be obtained by examining historical trends in production and proven reserves. Figure 7.8 suggests that the peak of

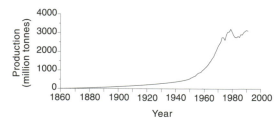

Fig. 7.8 World oil production, 1860–1992. (*Source:* compiled from data in BP, 1993, and miscellaneous sources.)

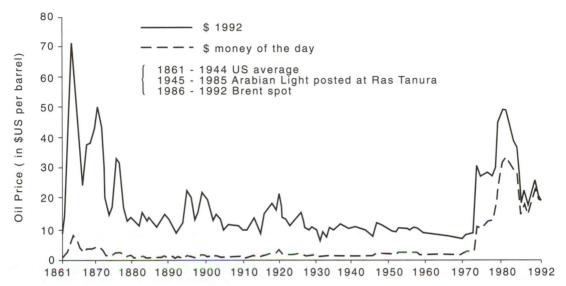

Fig. 7.9 Oil prices, 1861–1992. (*Source:* BP, 1993.)

world oil production may already have been reached as the rapid growth of the 1950s and 1960s was replaced by a sharp fall at the beginning of the 1980s followed by a gradual recovery. However, the model indicates that this peak should have been preceded by a steady decline in proven reserves (Fig. 7.6). There is no evidence of such a decline (Fig. 7.7) and the recent stagnation of world oil production cannot be attributed to physical scarcity.

Oil: geopolitical scarcity and response

It was noted in Chapter 1 that 'scarcity' carries several different meanings in the context of resources and that social, economic and political factors may be more important influences upon decision-making than physical limits. Coincidentally, Odell's rejoinder to Warman was published within months of a major jump in the international price of oil at the end of 1973. This event, which was related to a struggle for control of the resource, has had a much more dramatic effect upon the subsequent development of the industry and upon patterns of energy consumption than any fears about the ultimate exhaustion of the resource base.

The world oil industry was traditionally dominated by the United States. This reflected the origin of the industry in the United States and the early importance of oil in the country's overall pattern of energy consumption (Fig. 7.5). Despite periodic fears about

oil shortages and the initiation in the 1920s of an international search for new sources of supply, the United States was, by the end of the Second World War, not only meeting its own requirements, but also exporting to other countries. This position could not be sustained in the face of rapid increases in oil consumption in Western Europe and Japan. These increases were overwhelmingly met by suppliers from the Middle East and North Africa. By 1972, imported oil was meeting well over half of Western Europe's primary energy needs and the United States was also no longer self-sufficient. The political risks of this situation were acknowledged in various reports and emphasized by events such as the Suez crisis in 1956 and the Six Day War involving Israel and its Arab neighbours in 1967. The worst fears of the oil-consuming countries were realized in 1973 when the members of the Organization of Petroleum Exporting Countries (OPEC) succeeded in reconciling their internal differences and imposed dramatic increases in price upon the multinational oil companies and also expropriated most of their assets within the producing countries. Figure 7.9 emphasizes the significance of these and a second phase of OPEC-engineered price increases in 1979 in the context of the historical evolution of oil prices. These 'shocks', and earlier fluctuations associated with events in the United States, suggest that physical availability has been less important in determining oil prices than political factors relating to control of the resource (Suárez, 1990).

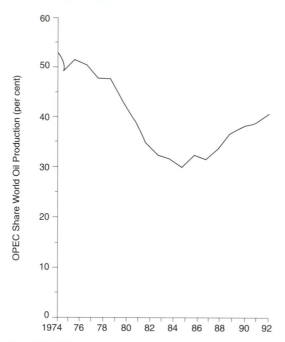

Fig. 7.10 OPEC share of world oil production, 1974–92. (*Source:* compiled from data in BP, 1993.)

International oil prices reached a peak exceeding US $35 per barrel in 1981 and subsequently declined to significantly lower levels (Fig. 7.9). The fluctuation of oil prices since 1973 provides a good illustration of the feedback mechanisms affecting the perceived scarcity and abundance of a resource as regulated by the interactions of supply, demand and technological change (Fig. 1.6). These mechanisms emphasize that consumers are not passive victims of scarcity and they introduce a dynamic element into the estimation of reserves. In the case of oil, the first price shock in 1973 not only stimulated short-term responses such as panic buying and rationing, but also set in motion longer-term adjustments on the part of governments, oil companies and individual consumers. These adjustments made it impossible for OPEC to maintain prices at the 1981 level. On the supply side, higher oil prices relative to the pre-1973 situation improved the economics of previously marginal fields and, more importantly, encouraged the search for oil outside the OPEC countries. The rapid development of the North Sea as a major oil-producing province was the most spectacular result of geographical diversification and Fig. 7.10 indicates a sharp decline in the OPEC share of world oil production between 1974 and 1984. The subsequent recovery of this share

mirrors the decline in international oil prices in the 1980s rather than a re-assertion of OPEC leverage. Any attempt by OPEC to raise prices would result in an immediate fall in its share as a result of the greater flexibility established by the consuming countries since 1973. Part of this flexibility is related to developments on the demand side where higher oil prices encouraged greater economy of use and, where possible, a switch to alternative fuels. It is this price-induced reduction in demand rather than any physical shortage which accounts for the stagnation of world oil production in the 1980s. Technological advance is a central link in the sequence of responses to perceived scarcity (Fig. 1.6). In the case of oil, the effects of innovation have been both direct, such as advances in the technology of exploration and production or improvements in the fuel efficiency of motor vehicles (see below), or indirect such as the impetus to research and development into alternative sources of energy.

The nuclear option

Nuclear power received a significant boost in the aftermath of the oil price shocks. Several countries, especially France and Japan, accelerated their nuclear programmes as a result of heightened fears about the security of their oil imports and the possibility of further price increases. Nuclear power was a spin-off from the Manhattan Project which created the atomic bombs dropped on Japan and military rather than civilian uses motivated post-war research into this technology. By the mid-1960s, however, nuclear power was regarded as a potentially important source of energy for peaceful purposes. This optimism contrasts sharply with the situation a generation later and the contribution of nuclear power to commercial energy consumption in countries such as the United States (Fig. 7.5) falls far short of predictions made 25 years ago.

The nuclear cycle

The physical basis of nuclear power was outlined at the beginning of the chapter and existing commercial uses involve fission which is based on ^{235}U derived from non-renewable supplies of uranium. The energy locked up is released by breaking up large nuclei to create heat which is successively converted into steam and electricity. Much of the energy contained in other

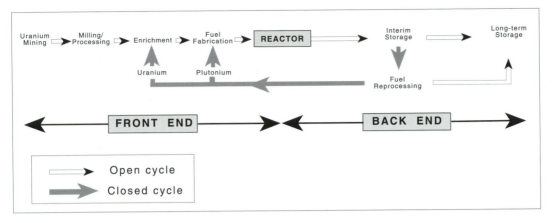

Fig. 7.11 The nuclear fuel cycle.

fossil fuels such as coal is converted to electricity in power stations, but there are fundamental differences between nuclear and other non-renewable sources of energy. These are reflected in the concept of the nuclear fuel cycle, which acknowledges the uniquely toxic nature of the residuals associated with nuclear power, and in the commitment which must be made to their storage and/or reprocessing.

Figure 7.11 identifies the stages of the cycle. Many of these are widely separated and transportation linkages are essential to the functioning of the system. The volume of movement is, however, limited by comparison with many other fuels because nuclear power meets such a small proportion of the world's energy needs. The starting point in the cycle is the mining of uranium which is contained, at very low concentrations, in a variety of granitic and meta-morphic rocks. Processing facilities are required to produce concentrated uranium oxide (i.e. yellow-cake). This in turn is refined or purified to produce uranium hexafluoride. Enrichment to increase the proportion of the isotope ^{235}U, which is critical to nuclear fission, is the next stage. The enriched fuel is then converted into pellets of uranium dioxide which are assembled into alloy tubes to form the fuel rods used in nuclear reactors. This sequence of input or 'front-end' operations is followed by the output or 'back-end' activities dealing with the radioactive residuals from the reactors. Spent fuel is periodically removed and stored on-site in specially constructed containers to permit the high levels of radioactivity to decline. This is at best an interim solution and there are two alternatives beyond this stage. The first is to find sites, possibly in deep geological formations, for the long-term storage of both high- and low-level radioactive wastes. The second is to 'close' the nuclear fuel cycle by reprocessing spent fuel. Although reprocessing has the dual attraction of reducing the demand for new supplies of uranium and of minimizing the long-term storage problem, only France and the United Kingdom have such facilities and the closure of the nuclear fuel cycle is unlikely to be achieved to any significant extent because of serious doubts about the commercial viability of reprocessing (see Box 7.1).

Reprocessing seemed highly desirable in the early days of nuclear power when the apparent rarity of the isotope ^{235}U, which accounts for only 0.7 per cent compared with 99.3 per cent of non-fissionable ^{238}U occurring in uranium ore, prompted fears about the adequacy of the resource base. At the height of the cold war in the 1950s and 1960s, these fears were driven more by concerns related to the military rather than the civilian uses of nuclear power and Hubbert (1971: 38) reflected contemporary opinion in predict-ing '...an acute shortage of low-cost ores...before the end of the century'. By the early 1980s it was clear that no such shortage was in prospect and the case of uranium is another illustration of the 'flexible' nature of reserve estimates. Several factors have contributed to this revision of opinion, especially the slow-down of investment in nuclear power. Many countries, such as Denmark and Australia, have abandoned plans to introduce nuclear power, the industry has 'collapsed' in the United States (Campbell, 1988) and most countries in Western Europe (with the conspicuous exceptions of France and the United Kingdom) have imposed moratoriums on further investment. Figure 7.12 emphasizes the very rapid growth of nuclear generating capacity in the United States during the

Box 7.1 THORP

British Nuclear Fuels commissioned its Thermal Oxide Reprocessing Plant (THORP) at Sellafield in Cumbria in 1994. The plant is designed to take used fuel rods from nuclear power stations and separate out the re-usable plutonium and uranium, leaving a residue of radioactive waste. The project has been controversial ever since it was first proposed in 1974.

Supporters argue that:

- it will reduce the level of nuclear waste by 'closing' the nuclear cycle in the United Kingdom;
- it will earn substantial export earnings by reprocessing waste from overseas power utilities (mainly from Germany and Japan);
- it will ensure that the United Kingdom remains at the leading edge of nuclear technology.

Opponents argue that

- the need for reprocessing has disappeared because there is no longer any shortage of uranium;
- reprocessing is more expensive than other methods of treating used nuclear fuel such as 'dry storage';
- the health risks arising from radiation exposure for those working in or living near the plant are unacceptable;
- the transportation of nuclear waste to and from the plant is inherently risky;
- the United Kingdom may become a 'nuclear dustbin' because foreign customers may renege on contracts requiring them to take back all of the products of reprocessing.

With the benefit of hindsight, it is clear that THORP should not have been built. The prospects for nuclear power look very different in the 1990s than they did when the project was conceived in the 1970s. The story of THORP illustrates the fact that large-scale energy projects, especially relating to nuclear power, achieve great momentum and, once initiated, cannot be readily halted or modified.

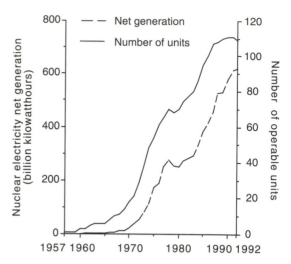

Fig. 7.12 Nuclear electricity generation in the United States, 1957–92. (*Source:* compiled from data in US Department of Energy, 1993.)

since the short history of nuclear power illustrates the sensitivity of our perceptions of natural resources to changing circumstances.

Technology and economics

It is generally accepted that the generation of electricity from nuclear power has proved more expensive than anticipated. Estimates of costs are, however, frequently disputed by the advocates and opponents of nuclear power (see Jeffrey, 1987; Sweet, 1990). Indeed, the economics of nuclear power are notoriously difficult to disentangle. There are many reasons for this. The mixed motives underlying the early development of nuclear power encouraged various government subsidies which were not always transparent. Different methods of accounting produce different financial outcomes and public inquiries into specific projects have revealed that the two sides of the debate usually start from different assumptions. For example, the case will appear more favourable if estimates of costs are limited to those involved in the construction and operation of a single power station; they will appear less favourable if account is taken of the additional costs implicit in the various operations associated with the fuel cycle that are required to support such a plant.

Even if these wider issues are disregarded, there is little dispute that the costs of building and operating a nuclear power station have escalated dramatically. The technology is complex and seems to have

1970s, when investment in new reactors reached its peak, and the relative stagnation since the mid-1980s. The US experience is typical of the reversal of policy which has occurred in most developed countries and it is interesting to consider some of the reasons for this stretched the capabilities of the nuclear engineering industry to the limit, resulting in massive over-runs in construction times. In both the United States and United Kingdom, the lack of agreement on a standard reactor design aggravated these difficulties because experience acquired on one project was not necessarily applicable to another. Delays were not only related to engineering difficulties. Concerns over safety (see below) led to a steady increase in the period from initial planning to project approval. This was partly due to the involvement of official regulatory agencies such as the Nuclear Regulatory Commission (NRC) in the United States. The review time for construction permit applications to the NRC (and its predecessor the Atomic Energy Commission) rarely exceeded 12 months before 1967, but was averaging 45 months 10 years later (Campbell, 1988: 41). These agencies were themselves responding to public pressure and many delays were attributable to grass-roots opposition to specific projects. It has been suggested that in Japan, for example, '... the average lead times necessary to gain public acceptance of nuclear projects have increased from 2 to 3 years in the 1960s to 14–15 years in the 1980s' (Lesbirel, 1990: 268). The soaring capital costs of nuclear power stations make it all the more important that they are reliable in operation. Although there have been wide variations in performance between reactor designs and between countries, experience has fallen short of expectations with load factors as low as 50–60 per cent of capacity not unusual.

Nuclear power and the environment

Many of the economic problems of nuclear power stem from fears about its safety. The technology has found it difficult to 'shake off' its military origins. The fact that early nuclear plants in the United States and United Kingdom were geared to the production of plutonium for weapons sets a bad example for others to follow. Considerable efforts have been made by existing members of the 'nuclear club' to discourage such ambitions. Economic constraints upon the adoption of the technology have, therefore, been reinforced by political ones designed to limit its proliferation at the international scale. Attempts to limit its adoption within individual countries, as reflected in organized opposition to specific proposals, have, above all, focused upon the potentially catastrophic consequences of a nuclear accident. The uniquely pervasive impact of major releases of radioactivity into the environment has meant that

nuclear power installations have been properly required to meet very demanding safety standards extending across the whole spectrum of activities from the choice of site, through the design of equipment to the regulation of operating procedures. These standards have certainly added to the costs and defenders of nuclear power would argue that its safety record compares very favourably with, for example, coal mining. Nevertheless, the incidents at Three Mile Island in Pennsylvania in 1979 and at Chernobyl in the former Soviet Union in 1986 emphasized that there are no guarantees against human error. Chernobyl in particular confirmed the worst fears of a nuclear accident, rendering a wide area in the vicinity of the plant uninhabitable and releasing radioactive debris which was subsequently detected over much of northern Europe, The long-term health effects of Three Mile Island and Chernobyl are uncertain, but these events had an immediate impact upon public confidence in nuclear power. No new orders for reactors have been placed in the United States since Three Mile Island and many projects have been cancelled. Chernobyl only served to reinforce the effect of Three Mile Island in promoting similar re-appraisals of the nuclear option in numerous countries.

The externalities associated with nuclear power are not limited to accidental releases; they are also programmed into the technology. All forms of energy use involving combustion generate by-products which are released into the environment, but none are anything like as toxic as the radioactive wastes produced at various stages in the nuclear fuel cycle. A distinction is usually made between low- and high-level wastes. The former give off small amounts of ionizing radiation, usually for a short time; the latter remain highly dangerous for periods which are virtually infinite on a human time scale. Low-level wastes have been dumped at sea in steel drums by many countries, but are now usually placed in landfills. Leakages from containers have occurred on many occasions, leading to attempts to regulate these practices better. The major controversies have focused on high-level wastes. Nobody welcomes the prospect of a nuclear dump on (or under) their doorstep, whatever assurances are provided about the reliability of storage procedures. Despite numerous technical studies and considerable political effort, no country has solved this problem by establishing a permanent storage facility for high-level waste. Yet another concern is the de-commissioning of nuclear reactors. These, like any other capital equipment, wear out. This is a pressing issue in the United Kingdom where the early Magnox reactors are already beyond the end of their original design life. Such facilities cannot be demolished in the same way as a conventional power station because of the risks of radioactive contamination and there is considerable uncertainty surrounding future de-commissioning strategies (see Pasqualetti, 1989).

Overall, it is clear that the future of nuclear power looks very different in the 1990s, compared with just 20 years earlier. This change is not due to any constraint imposed by the natural resource base since uranium appears much more abundant than it did when the first nuclear power stations were commissioned in the 1950s. Rather it reflects a complex amalgam of social, political and economic factors which have combined to undermine confidence in the technology. In a sense, nuclear power has become less acceptable in environmental terms and this has diminished the perceived value of the resource (i.e. uranium) upon which it is based.

Location, mobility and energy resources

The problems associated with the development of nuclear power illustrate some of the limitations of technology. However, it is important to place the relatively short and troubled history of nuclear power into historical perspective. Many innovations related to the exploitation of energy sources have dramatically improved the quality of human life. Some of the most important have been associated with the mobility of energy, which is closely related to the issue of scarcity. A basic response to resource scarcity is to seek new supplies by widening the spatial range of resource exploitation (see Chapter 3). Improvements in the mobility of energy, therefore, have a significance which extends beyond the matters of convenience or flexibility. This is evident in the link between the enhanced mobility of energy in its various forms and the previously noted transformation from low- to high-energy societies. This transformation has been accompanied by a progression from essentially local to international energy supply systems.

Local supply systems

Low-energy societies are typically self-sufficient in energy terms. The principal inputs are the solar energy required for cultivation, human and animal labour, renewable biomass fuels and the recycled energy generated by food consumption (Rappaport, 1971). The only significant energy movements are those associated with routine agricultural activities and with the collection of firewood. There is no requirement to seek out and 'import' energy from external sources. Such societies are becoming relatively rare, to the extent that even the most basic agricultural community often uses, for example, petroleum-derived products such as kerosene for heat and light. Furthermore, Chapter 5 has already drawn attention to the contemporary 'fuelwood crisis' in many parts of the developing world (Fig. 5.6). An increase in collection times and an extension of the radius of search around the focal settlements are important manifestations of this 'crisis'. Table 7.1 summarizes the results of various studies describing the problem.

National supply systems

As levels of energy consumption increase it becomes more difficult to meet these demands from local sources. Urban life breaks the link with the land and necessitates an ever widening search for energy sources. For example, the availability of fuelwood for London was causing concern as early as the fifteenth century and the development of the coastal coal trade from Newcastle was a direct response to this problem. This trade was an early illustration of the feasibility of moving fossil fuels from the point of production to the point of consumption. This approach was, however, relatively unusual prior to the twentieth century and the most common response to the uneven distribution of energy resources was to locate economic activities in close proximity. This was evident in the distribution of water-powered woollen mills in England by the sixteenth century but it had its most durable impact as a consequence of the 'pull' of the coalfields during the Industrial Revolution. The attractions of the coalfields were ultimately based upon the energy-intensive nature of many of the economic activities associated with the Industrial Revolution, notably the manufacture of iron and steel, and upon the high costs of moving coal in sufficient quantities to sustain them. Although the availability of abundant, low-cost energy continues to attract certain types of economic activity such as mineral smelting to, for example, sources of hydro-electricity or natural gas, the real costs of transporting energy have declined, leading to a weakening of the

Table 7.1 Time taken and distance travelled for firewood collection

Place/date	Firewood collection	
	Time taken	Distance travelled
Nepal		
Tinan (hills) 1978	3 hours/day	–
Pangua (hills) late 1970s	4–5 hours/bundle	–
India		
Chamoli (hills) 1982	4–5 hours/day	3–5 km
Gujerat (plains) 1980		
(a) Forested	Every 4 days	–
(b) Depleted	Every 2 days	4–5 km
(c) Severely depleted	4–5 hours/day	–
Madhya Pradesh 1980	1–2 times/week	5 km
Kumaon Hills 1982	3 days/week	5–7 km
Karnataka	1 hour/day	3 km
Gashwal (hills)	5 hours/day	10 km
Africa		
Sahel 1977	3 hours/day	10 km
1981	3–4 hours/day	–
Niger 1977	4 hours/day	–

Source: Agarwal (1986).

ties which created the concentrations of mining, industry and population in areas as diverse as Pennsylvania and the Ruhr.

Despite a massive dependence upon very unevenly distributed fossil fuels, domestic and industrial consumers within modern economies generally assume that their energy requirements will be met at almost any location. The existence of efficient and comprehensive energy distribution systems is taken for granted. Many technical advances have contributed to this situation, but the most significant single development has been the creation of electricity transmission networks. The public supply of electricity was introduced in the United Kingdom in 1881. It was initially restricted to the major cities, but it has become a virtually ubiquitous source of energy within developed economies.

International supply systems

The movement of commodities such as coal and oil between countries and continents represents the ultimate expression of the trend towards the increasing mobility of energy. This trend reflects a progressive decline in the real costs of transport. Although international movements take place by

pipeline (oil and gas) and by transmission wire (electricity), ships are by far the most important mode of transport at this scale. Advances in the design, construction and operation of merchant ships have been responsible for the fall in transport costs. For example, the launch of the first genuine oil tanker in 1885 was significant because it heralded the bulk transportation of oil which had previously been shipped in individual barrels. Similarly, bulk transportation acquired a new meaning with the advent of the 'supertanker' in the early 1960s and the economies of scale associated with these vessels dramatically reduced the unit costs of moving oil by sea (although the environmental costs associated with such shipments are high when accidents involving large vessels such as the *Exxon Valdez* occur).

Technical change has not only reduced the costs of energy transport, but has also created entirely new opportunities. The development of cryogenic techniques has, for example, led to the emergence of an international trade in liquefied natural gas (LNG) in refrigerated tankers. Such maritime transport of natural gas was not possible before 1960 and it has allowed countries such as Indonesia and Algeria to export natural gas to Japan and France respectively. In the case of Japan, these shipments are just one element in a strategy of securing access to overseas energy sources which may be traced back to its invasion of Manchuria in the 1930s. This strategy is apparent in other resource sectors such as industrial wood (Chapter 5) and minerals (Chapter 8). It is a response to Japan's relative lack of indigenous energy resources and has been described by Forbes (1982) as 'energy imperialism'.

It has already been noted in Chapter 3 that developed countries are not necessarily constrained by the limitations of their domestic resource base. The international trade in oil is the clearest example of this. The scale of this trade increased dramatically during the 1950s and 1960s as oil became the principal single source of energy in Western Europe and Japan as well as the United States. The geography of the trade is represented in Fig. 7.13. Although the pattern has become more complex as a result of the increasing dependence of the United States upon imports and the introduction of new sources of external supply to Europe, such as Russia, the focal position of the Middle East has been evident for more than 40 years. This reflects its status as the principal centre of oil production, but, more important in the long term, its overwhelming share of proven reserves.

The risks of dependence upon Middle Eastern oil

Plate 7.2 The importance of coal as the principal energy source of the Industrial Revolution created new communities in places such as South Wales. The subsequent decline of coal has led to the closure of every deep mine in South Wales and to the economic collapse of the communities based on this activity. (*Source:* K. Chapman)

have already been discussed and the events that led to a stagnation in the expanding volume of international oil movements in the 1980s had exactly the opposite effect on the corresponding trade in coal. International movements of coal are known to have taken place as early as the fourteenth century with shipments from north-east England to France. Indeed, the United Kingdom remained the world's major exporter until the 1920s. After the Second World War, the United States became dominant, but the quantities involved were small by comparison with oil and the volume of coal entering international trade declined between 1950 and 1970. The oil price shocks improved the relative position of coal and prompted a revival of the industry in established coal producers such as the United States and the United Kingdom (see Spooner, 1981b). It also encouraged other countries such as Venezuela and Colombia to establish an export capability based on indigenous resources. The desire of energy-importing countries to diversify both the nature and the geographical sources of their supplies following the oil price shocks increased their interest in coal, especially in view of the fact that the principal exporters such as the United States and Australia were perceived as more

'secure' than the Middle East. In these circumstances, the volume of coal entering international trade increased from 193 million tonnes in 1975 to 334.9 million tonnes in 1985 (Owen, 1988). This trade has become very competitive, however, as the oil price decline from its 1981 peak and the availability of increasing quantities of low-cost, open-cast coal from overseas has contributed to the collapse of the UK coal-mining industry. Indeed, in view of the connection between the expansion of this trade and the security fears engendered by dependence on Middle East oil, it is ironic that it has been partly responsible for weakening the United Kingdom's capacity to maintain a diversified indigenous energy resource base.

Energy policy

Policy objectives

The link between trends in international oil prices and the fortunes of the UK coal industry is just one illustration of the interdependencies arising from the

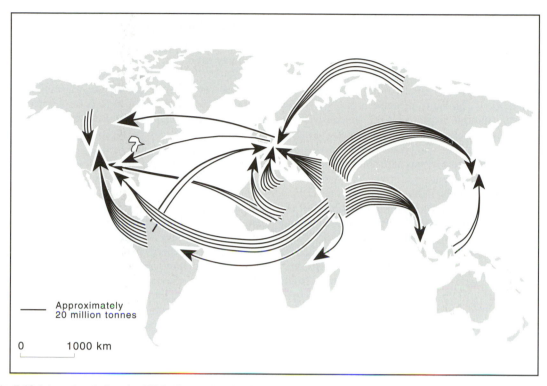

Fig. 7.13 International oil trade, 1992. (*Source:* BP, 1993.)

central role of energy in modern economies. A study of Canadian energy prospects, for example, was given the title *Connections* in an explicit acknowledgement of this interdependence (Economic Council of Canada, 1985). The same interdependence has also encouraged governments in many countries to declare their commitments to national energy policies. Such commitments are especially significant in countries such as the United States which have a traditional antipathy towards government intervention in economic affairs. In practice, the effectiveness of energy policy is variable. President Carter's National Energy Plan published in 1977, for example, had limited impact, partly as a result of the very different philosophy of his successor (President Reagan) in the White House. Nevertheless, it is possible to identify certain common themes in energy policy statements made by different governments at different times. The basic objective is to guarantee an uninterrupted supply of low-cost energy in sufficient quantity to meet demand. Typical supply-side measures seek to promote indigenous sources of energy, the geogra-

phical diversification of any imports and a balanced fuel 'mix'. These may be complemented on the demand side by instruments to encourage more efficient use of energy (pp. 156–7).

Policy problems

Energy policy must not only be internally consistent, but must also take account of other areas of policy concern. Attempts to encourage more efficient use of energy by the imposition of taxes or by the removal of subsidies may, by raising costs to the consumer, depress the rate of economic growth when other policies are designed to have exactly the opposite effect. Difficulties also arise because national, regional and local interests frequently diverge. Major tensions occurred between Alberta and the federal government in Canada following the publication of the National Energy Programme in 1980. These tensions were based upon Alberta's desire to maximize the income from its oil resources and the federal government's wish to limit the impact of rising international prices upon the Canadian economy.

Some of the most obvious conflicts are associated with the environmental effects of energy developments. It is often local communities which experience the negative effects while the benefits are directed at national level. Successive UK governments have attached high priority to the rapid exploitation of North Sea oil. This led to pressure in the early 1970s to identify suitable coastal sites for the construction of production platforms. These sites were concentrated in Scotland and considerable permanent environmental damage was inflicted on some high amenity areas in preparing sites which proved to have very short useful lives. Indeed, one site was never used for its intended purpose. Geographical scale is not the only dimension of such conflicts between amenity and energy developments and direct clashes may occur between energy and environmental policy objectives at national level. There have been several disputes surrounding the exploitation of coal in the western United States. Large deposits suitable for strip mining are found from Montana in the north to New Mexico in the south. These are commercially attractive and their development is consistent with the energy policy objectives of promoting indigenous energy sources. On the other hand, many of these deposits are in areas heavily protected by various aspects of federal environmental legislation.

Reconciling conflicting objectives is not the only problem associated with the implementation of energy policy. Another set of difficulties stems from the uncertainties of energy planning. It is easy, with the benefit of hindsight, to identify some spectacular forecasting errors. The experience and technical sophistication of the oil companies did not protect them from grossly over-estimating future levels of oil demand in projections made at the beginning of the 1970s, and public disputes over particular projects often reveal very different forecasts of energy demand from 'expert' witnesses on opposite sides of the argument. The significance of differences is magnified by the very long lead times associated with major energy projects (Box 7.1). Uncertainty and the essentially technical aspects of forecasting are not the only difficulties involved in framing energy policy. The institutional framework within which many crucial decisions are taken does not help. The electoral cycle within democratic societies is much shorter than the time scale appropriate to sensible energy planning. Controversial and unpopular decisions are difficult to make and short-term political expediency is not consistent with the vision required to influence the direction of policy in the medium to long term.

Energy efficiency

Energy ratio

The link between levels of energy consumption and economic development has already been noted and is implicit in the reference to low- and high-energy societies. It has also been shown that contemporary high-energy societies are heavily dependent upon non-renewable energy sources. This combination of rapid increases in energy consumption and reliance upon fossil fuels raises obvious questions about the sustainability of existing trends. Focusing upon the demand side, recent experience in the developed economies provides some grounds for optimism as present levels of energy consumption have fallen well short of predictions made in the early 1970s. This partly reflects the much lower rates of economic growth relative to the experience and expectations of the 1960s. More significantly, however, it also reflects a change in the nature of the link between energy consumption and economic growth within the developed economies. This link is defined by the energy ratio:

$$\text{Energy ratio} = \frac{\begin{array}{c}\text{Rate of change in}\\\text{energy consumption}\end{array}}{\begin{array}{c}\text{Rate of change in}\\\text{economic growth (GDP)}\end{array}}$$

A value greater than 1 indicates that the amount of energy required to create an additional unit of gross domestic product (GDP) is increasing; a value less than 1 suggests the reverse. Expressed another way, a declining energy ratio suggests greater efficiency in the use of energy to create the goods and services which constitute GDP. Figure 7.14 shows that the ratio changes in a systematic way that is linked to the progression from subsistence agricultural to post-industrial societies. Rates of increase in energy use typically outstrip the incremental growth in economic output during the phase of rapid industrialization. This is related to the energy-intensive nature of heavy industries such as iron and steel associated with this phase. The relative importance of these industries tends to decline in the post-industrial phase as services and information-handling activities, which require lower energy inputs, become dominant. The effect of such changes in the structure of economies upon energy consumption is evident in the flattening out of the logistic curve representing change in per capita energy consumption.

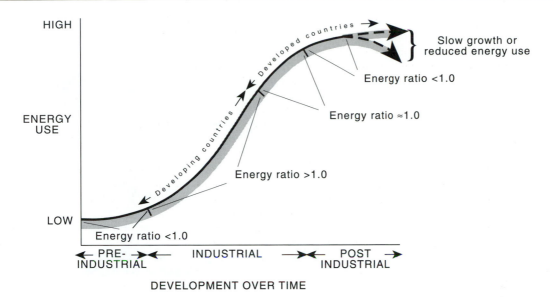

Fig. 7.14 Energy ratio and economic development. (*Source:* Odell, 1989.)

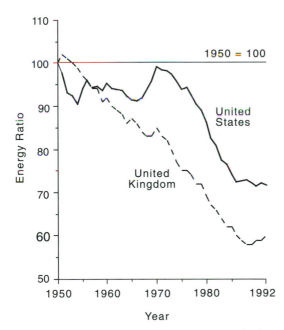

Fig. 7.15 Energy ratio in the United Kingdom and the United States, 1950–92. (*Source:* compiled from data in Department of Trade and Industry, 1993, and US Department of Energy, 1993.)

Figure 7.14 is an over-simplication and detailed studies of individual countries have emphasized the sensitivity of such measures of energy efficiency to short-term events such as oil price changes and economic recession (see Edmonds and Reilly, 1985: 46–56; Chern and James, 1988). Nevertheless, Fig. 7.15 plots a steady downward trend in the energy ratio of both the United Kingdom and the United States since 1950. This is consistent with the model (Fig. 7.14) and is typical of the experience of most developed countries over the same period. However, it is important to recognize that the bulk of the world's population lives in countries which are placed somewhere along the steepest section of the curve in Fig. 7.14. If they are to follow the path of the developed economies, their aspirations can only be achieved by substantial increases in energy consumption. Realistically, these aspirations must fall far short of parity, in terms of living standards with the United States. Even the most modest objectives, such as bringing the convenience of electricity to their rural populations, can only be achieved at the expense of a deteriorating energy ratio. Time-series data for developing countries comparable to the information in Fig. 7.15 for the United Kingdom and the United States are difficult to obtain, but it is safe to say that most will display exactly the opposite trend in their energy ratio. However, many predictions of further growth in world energy demands are based on the assumption that the reversal of this trend will be achieved more quickly as a result of knowledge gained within the developed economies (World Energy Council, 1993). This implies that the point of inflexion in Fig. 7.14 will shift to a lower point on

Plate 7.3 The 'smog' problems of US cities such as Houston have been aggravated by high levels of car ownership. Legislation has improved the energy efficiency of vehicles and imposed tougher exhaust emission standards. (*Source:* K. Chapman)

the curve. Nevertheless, viewed from the 'South' it is difficult to accept appeals from the 'North' to a sense of global responsibility as a good reason for restraint. The legitimacy of such appeals is dubious in view of the massive imbalances in per capita energy consumption between countries (Fig. 7.4). The moral responsibility for making more efficient use of energy lies with the 'North'.

Promoting energy efficiency

Discharging this responsibility will require more than a passive strategy based on the assumption that the transformation to a post-industrial society will automatically reduce levels of energy consumption. The greatest stimulus to action is provided by perceptions of scarcity (Fig. 1.6). Energy policies before 1973 were directed towards minimizing costs to consumers and promoting security of supply. The oil price increases encouraged a shift away from this preoccupation with the supply side and greater efficiency of energy use became a more important consideration. For example, the demise of the 'gas guzzler', which was almost symbolic of the 'American way of life' in the 1960s, was largely a response to consumer choice as smaller, more energy-efficient, imported cars (mainly from Japan) became more

attractive at a time of rising gasoline prices. The appeal of these vehicles was reinforced in the 1970s by the prospect of fuel rationing. The US car-makers were relatively slow to react, but the federal government responded in 1975 by imposing the first mandatory fuel economy targets for new vehicles. Further pressure was applied by the introduction of purchase taxes directly linked to fuel efficiency which meant that the 'gas guzzler' became more expensive to buy as well as to run. Figure 7.16 indicates the results of these policies. The average annual mileage per vehicle has shown little change (although the absolute number of cars on the road has increased), but the economy of fuel use has improved significantly. The achievement is, perhaps, less impressive than it seems. It partly reflects the gross inefficiency of the typical American car in the 1960s. It is also worth remembering that the internal combustion engine is a very poor energy converter, with approximately 90 per cent of the total energy contained in the crude oil which is its ultimate source of fuel 'lost' as waste heat and combustion products. The significance of this characteristic is emphasized by the massive growth in the size of the world's motor vehicle fleet which means that an increasing proportion of total energy consumption is accounted for by this end-use. For example, the proportion of energy delivered for use in

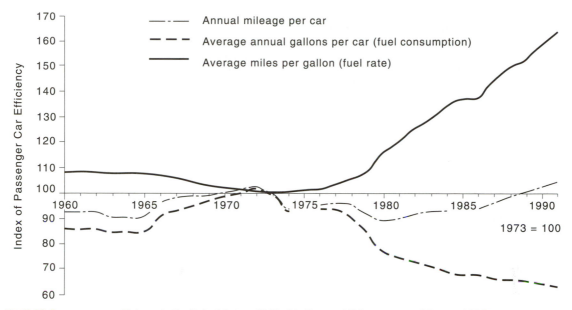

Fig. 7.16 Passenger car efficiency in the United States, 1960–91. (*Source:* US Department of Energy, 1993.)

the United Kingdom that is directed towards transportation (mainly private cars) increased from 17.4 per cent in 1960 to 32.4 per cent by 1992 (Department of Trade and Industry, 1993). The potential impact of further improvements in energy efficiency in this sector is, therefore, considerable.

Even greater opportunities exist within the total energy systems of the developed countries. It has been suggested that per capita energy use in these countries may be reduced by 50 per cent by 2020 while maintaining strong economic growth and high standards of living (World Resources Institute, 1990: 147). In theory, these savings could be achieved without any radical change in lifestyles by such measures as improving the design and construction of buildings; the adoption of more efficient energy-using practices in industry; and the widespread introduction of co-generation. The latter refers to the multiple use of steam raised in power stations. Most of this heat is presently released to the environment as warm water to rivers and steam from cooling towers. Denmark and Finland have pioneered the development of schemes which direct this energy in the form of piped hot water to buildings in the vicinity of power stations. Similarly, power stations located on major industrial sites such as chemical complexes provide not only electricity, but also heat for various manufacturing operations. The primary incentive for decision makers, from government agencies to private

sector corporations and individual households, to implement energy-saving measures has always been economic and various studies have shown that the overall improvements achieved in the immediate aftermath of the oil price shocks were not maintained as energy prices (especially that of oil) stabilized at a lower level in the 1980s. There is, however, another, ultimately more powerful, reason for using energy more efficiently. This is a growing awareness that, as a consequence of the constraints imposed by the First and Second Laws of Thermodynamics (see Box 7.2), a sustainable energy future may depend more upon controlling the by-products of energy use than upon the limitations of a finite resource base.

Energy consumption and environmental externalities

Human use of energy is based on supply systems which involve a sequence of operations from the finding of, for example, oil, through its production, transportation, refining, distribution and consumption. Other sources of energy may involve less sophisticated and extensive supply systems, but the principle remains valid. All of these various activities are responsible for impacts on the environment, but the most significant, in global terms, are associated

Box 7.2: The First and Second Laws of Thermodynamics

Energy is defined as the capacity for doing work. Work is necessary to grow food, to move things from place to place and to change matter from one physical or chemical form to another. Human use of energy resources rests upon two fundamental physical laws which are universal in their application and of considerable practical significance.

The First Law of Thermodynamics states that energy is neither created nor destroyed. Thus in carrying out work, such as the generation of electricity in a coal-fired power station, energy is converted from one form to another. In this case, the energy stored in coal deposits was originally derived from solar radiation in the geologic past as a result of the creation of vegetable matter by photosynthesis. At face-value, the proposition that energy cannot be destroyed suggests that there should be no shortage of supply. This conclusion is not consistent with periodic and well-publicized fears of an 'energy crisis'. The paradox may be resolved by recognizing that energy varies in quality as measured by the ability to do useful work. Returning to the example of a coal-fired power station, much of the energy originally stored in the coal used to fire the boilers to raise the steam for the turbines which in turn drive the generators to produce electricity is dissipated as waste heat. This is released to the atmosphere via chimneys and also to any adjacent river or lake which may receive cooling water from the power station. Such waste heat is evidence of the operation of the Second Law of Thermodynamics which states that any conversion from one form of energy to another involves a decrease in the amount of useful energy. The efficiency of this conversion process is measured by the ratio between the energy input and the output of useful work. In the case of a power station, the objective is to maximize the output of electricity and to minimize the 'loss' of energy associated with various forms of waste heat.

Taken together, the First and Second Laws of Thermodynamics have important practical consequences. Although the First Law suggests that energy cannot be destroyed, the Second Law states that it is degraded into less useful forms. Effectively this means that in burning fuels such as coal or oil, a significant proportion of high-quality, concentrated energy is degraded into dispersed, low quality heat. Advances in technology may increase the possibilities of using such diffuse heat, but much can never be recovered in an energy-efficient way. Such heat not only represents a 'loss' of energy in human terms, but also contributes to the pollution of the environment by producing undesirable side-effects ranging from adverse impacts upon fish-life within individual rivers to warming of the global atmosphere. The fundamental principles embodied within the First and Second Laws of Thermodynamics must determine the approach towards any conservation-oriented strategy of energy use.

with the final stage of consumption. The chemical energy contained in the fossil fuels (and renewable biomass) which dominate existing patterns of commercial energy consumption is released by combustion. This process also releases a range of by-products or residuals to the atmosphere including carbon dioxide, nitrogen oxides, unburnt hydrocarbons and sulphur dioxide. The relative importance of these by-products varies between fuels. Coal is a 'dirty fuel', partly because of the impact of mining operations on the landscape (see Chapter 8), but more especially because of the effects of its combustion upon the atmosphere. It releases approximately 20 per cent more carbon dioxide per unit of energy than oil, while various impurities contained within coal, including sulphur, clays and carbonates, are also produced as visible particulates and invisible gases. Generally speaking, natural gas contains fewer impurities than either coal or oil, releases even less carbon dioxide per unit of energy and is regarded as a comparatively 'clean' fuel. Nevertheless, atmospheric pollution is an inevitable consequence of combustion processes whatever fuel is involved and it exemplifies the problems posed by *environmental externalities*.

Externalities arise when an economic activity imposes costs upon others without compensation. In the case of atmospheric pollution, these costs may be experienced at a range of geographical scales from the local to the global. Soot deposited on clothes on a washing-line downwind of a factory represents one end of the spectrum; the impact of the release of 'greenhouse gases' upon world climate represents the other. The local problem may be addressed by national legislation. The Clean Air Acts of 1956 and 1968 introduced measures which contributed to significant air quality improvements within the major industrial cities of the United Kingdom where the smog, associated particularly with the impact of coal combustion, that was a common winter event as late as the 1950s has become a distant folk-memory. This problem has, to some extent, been replaced by another arising from motor vehicle emissions and efforts to deal with this are evident in, for example, the requirement to fit catalytic converters to new cars. The considerable problems of dealing with atmospheric pollution at the urban scale appear trivial by comparison with those extending across international boundaries. Problems of international environmental management are not restricted to air pollution, but two of the most important ones – acid rain and global warming – are related to the atmospheric by-products of energy use.

Acid rain

Acid rain, properly termed acid deposition, results from chemical reactions involving sulphur dioxide and nitrogen oxide emissions, mainly from coal-burning power stations. These reactions form secondary pollutants such as nitrogen dioxide, nitric acid vapour and droplets containing solutions of sulphuric acid. These pollutants are deposited in dry form as gases, aerosols and fine particles and in wet form as rain or snow. Dry deposition usually occurs relatively close to the source and most attention has focused upon wet deposition which may return to the ground several days later in another country 1000–2000 km downwind. It is now generally accepted that acid rain adversely affects freshwater lakes and river systems as well as trees. These effects are seen in reduced species diversity in 'acidified' lakes and in tree damage and death (see Park, 1991). The geographical extent of the problem may be monitored by measuring rainfall on the pH scale of acidity which ranges from 0–14. A neutral solution has a pH of 7; lesser values indicate increasing acidity. 'Clean' rainfall varies in acidity about an average of 5.6. Values below this are not necessarily 'proof' of human impact, but the isopleths in Fig. 7.17 provide strong circumstantial evidence of such an impact in Western Europe. Indeed, the phenomenon first became a public issue as a result of studies conducted in Scandinavia in the late 1960s. It was noted that, by building taller chimneys to reduce local air pollution problems in British cities, emissions from power stations were given a flying start on their downwind journey across the North Sea. Canada has similarly charged the United States with 'exporting' acid rain as prevailing south westerly winds carry combustion products from the power

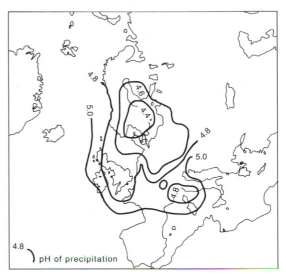

Fig. 7.17 'Acid rain' over Europe, 1986. (*Source:* UN Environmental Programme 1989/90.)

stations of the Ohio valley across the border. The occasionally acrimonious debate over this problem emphasizes the special difficulties posed by trans-frontier pollution. Despite pressure from their neighbours, both the United States and United Kingdom have been reluctant to take steps to reduce emissions from their power stations. Major improvements could be achieved by fitting flue-gas de-sulphurization (FGD) equipment to new and existing stations. Such equipment is very expensive, however, and a major project to install FGD at Drax power station, the first stage of which was completed in 1993, is unlikely to be repeated at other coal-fired power stations by the privatized electricity supply industry in the United Kingdom. The financial imperatives of corporate balance-sheets often take precedence over environmental 'costs' borne by others. In theory, the United Kingdom is subject to supra-national pressure from the European Commission; in practice, it is not willing to meet the same targets for sulphur dioxide reduction by 2003 which have been agreed by most other EU member states.

Global warming

The problem of controlling acid rain pales into insignificance when compared with the issue of global warming. The Earth's atmosphere regulates the flow of solar energy. Much out-going radiation is

Plate 7.4 Tall chimneys, exemplified by this smelter at Sudbury, Ontario, do not solve the problem of air pollution, but rather transfer it further downwind. (*Source:* A. Mather)

temporarily trapped by 'greenhouse gases' (i.e. carbon dioxide, methane and chlorofluorocarbons (CFCs)) which are essential in maintaining the ambient temperature conditions which sustain life on the planet. The proportion of carbon dioxide in the atmosphere (currently 0.035 per cent) is known to have fluctuated in geological time; it is also known to have increased by 25 per cent in the last 100 years (World Resources Institute, 1990: 11–31). This recent trend probably reflects human activity. Deforestation has been important (see Chapter 5), but more than half of the carbon dioxide released to the atmosphere since 1860 is estimated to be a consequence of burning fossil fuels. Global surface temperature seems to have been rising during the same period (Fig. 7.18). It is impossible with existing scientific knowledge to prove that this rise is human-induced or even that it is a consequence of increasing atmospheric concentrations of greenhouse gases (Jones *et al.*, 1986). The

effects of further rises in global mean surface temperature upon climate and natural ecosystems are even more uncertain. Despite this uncertainty (or perhaps because of it), there is an almost universal scientific consensus favouring attempts to control 'greenhouse gas' emissions.

The translation of this consensus into action is problematic. Rees (1991b: 292) observes that '...the political decision making process appears to be far slower than the process of environmental change'. This is especially true at the international scale where the rhetoric of sustainable development and global cooperation tends to depart from the reality of national self-interest. Bearing in mind the link between energy consumption and economic development, it is very difficult to believe that Third World countries will limit their emissions of 'greenhouse gases'. Energy-related carbon dioxide emissions, for example, increased by 82 per cent in the developing countries between 1970 and 1992 compared with a much slower rate of 28 per cent in the developed countries (US Department of Energy, 1994). Table 7.2 indicates the rank and per cent share of the global warming potential attributable to the countries that emitted the most 'greenhouse gases' in 1989. These gases include not only carbon dioxide, but also methane and the industrial chemicals chlorofluoro- carbons (CFCs), both of which are especially power- ful infra-red absorbing gases. The table emphasizes that the United States is the most important contributor to these emissions, but it also shows that several developing countries are in the top ten, mainly as a result of their very large populations. Expressed in per capita terms, the contributions of the developed economies are generally higher and countries such as China and India may justifiably question the equity of seeking to stabilize an existing situation which perpetuates huge disparities in levels of national energy consumption. The implications of this position, however, are profound and it is estimated that if countries outside the Organization for Economic Cooperation and Development (OECD), which includes the 24 major 'market' economies, had emitted carbon at OECD per capita rates in 1990, world fossil fuel consumption and carbon emissions would have been *three times* higher (US Department of Energy, 1994).

The best prospect for practical policy intervention to reduce 'greenhouse gas' emissions lies at the national level. Leaving aside the important option of slowing down the rate of deforestation in the Tropics (see Chapter 5), there are three obvious

Table 7.2 Countries with highest greenhouse gas emissions, 1989

Rank	Country	Carbon Dioxide (CO$_2$)	Methane (CH$_4$[1])	CFC's[1]	Total	% of World total	Per capita (tonnes)
		(000 tonnes)					
1	USA	4 891 005	777 000	763 490	6 431 495	17.2	25.8
2	USSR	3 804 001	714 000	393 491	4 911 492	13.2	17.1
3	China	2 388 613	840 000	70 476	3 299 089	8.8	2.9
4	Japan	1 040 554	86 100	557 935	1 684 589	4.5	13.7
5	India	771 936	756 000	23 492	1 551 428	4.2	1.9
6	Brazil	1 156 957	184 800	35 238	1 376 995	3.7	9.3
7	Indonesia	1 007 726	136 500	5 873	1 150 099	3.1	6.4
8	Germany	641 398	63 000	158 571	862 969	2.3	11.0
9	UK	568 451	81 900	146 825	797 176	2.1	13.9
10	Mexico	519 702	48 300	29 365	597 367	1.6	6.8
	World	28 263 088	5 670 000	3 406 340	37 339 428	100.0	7.2

Sources: Compiled from data in World Resources Institute (1992), tables 24.1 and 24.2 (greenhouse gas emissions in 000 MT), with population figures from FAO (1990) and conversion factor for CH$_4$ and CFC's from Intergovernmental Panel on Climate Change (1990).
Note:[1] The estimated emissions by weight of methane and CFCs are significantly lower than indicated. Conversion factors are used to express their much greater radiative effect (molecule for molecule) than carbon dioxide.

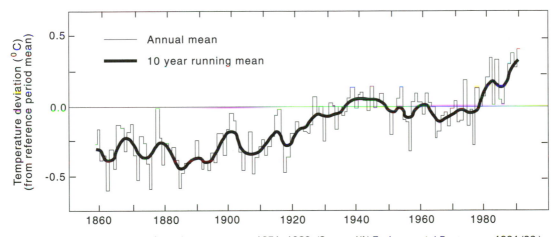

Fig. 7.18 Global mean annual surface air temperatures, 1851–1989. (*Source:* UN Environmental Programme 1991/92.)

energy-related approaches to the problem. First, improvements in energy efficiency will reduce the overall consumption of energy and, therefore, the discharge of residuals to the environment. If attention is focused on the developed economies, there are wide variations in energy efficiency which suggest that there is considerable scope for improvement in, for example, the United States to bring it up to Japanese standards. A second strategy involves switching the energy 'mix' towards cleaner fuels. Natural gas is attractive from this perspective and many forecasters predict a growing share for this fuel within most developed economies (see Odell, 1988). Such a switch depends, of course, upon resource availability. In the longer term this suggests a third option of converting to carbon-free energy sources. These include the various renewables considered earlier in the chapter as well as nuclear power. We have already seen that nuclear power creates a different set of environmental problems, while the current contribution of renewables within developed economies is minimal. Furthermore, the research and development commitment to renewable energy by both government and the private sector is limited and improvements in the efficiency of energy converters based on fossil fuels seem the most likely route to the reduction of 'greenhouse gas' emissions by the developed countries – a route made attractive by potential cost-savings rather than by any acknowledgement of a wider responsibility.

This chapter has emphasized the significance of energy use in the long-term evolution of human society. It has also identified some of the factors affecting contemporary decisions relating to energy resources. There is no dispute that the existing dependence on fossil fuels cannot be sustained, but there is considerable disagreement about the urgency of any shift from non-renewable to renewable sources of energy and there is little evidence of such a change. Indeed, reflecting upon the world energy scene during the last 20 years, it is difficult to avoid the conclusion that the perceptions of individuals and the policies of governments are based on short-term expediency rather than any clear vision. Official enthusiasm for alternatives to fossil fuels waned during the 1980s as a result of the combined effects of recession upon consumption and an apparent abundance of supply in relaxing the pressures imposed by the oil price shocks of the 1970s. It seems that further shocks, such as a major environmental catastrophe unambiguously attributed to global warming, will be required to focus the minds of policy-makers upon more distant energy futures.

Further reading

Blunden, J. and Reddish, A. (1991) *Energy, resources and environment*. London: The Open University.

Chapman, J.D. (1989) *Geography and energy*. Harlow: Longman.

Cook, E. (1976) *Man, energy, society*. San Francisco: W.H. Freeman.

Edmonds, J. and Reilly, J.M. (1985) *Global energy: assessing the future*. New York: Oxford University Press.

Smil, V. (1987) *Energy, food, environment*. Oxford: Clarendon Press, pp. 22–96.

CHAPTER 8

Mineral resources

The significance of minerals is reflected in the classification of major periods of human history in terms of the links between dominant technologies and specific resources. Copper, tin and iron were essential to the metal-using cultures of the Bronze and Iron Ages. The manufacture of metals of various types continues to be the most important application for mineral resources, but they are also used in construction and in the chemicals industry. Indeed much that we take for granted in modern society depends upon the exploitation of mineral resources.

Scarcity and abundance

Nature of mineral resources

A 'mineral' is any naturally occurring inorganic substance; a 'mineral resource' is a 'mineral' which is recognized as having utility. 'Minerals' are created by natural processes; they are transformed into 'mineral resources' by cultural processes such as advances in knowledge. For example, flint was a mineral resource during the Neolithic, but iron did not 'become' a resource until much later. Thus the range of inorganic substances contained within the Earth's crust that are defined as mineral resources has increased through time as humans have developed the means of exploiting them. A distinction is usually made between the fuel (i.e. coal, oil, natural gas, etc.) and non-fuel mineral resources and the latter are further subdivided into metalliferous (i.e. gold, silver, iron, etc.) and non-metalliferous (i.e. limestone, sulphur, potash etc.) categories. It is beyond the scope of this book to provide a systematic review of the occurrence and applications of contemporary mineral resources (see Blunden, 1985), but an awareness of certain general characteristics is important to an understanding of patterns of exploitation.

Most minerals are compounds of the elements which make up the Earth's crust although a few, such as gold and silver, occur as free elements. There are wide variations in the crustal abundance of these elements and, therefore, in the minerals derived from them. Table 8.1 indicates that aluminium is the most abundant metallic mineral, accounting for an estimated 8 per cent by weight of the Earth's crust. This global average conceals wide geographical variations in the concentration of bauxite from which aluminium is extracted. Table 8.1 suggests that an aluminium content of at least 25 per cent is required to justify mining. The corresponding thresholds for the less abundant minerals such as gold are, not surprisingly, much lower.

The definitions of minimum ore grade in Table 8.1 are approximations which vary with economic circumstances and scarcity will, in theory, encourage a trend towards lower-grade ores. Figure 8.1 indicates a general decline in the average grade of copper ore mined in the United States over a period of almost 80 years. This is consistent with the interpretation of a long-term fall in the quality threshold linked to increasing scarcity, but the trend has been reversed since 1987. This probably reflects falling copper prices which have effectively raised the cut-off point for economic operation and resulted in the closure of mines with ores below this threshold. The postulated link between scarcity and minimum ore grade is, therefore, not a simple one and a detailed study covering 50 years of production history in Canada for several minerals revealed no evidence of a consistent downward trend (Martin and Lo Sun-Jen, 1988).

Various geological processes are responsible for the accumulation of mineral deposits in sufficient quantities and of sufficient quality (i.e. concentra-

Table 8.1: Crustal abundance and minimum ore grade for selected metals

	Crustal abundance (%)	Minimum ore grade (%)
Aluminium	8.0	25
Iron	5.8	25
Manganese	0.1	15
Chromium	9.6×10^{-3}	15
Zinc	8.2×10^{-3}	2.5
Nickel	7.2×10^{-3}	1.0
Copper	5.8×10^{-3}	0.5
Lead	1.0×10^{-3}	2.0
Uranium	1.6×10^{-4}	0.18
Tin	1.5×10^{-4}	0.2
Molybdenum	1.2×10^{-4}	0.25
Tungsten	1.0×10^{-4}	1.35
Silver	8.0×10^{-6}	0.01
Platinum	5.0×10^{-7}	0.000 3
Gold	2.0×10^{-7}	0.000 8
Mercury	2.0×10^{-6}	0.20

Source: Chapman (1983) p. 56

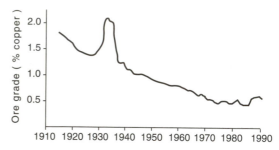

Fig. 8.1 Average grade of copper ores mined in the United States, 1911–91. (*Source:* US Bureau of Mines, annual.)

tion) to justify their extraction. Some of these processes are more common than others. Sedimentary rocks such as chalk and limestone as used in construction or road building are mineral resources. These rocks are not ubiquitous, but they are found over wide areas. The majority of mineral resources, however, are anything but ubiquitous and their spatial distribution is determined by complex relationships between the abundance of the elements which provide the raw materials for their creation and the chance combinations of geological circumstances which result in their formation or concentration in particular places. These places are more numerous in the case of the abundant minerals such as aluminium and iron; they may be very infrequent for rare minerals. There are also differences in size of individual deposits. The abundant minerals are found in large quantities, but even the most favoured occurrences of the rare minerals contain limited quantities.

Life indices for mineral resources

The 'quantity' of mineral resources is obviously most important at the global scale because they are finite or *stock* resources which, in the absence of recycling, are effectively 'used up'. Some of the most alarming projections contained in *Limits to Growth* (Meadows *et al.*, 1972) related to mineral resources. These were based on the calculation of the so-called exponential index which indicated the number of years that proven global reserves would last with consumption growing exponentially at the prevailing average annual rate of growth. The exponential index suggested that the world would, for example, run out of gold by 1979, silver by 1983 and tin by 1985. The exponential index was not the only basis for projection contained in *Limits to Growth*, but it was the one which attracted the greatest attention because of its doomsday scenario. In fact, it was the most unrealistic basis for forecasting. The rates of economic growth experienced in the developed economies during the 1950s and 1960s were exceptionally high by both historic and current standards. Furthermore, little or no allowance was made for the growth of recycling and other improvements leading to more economical use of mineral resources. Simple extrapolation leading to sustained exponential growth in the consumption of resources was, therefore, simplistic and pessimistic.

The apparent exhaustion of a finite resource is not only defined by levels of consumption. It has already been noted with reference to energy resources (Chapter 8) that reserve estimates change through time and that proven reserves usually represent only a relatively small proportion of the ultimate resource base (Fig. 1.7). Thus the proven reserves of 1970 which were used to calculate the exponential index in *Limits to Growth* were not the same as current estimates of proven reserves. Table 8.2 emphasizes some of the implications for projections of resource 'life' of these complex interactions between levels of consumption and reserve estimates. The methods of calculating the respective 'life indices' in 1970 and 1990 are broadly comparable. The 1970 forecasts are the most optimistic that were contained in *Limits of Growth* and assume constant rather than increasing levels of annual consumption. Even so, it was projected that mercury, tin and zinc would have

Plate 8.1 The finite nature of mineral resources is evident in the 'boom and bust' syndrome associated with many mining towns. The main street of Clyde in Otago, New Zealand is quieter than it was when the town was established during a late-nineteenth-century 'gold rush'. (*Source:* A. Mather)

been exhausted by 1983, 1987 and 1993 respectively, with lead and copper suffering the same fate by 1996 and 2006. Comparison with the 1990 forecasts suggests that these 'exhaustion dates' have all been pushed back into the next century. It also emphasizes that the future for most of the minerals identified in the table looked brighter in 1990 than it did 20 years earlier. It may be argued that we have no reason to be sanguine when the revised exhaustion dates for several minerals are less than a generation away. However, experience suggests that the rolling upward revision evident in Table 8.2 is typical (see Chapter 7) and that the ultimate resource base for most minerals is sufficiently large to allow this process to continue for the foreseeable future.

Finding mineral resources

Economics of exploration

The discovery of a new deposit is an obvious way of extending the life of a mineral resource. The image of the lone prospector armed with experience, intuition and a geological hammer is an appealing one and

many major discoveries have been made by such individuals. However, many of the techniques now used to find mineral deposits, including remote sensing from satellites and aircraft, are expensive and sophisticated and can only be carried out by large organizations. Opportunities remain for individuals and small, specialized exploration companies, but multinational corporations are the principal agents in exploration activity. The intervention of governments, especially in Third World countries, has weakened the dominance of these corporations in recent years by promoting the growth of state-owned enterprises. Nevertheless, the mining multinationals are well represented in listings of the world's largest companies and include Aluminium Company of America (US), RTZ Corporation plc (UK), Noranda Inc. (Canada) and Broken Hill Proprietary (Australia). These companies typically operate in many countries and their activities extend from exploration, through mining to processing and the manufacture of metal and other mineral-based products.

As might be expected, such companies regard exploration as an economic activity motivated by the prospect of financial returns. Its overwhelming characteristic is the combination of high levels of

Table 8.2 Estimated Life Indices of selected mineral resources

	1970 Projections[1]		1990 Projections[2]	
	Life Index (Years)	'Exhaustion' Date	Life Index (Years)	'Exhaustion' Date
Aluminium	100	2070	200	2190
Cobalt	110	2080	109	2098
Copper	36	2006	36[3]	2026[3]
Iron	240	2210	175	2165
Lead	26	1996	21	2011
Mercury	13	1983	22	2012
Molybdenum	79	2049	67[3]	2056[3]
Nickel	150	2120	52	2042
Platinum Group	130	2200	2 25[3]	2215[3]
Tin	17	1987	27	2017
Zinc	23	1993	20	2010

Sources: [1]Meadows *et al.* (1972)
[2]World Resources Institute (1992)
[3]Frosch and Gallapoulos (1989) – reserves in 1989
Notes: Projections from all sources were obtained by dividing proven recoverable reserves by annual rates of consumption. Reserves estimates and consumption rates contemporary with the date of projections.

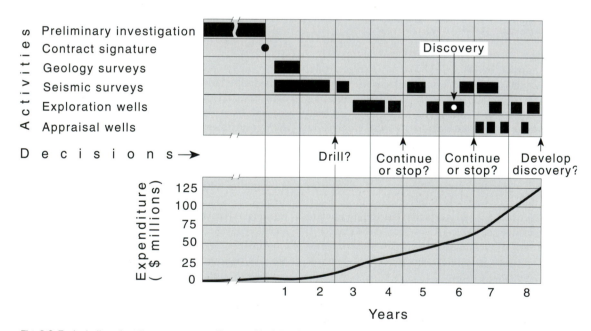

Fig. 8.2 Typical oil exploration programme. (*Source:* Shell Briefing Service, 1994.)

risk and uncertainty. Large companies are better equipped to cope with these problems, which have, therefore, encouraged the domination of the world mining industry by major multinationals. Despite significant advances in the technology available for finding new deposits, most exploration projects are unsuccessful and do not even recover their costs, let alone secure an acceptable return on the initial capital investment. Earnings are typically generated by a small number of very lucrative finds and, in this

Plate 8.2 Economic and political considerations have focused mineral exploration upon the developed countries. Mineral resources are, however, vital to the economic prospects of several developing countries. Brazil has an especially favourable mineral resource base with substantial reserves of bauxite, iron ore, nickel and tin. Alluvial tin mining at Pitinga in Amazonia. (*Source:* L. Joels)

respect, exploration has more in common with research and development than with manufacturing as an economic activity. Figure 8.2 summarizes the activities, stages, decision points and expenditures associated with a search for oil. Similar steps are involved in any mining exploration programme and the diagram suggests that eight years may elapse between preliminary investigations and the decision to develop a discovery into a producing field. Expenditure rises steadily, but pales into insignificance by comparison with the development costs involved in bringing an oilfield or a mine into production. This transition is not only costly, but also time-consuming and may take from 5 to 10 years for a large project in a remote location.

Various economic techniques are used in making development decisions, but the very long lead-times mean that such decisions are inevitably based on assumptions about unpredictable or unknown future events. Cost over-runs are common and anticipated revenues may be affected by changes in price and levels of demand which are themselves related to macroeconomic conditions. Prices are not only relevant to decisions on specific projects, but also to general levels of exploration activity. Indeed, given that future prices will ultimately determine the commercial fortunes of a mining project, exploration is surprisingly sensitive to current prices and expenditure on exploration tends to reflect short-term fluctuations in economic circumstances (Crowson, 1988). For example, the slump in the world oil price in 1986 (Fig. 7.9) resulted in drastic reductions in exploration budgets and the 'mothballing' of large numbers of drilling rigs. Attitudes towards exploration are influenced by corporate perceptions of political as well as economic futures. The large scale and economic importance of mining projects ensure that they attract the attention of governments and policies relating to taxation and foreign investment, for example, are important variables affecting the exploration and development decisions of multinational companies. These variables are additional sources of risk and uncertainty and their significance is increased by the volatility of political events relative to the lengthy time horizons involved in finding and developing mineral deposits.

Geography of exploration

The distinction between reserves and resources has already been shown to be important in the context of debates about the 'life' of non-renewable resources. It is also relevant to an understanding of their geography. Thus while the overall distribution of a mineral resource within the Earth's crust is physically determined, our knowledge of this distribution is incomplete and we can only exploit deposits which have been identified as reserves. Viewed in global terms we have much more information about geological conditions in some places than in others. Odell (1989) emphasizes that, compared with the United States, very few exploration and development wells have been drilled in the search for oil elsewhere in the world, especially in China, Africa, Latin America and South-East Asia. Crowson (1988) notes a similar pattern for all other minerals and suggests that total annual exploration expenditures in developing countries have never exceeded 28 per cent of the world total since the 1960s and have sometimes fallen to less than half of this figure. Thus while patterns of exploration within favoured areas have been shaped by judgements based on geological factors, other, essentially cultural, variables, have defined the favourable areas themselves leading to a highly selective search.

The relative neglect of the developing countries is not surprising in view of the historic pattern of industrialization. The demand for mineral and energy resources increased sharply as a result of the changes accompanying the Industrial Revolution and it was natural to try and satisfy this demand by looking for deposits in the countries directly affected by these events. This geographical concentration of exploration within a few countries tended to be self-perpetuating because it led to the selective accumulation of geological information. There is a strong 'bandwagon effect' in mineral exploration as significant discoveries within an area stimulate efforts to repeat earlier successes. Viewed over the long term, this pattern of behaviour has reinforced the effect of economic arguments promoting exploration in the developed rather than the developing countries. Political considerations have also contributed to this situation. Operations within developing countries are regarded as more risky from the point of view of the mining multinationals. Fears of political changes leading to the abrogation of earlier agreements and the possible expropriation of assets (see p. 174) influence the thinking of these corporations. Attractive geological prospects in a particular country may be more than offset by perceptions of 'political' risk and such judgements certainly influence patterns of mineral exploration (Johnson, 1990).

Exploiting mineral resources

Economics of production

The discovery of a mineral deposit is no guarantee of its development which will depend upon a complex process of resource appraisal embracing geological, economic, political and other considerations. It is the interaction of geological and economic factors as influences upon profitability which is the initial consideration in assessing the feasibility of a mining project, although political and other factors may be decisive in the final analysis. The relative importance of these factors will vary from one company to another and the priorities of state-owned enterprises are not necessarily the same as those of private-sector corporations. Nevertheless, projections of costs and revenues are required in the preliminary appraisal of any discovery.

Production costs reflect various attributes of the deposit including its size, quality and accessibility. The high capital costs of mining projects ensure that large deposits are more attractive than small ones since a long producing life is likely to generate a better return on investment. The quality of the deposit, as defined by its mineral content, is also important. It has already been noted that for any mineral there is a minimum concentration below which exploitation is not economic (Table 8.1). Although this threshold shifts over time in response to the interacting influences of advances in technology and changes in costs and prices, a company will, other things being equal, choose to exploit a higher- before a lower-grade deposit. Accessibility is another factor influencing costs. Generally speaking, a deposit which lies at or close to the surface is easier to develop than one at greater depth. Extraction costs are affected by other aspects of geological accessibility including the thickness of seams or veins, the degree of faulting and, in the case of oil and salt, the porosity and permeability of the surrounding rocks.

Geography of production

Accessibility has a spatial dimension related to transport costs. Minerals only have commercial value if they can be sold in a market. Although coalfields have attracted coal-using economic activities to the resource to a significant extent, this is unusual and mining operations are often remote from the principal centres of consumption. In these

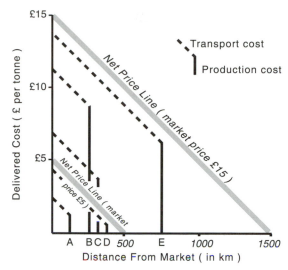

Fig. 8.3 Transport and production costs in the exploitation of mineral deposits. (*Source:* modified after Hay, 1976.)

circumstances, differences in transport costs may determine development decisions. Given two newly discovered deposits of a particular mineral with equal production costs, the one nearer the market is likely to be developed first. Various attempts have been made to demonstrate the nature of the relationships between production and transport costs in mining. Studies of individual coalfields, for example, have drawn attention to patterns of development in which the focus of operations progressively shifts towards locations which are less accessible in both a geological and a geographical sense (see Wilson, 1968; Millward, 1985).

Surface patterns of exploitation are often easy to interpret in the case of coal because they partially reflect sequential changes in underground conditions such as a steadily dipping seam. However, the distribution of most mineral resources is discontinuous and irregular even at the regional scale. Figure 8.3 represents this more complex situation (Hay, 1976). The location of several deposits relative to a market is given. Production costs are assumed to vary and are represented as vertical bars of differing height. The intersection of the bars with the base line defines the distance of each deposit from the market. Assuming a freight rate of 1p per tonne km, total transport costs rise with increasing distance from the market. The cost of transport must be deducted from the prevailing market price to yield a net price per tonne at the mine which therefore declines with

distance at the rate of 1p per km or £1 per 100 km. Given a market price of £5 per tonne, the net price declines to zero at 500 km from the market. Within this spatial margin of viability, however, it is clear that distance from the market does not exclusively determine the sequence of development. Assuming simultaneous knowledge of all deposits, A will be exploited first, followed by D. Although nearer to the market than D, B and C will not be developed because their higher production costs mean that their output cannot be delivered below the market price.

Figure 8.3 emphasizes that the decision to exploit a mineral deposit must take account of prices as well as costs. Indeed, 'it is the price...that ultimately determines whether a deposit...is a mineable ore or just so much rock' (Aschmann, 1970: 174). Thus all deposits in Fig. 8.3 other than A and D are, in Aschmann's term, '...so much rock' at a market price of £5. The position is very different when the price rises to £15 making the deposits at B, C and E commercially viable. The situation depicted in Fig. 8.3 incorporates several simplifying assumptions such as a linear increase in transport costs with distance, the homogeneity of product between different mines and a rather naive view of the relationship between supply and demand at different price levels. Nevertheless, Fig. 8.3 illustrates a real problem in the mining industry. Mineral prices are, for a variety of reasons, notoriously volatile and this characteristic seems to have become even more pronounced in recent years (Daniel, 1992). Major investments are difficult to make in these circumstances. Price changes are also significant influences upon the fortunes of existing mines. Many close not because their reserves are exhausted, but because falling prices make it impossible to extract them at a profit. The recent history of tin mining in Cornwall illustrates the point. Various attempts were made to revive the industry in the 1970s, approximately 100 years after it reached its zenith. These efforts were stimulated by a combination of political and economic considerations as the industry seemed to offer the possibility of new jobs in a depressed area (Spooner, 1981a). The final mine closed in 1991. Several factors contributed to this decision, but the collapse of the international tin price from £8900 per tonne in 1986 to £2865 just 5 years later was certainly one of them. It is not inconceivable that future moves in the opposite direction may encourage yet another revival of tin mining in Cornwall, but technical problems such as flooding often prevent the re-opening of deep mines.

Technical change

It is not only changes in price which affect the perception of mineral resources and the definition of reserves. Technical change, which is an essentially cultural phenomenon, is also important in both 'creating' and 'destroying' mineral resources. The flint mines of the Neolithic are now abandoned because humans have discovered better materials for the manufacture of tools. One of these is tungsten which is contained in scheelite and wolframite found in granitic rocks. Tungsten is exceptionally hard and is used in cutting- and machine-tools and in the production of tough steel alloys. It is now regarded as an important material, but these applications have only developed to a significant extent since the Second World War and, certainly, no practical uses had been found for tungsten at the beginning of the twentieth century. More gradual effects upon the demand for particular mineral resources may arise as a result of technical changes promoting conservation or substitution. Plastics have replaced metals in many applications over the last 30–40 years and continuing advances in the development of high-performance polymers are, for example, bringing the idea of an all-plastic car closer to reality. Conservation pressures are encouraging the adoption of recycling schemes. Scrap has been used in the manufacture of steel for a long time and accounts for a substantial proportion of the input to furnaces. Similar practices are being adopted for other metals such as aluminium. Such recycling is not only motivated by the desire to conserve finite resources, but also by a growing awareness of the environmental risks associated with the dumping of waste containing toxic metals such as lead and mercury.

Technical change affects the supply of, as well as the demand for, minerals. Various advances have made it possible to exploit lower-grade ores. These have affected both methods of extraction and the preliminary processing of ores. Ever larger machines have, for example, accelerated the rate at which rock may be removed and techniques for the concentration of ores, such as pelletization, have improved the economics of lower-grade deposits by allowing preliminary on-site up-grading before onward shipment. Such techniques reduce the proportion of waste material contained in a cargo and, therefore, the unit costs of transportation. This is significant because transport costs have exerted a strong influence upon patterns of mineral exploitation (Fig. 8.3). The costs of transporting coal, for example, were very high in the eighteenth century, but it is now an important commodity in world trade (see Chapter 7). Indeed the bulk ocean carriers which have facilitated the growth of this trade are also used to move the other key resource of the Industrial Revolution – iron ore (Bradbury, 1982). The geographic association of iron ore and coal was largely responsible for the creation of communities such as Middlesborough in the mid-nineteenth century. The last remaining steelworks in north-east England is sited downstream from Middlesborough at the mouth of the River Tees and depends upon iron ore imported from as far afield as Australia. This example is typical of a general trend towards the increasing geographical separation of the points of mineral production and consumption which has been made possible by advances in the technology of transportation (Chisholm, 1990). As already noted in Chapter 3, these innovations have promoted the growth of a progressively more complex global system of resource-based interdependencies. The operation of this system raises many political and ethical questions, not least the relative positions of the developed and developing countries as beneficiaries from the trade in commodities such as minerals.

Mineral resources and international relations

Strategic minerals

Earlier references to the 'life' of mineral resources emphasized their physical availability, but political restrictions in supply have often been of more immediate concern. Hitler regarded self-sufficiency in mineral and other raw materials as essential to the security and self-esteem of the Third Reich. Following its defeat in the First World War, Germany surrendered the iron ore of Lorraine to France and became largely dependent upon imports from Sweden in the inter-war years. Hitler's fears may be dismissed as the paranoia of a demagogue, but many other political leaders have displayed similar concerns since the Second World War, especially at times of international tension or conflict such as the Korean or Viet Nam Wars (see Mandel, 1988; Haglund, 1989). These fears have focused on the 'strategic' minerals. There is no unambiguous definition of this term, but the essential characteristics of strategic minerals rest upon a combination of economic importance, especially for military industries, and accessibility. They are defined from a particular national perspective and a mineral that is 'strategic'

Table 8.3: Production and proven reserves of selected metallic minerals, 1990

		Production		Reserves[2]	
		(in 000 tonnes)	% of world total	(in million tonnes)	% of world total
Copper	Chile	1 603.2	18.2	85.00	26.5
	United States	1 587.2	18.0	55.00	17.1
	Canada	779.6	8.8	12.00	3.7
	USSR	600.0	6.8	37.00	11.5
	Zambia	445.0	5.0	12.00	3.7
Lead	Australia	563.0	16.7	14.00	19.9
	United States	495.2	14.7	11.00	15.6
	USSR	450.0	13.4	9.00	12.8
	China	315.0	9.3	6.00	8.5
	Canada	236.2	7.0	7.00	9.9
Zinc	Canada	1 177.0	16.1	21.00	14.6
	Australia	937.0	12.8	19.00	13.2
	USSR	750.0	10.2	10.00	6.9
	China	619.0	8.4	5.00	3.5
	Peru	576.8	7.9	7.00	4.9
Iron ore	USSR	236 000.0	27.3	23 500.00	36.3
	Brazil	150 000.0	17.3	6 500.00	10.0
	China	118 000.0	13.6	3 500.00	5.4
	Australia	110 000.0	12.7	10 200.00	15.8
	United States	59 032.0	6.8	3 800.00	5.9
Nickel	USSR	259.0	27.6	6.62	13.6
	Canada	201.9	21.5	8.13	16.7
	New Caledonia	88.0	9.4	4.54	9.3
	Australia	70.0	7.5	1.27	2.6
	Indonesia	58.0	6.2	3.20	6.6
Bauxite[1]	Australia	40 697.0	37.3	4 400.00	20.4
	Guinea	16 500.0	15.1	5 600.00	26.0
	Jamaica	10 921.0	10.0	2 000.00	9.3
	Brazil	8 750.0	8.0	2 800.00	13.0
	India	5 000.0	4.6	1 000.00	4.6

Source: World Resources Institute (1992), tables 21.5 and 21.6
Notes: [1]Dry weight
[2]Mineral reserves are those deposits whose quantity and grade have been determined by samples and measurements and can be profitably recovered at the time of the assessment. Changes in geologic information, technology, costs of extraction and production and prices of mineral products can affect the reserve estimates. Reserves do not signify that extraction facilities are in place and operative.

for one country which must rely on imports to meet its needs is not 'strategic' for another with a domestic source of supply. Minerals which are found in relatively few places are, therefore, more likely to be perceived as 'strategic'. Table 8.3 identifies the principal sources of selected minerals in terms of their shares of world production and of proven reserves in 1990. The production figures emphasize the dominance of relatively few countries, notably the former USSR, Australia, Canada and the United States, although individual developing countries are important for specific minerals. The distribution of proven reserves is even more uneven in some cases with a single country accounting for more than 40 per cent of the world total (e.g. South Africa – chromium and manganese, US – molybdenum, Zaire – cobalt, China – tungsten).

The concept of strategic minerals is defined with reference to the interests of the developed countries since they consume the overwhelming proportion of the world's mineral resources. For example, Japan was the world's second largest (after the former USSR) steel producer in 1991, but its domestic production of iron ore was negligible. Indeed, Japan

Table 8.4: Percentage shares of total mine production (tons of contained metal) for selected minerals by world economic regions

	Industrial			Developing			Other[1]		
	1965	1980	1990	1965	1980	1990	1965	1980	1990
Copper	39	31	40	45	49	45	16	20	15
Lead	46	45	51	34	33	26	21	21	23
Zinc	57	50	55	27	28	27	16	22	18
Iron ore	44	34	31	29	38	41	27	28	28
Bauxite	16	34	44	67	55	49	17	11	7
Nickel	70	47	37	9	30	32	21	23	31

Sources: 1965 and 1980 data from Faber (1986) and 1990 data adapted from World Resources Institute (1992).
Note: [1] Mainly former and existing centrally-planned economies.

is especially dependent upon imported mineral (and energy) resources but it is US governments which have expressed the greatest concern about this issue (see Prestwich, 1975; Anderson, 1988; Mandel, 1988). The geographical origin of imports has an important bearing upon perceptions of political risk and the United States is probably less concerned about its heavy reliance upon Canada for supplies of nickel than it is about obtaining cobalt from Zaire or tungsten from China. The vulnerability of the United States and other developed countries to the disruption of supplies from the Third World has, in fact, been exaggerated in the minds of politicians. The success of OPEC in raising the price of oil in the 1970s (Chapter 7) suggested that similar successes might be achieved by producer cartels for other commodities. However, Table 8.3 emphasizes that the United States is itself one of the world's major mineral producers together with Australia and Canada. Furthermore, many developed countries have accumulated substantial stockpiles of minerals which they perceive to be 'strategic'. These provide insurance against supply disruptions as well as the opportunity to influence international prices in desired ways by selective releases on to the market. It is likely that the developing countries will increase their share of world mineral production, in the long term, but this trend seems likely to be much slower than many predicted in the 1970s (Tanzer, 1980). Indeed Table 8.4 shows that their share actually declined for several of the principal metallic minerals during the 1980s.

Export dependence

The developing countries' view of strategic minerals looks very different from that of the developed

nations. Minerals make a substantial contribution to GDP, government revenues and export earnings in several developing countries (Table 8.5). It is evident that some of the poorest countries in the world are mineral-based economies. Chile is the most developed of the countries identified in Table 8.5 with a per capita GDP that is 50 per cent of the world average and less than 10 per cent of the corresponding US figure. Most of the others lie well behind Chile on any measure of economic well-being. This is, at first sight, surprising since the possession of mineral deposits might be regarded as a source of comparative economic advantage and both economic theory and the experience of history in Europe and North America suggest that mineral resources can provide the initial impetus to development (Chisholm, 1982). The contrast between what Eyre (1978) called 'the real wealth of nations' as embodied in their resource endowments and the relative poverty of the mining economies as measured by conventional economic indicators is striking. There is no simple explanation and the situation is part of a much wider set of issues surrounding the core–periphery relationships between the developed and developing world (see Chapter 3).

The terms of trade measure the change in import prices relative to export prices and effectively define the rate of exchange. Simplifying considerably, the developing countries export primary commodities, including minerals, to the developed countries which supply manufactured goods in return. There is evidence that the rate of exchange associated with this pattern has steadily shifted in favour of the developed countries. The mineral exporters have been especially hard-hit (although it is important to remember that these include countries such as

Table 8.5: Ores and metals as percentage of total value of exports in selected countries

Country	Year			Per capita GDP, 1990 (in US $)	Per capita GDP, 1990 (% of world average)
	1970	1980	1990		
Chile	88.1	64.1	53.1	2 110	50.0
Liberia	73.6	57.6	59.7[1]	333	7.9
Mauritania	88.3	78.3	48.6	520	12.3
Niger	0.2	84.7	86.0[1]	326	7.7
Papua New Guinea	0.8	49.1	41.3	827	19.6
Peru	48.4	42.3	42.6	1 691	40.1
Sierra Leone	18.9	34.6	40.9	217	5.1
Togo	24.9	40.3	44.7	459	10.9
Zaire	77.8	48.3	55.2	232	5.5
Zambia	99.1	93.4	83.4[2]	518	12.3

Source: adapted from UNCTAD (1993)
Notes: [1] 1986
        [2] 1989

Australia and Canada as well as developing nations). Despite short-term fluctuations and significant differences between individual minerals, there has been a secular decline in the real price of metals and minerals since 1950 (Fig. 8.4). The reversal of this and similar trends in the price of other primary commodities has been and remains an important political objective of the developing countries in international trade negotiations.

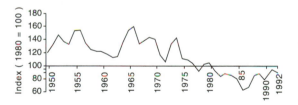

Fig. 8.4 Mineral/metals: real price trends, 1950–92. (*Source:* Auty, 1993.)

Ownership and control

Much of the international trade in minerals is internal to multinational companies with headquarters in the developed world. The trade is, therefore, externally controlled from the point of view of the developing countries and the economic interests of multinational companies and national governments will rarely coincide. For example, transfer pricing, in which one subsidiary of an integrated company is 'selling' to another allows profits to be made in the place which is most convenient to the company. This is frequently in the downstream manufacturing operations located outside the developing countries which, therefore, lose tax revenues on their mining activities. This practice is just one example of the tensions in the relationship between host governments and foreign multinationals – a relationship hinging upon the allocation of the resource rent which is the surplus earned by a mineral deposit over and above the minimum earnings required to attract the investment to develop it. This relationship is essentially unequal. The mining multinationals may have annual sales which exceed the total value of the GNP of the countries with which they are dealing. They also have the expertise which is conspicuously lacking in many developing countries and may be able to play one country off against another in negotiations in which they hold most of the 'cards' and, in particular, most of the information. In these circumstances, it is not surprising to find examples of deals over mining concessions which could only be described as the exploitation of the weak by the strong (see Lanning and Mueller, 1979).

Various attempts have been made to change this balance of power. Producer organizations have been established for bauxite, copper and iron ore, but they have been unable to achieve price increases comparable with the successes of OPEC in the 1970s. Several factors have contributed to their weakness, including a lack of internal cohesion and the fact that they have failed to recruit all of the major producers. Furthermore, the mining companies have, like the oil companies (Chapter 7), responded by diversifying

Plate 8.3 The derelict head-gear of former copper mines in Butte, Montana partly reflects a trend away from deep-mining towards more economic open-cast methods. (*Source:* K. Chapman)

their sources of supply. The success of this strategy is partly reflected in the previously noted failure of the developing countries to increase significantly their share of world mineral production (Table 8.4). Nationalization is a more direct approach to the problem and several governments in Africa and Latin America expropriated the assets of mining companies between the mid-1960s and the mid-1970s. In several cases, such as Zambia's acquisition in 1969 of a controlling interest in the copper mining operations of Anglo-American and Roan Selection Trust, these moves were regarded as a logical consequence of political independence, which also presented an opportunity for the renegotiation of agreements originally reached under colonial rule. Nevertheless, the additional leverage conferred by political independence does not obscure the fact that the mining multinationals dominate the international market and retain much of the expertise necessary for the achievement of the economic aspirations of the mineral-exporting countries. The Zambian copper mining industry, for example, continues to rely upon the technical and management skills of staff of Anglo-American and Roan Selection Trust. Similarly,

changing political attitudes in Chile have encouraged a reversal of its previous policy of excluding foreign multinationals from its mining industry and it entered into an agreement with a US–Canadian consortium in 1993 to develop a massive new copper mine at El Abra. The fact that Codelco, the state-owned copper company, is the minority partner in this project reinforces the point.

Expertise, or rather the lack of it, is also significant from another point of view. It would be misleading to attribute the difficulties of the mineral-dependent economies exclusively to the exploitative behaviour of the multinationals. Although there is no doubt that the options available to the governments of these countries are heavily circumscribed, there have been some spectacular examples of economic mismanagement with mineral revenues frittered away on prestige projects or diverted to the personal bank accounts of a corrupt ruling élite. Indeed it has been suggested that the possession of mineral resources has, in some cases, been a disadvantage or 'curse' because it has distorted the structure of national economies and discouraged the adoption of policies designed to secure long-term development (Auty, 1993).

Impacts of mining

Mining operations

Several of the issues surrounding the allocation of the benefits of mineral extraction at the international scale find parallels in debates over the impacts of mining at regional and local levels. These impacts ultimately depend upon the nature of mining operations. With the exception of solution mining techniques using boreholes which are used to recover potash, salt and sulphur, extraction involves either surface or underground mining. Several types of surface mining are used, including open pits or quarries and various methods of strip mining or open-cast working. Open-pit mining creates large holes in the ground which usually become a permanent feature of the landscape. Strip mining and open-cast, normally associated with the extraction of coal, involve the use of massive diggers, power shovels and drag-lines which are capable of removing overlying rock, soil and vegetation. These methods may lead to permanent scarring of the landscape over extensive areas, but environmental concerns have encouraged the adoption of techniques to limit this damage (see pp. 175–6). Many minerals may be extracted using surface or underground methods, but there is a clear trend towards the former. As late as the mid-1960s, underground mineral working was the more important in the United States; by 1980 over 80 per cent of the country's mineral wealth was extracted by open-pit methods (Blunden, 1985: 191). Generally speaking, advances in the technology of extraction have improved the economics of surface relative to deep-mining, a trend which has been reinforced in some cases by the exploitation of lower-grade ores which are better suited to surface removal.

Environmental conflicts

The disruptive consequences of open-pit mining, which is responsible for some of the largest human excavations in the world, are typical of the controversial character of all mining operations. Mining proposals often draw attention to differing perceptions of the environment and its resources. For example, plans to open quarries in environmentally protected areas produce a direct clash between commercial interests, sometimes supported by local communities seeking new jobs, and amenity groups anxious to preserve the aesthetic qualities of an

Box 8.1: Mining and aboriginal rights in Australia

Aboriginal perceptions of land have been expressed in the following terms:

'Land is not bound by geographical limitations . . . it is a living entity. It belongs to me. I belong to the land. I rest in it. I come from there'
(Dodeson, 1983, p.5)

Given this reverence for land, it is not surprising that modern large-scale mining operations are often resented by aboriginal communities. This resentment has been reinforced by a wider sense of injustice linked to the occupation of land that belonged exclusively to the indigenous peoples of Australia before the arrival of Captain Cook.

The exploitation of mineral deposits on aboriginal land has, in many cases, resulted in the subordination of community to corporate interests. The discovery of bauxite by a RTZ Corporation subsidiary on the Cape York peninsula in northern Queensland in the 1950s, for example, prompted the unilateral revocation of an earlier agreement reserving much of the peninsula for the use of its aboriginal inhabitants. This action was significant because it set a precedent that was followed in later projects associated with Australia's 'minerals boom' in the 1960s and 1970s (Howitt, 1992). Generally speaking, aboriginal communities have not only been unable to resist mining projects on land that was once theirs alone, but have also failed to reap significant economic benefits in return for the concessions imposed upon them.

The question of aboriginal land rights in Australia is a sensitive one and the Native Title Act, which became federal law at the beginning of 1994, is a significant and belated (compared to Canada and New Zealand) recognition of these rights. Under this legislation, title rights to land with which indigenous inhabitants can demonstrate 'a close and continuing association' may be granted. However, the validity of existing mining leases is also accepted with the proposal that native title may 'revive' with the expiry of such leases. Nevertheless, the mining industry is unhappy about the Native Title Act and the Australian Mining Industry Council estimates that it will cause new projects to be delayed by up to two years. While mining interests may be expected to take this position, there is clearly a national economic interest in sustaining a successful export-based mining industry in Australia. Despite the new legislation it may be difficult to reconcile this interest with the legitimate aspirations of the aboriginal population.

attractive landscape. These issues have been evident in disputes surrounding the quarrying of limestone in the Peak District National Park and the proposed extraction of china clay on the edge of Dartmoor National Park. Conflicts over resource use are even more difficult to resolve where they bring totally different cultures or value-systems into opposition. Australia, which is one of the world's largest mineral producers (Table 8.3), provides several examples of such conflicts (Connell and Howitt, 1991). Both federal and state (especially Western Australia) governments have made strenuous efforts to encourage the exploitation of the country's considerable mineral wealth as a means of stimulating economic development at national and regional levels. These efforts have provoked confrontation with aboriginal communities which have a spiritual relationship with their land that differs fundamentally from 'Western' concepts of property and ownership (Box 8.1). Such situations are not unique to Australia and similar confrontations have arisen in connection with oil developments in Alaska and copper mining in Papua New Guinea. The common thread is the question of land rights and there has been some belated recognition that the prior claims of indigenous peoples have a moral legitimacy that has been disregarded in the scramble for mineral resources. These peoples may have no use for bauxite or copper, but the land containing these minerals is, nevertheless, a highly valued natural resource.

Conflicts over land rights acquire even greater significance when it is remembered that any mine has a finite life and any assessment of the costs and benefits of a proposal must acknowledge the inevitability of closure, often within the span of a single generation. The ghost towns left behind provide a vivid reminder of this reality. The economic and social consequences of this boom and bust syndrome are most obvious in the immediate aftermath of a mine closure. Many former mining areas are identified as 'problem regions' characterized by high unemployment rates, out-migration and an ageing population. The prospects of inheriting these problems make it all the more important not only to maximize the economic benefits during the operation of a mine, but, where feasible, to find ways of securing a community future which extends beyond the life of the resource (Neil *et al.*, 1992).

Economic impacts

The efforts of governments to use mineral deposits to promote national economic growth have been matched by similar attempts to base the development of peripheral regions upon investments in mining. The lasting impact of the gold 'rushes' in South Africa and California demonstrate the possibilities (Warren, 1973: 209–25). The gold resources of the Witwatersrand were discovered in 1886 and were responsible for the rapid growth of Johannesburg which remains the economic core of South Africa. Similarly, the 'forty-niners' attracted by John Sutter's chance discovery were the catalyst in California's rise to economic importance and the population of San Fransisco rose from just 450 in 1847 to more than 55 000 by 1860. These experiences demonstrate the operation of processes of circular and cumulative causation based on the initial advantage provided by mineral resources. A review of these processes is beyond the scope of this book and fuller accounts may be found elsewhere (see Chapman and Walker, 1991: 170–77). The multiplier is the essential element in these processes. It rests upon the simple idea that investment introduced by, for example, a mining company will stimulate further investment by other businesses providing goods and services to and processing the output of the mine. Wages paid to employees that are spent locally will generate further demands, leading to the establishment of other economic activities such as shops serving the population rather than the original mining operation. The creation of a supporting social infrastructure including schools, hospitals and so on may transform a previously remote area and investment in transport facilities may improve accessibility to other centres of economic activity, further enhancing its prospects for development. Indeed, governments frequently regard mining projects as an opportunity to 'open-up' peripheral regions. The development of Cuidad Guayana in the eastern part of Venezuela began in 1960 as part of a deliberate strategy to create a new city to offset the dominance of the capital, Caracas. The choice of site was strongly influenced by the existence of an established iron ore mine and encouraging geological prospects for further mineral discoveries. Western Australia provides another example of substantial public investment in infrastructure in association with private-sector mining projects. However, the results of these investments have been disappointing (Alexander, 1988). Ambitious plans for manufacturing based on the iron ore and other mineral resources of the Pilbara region in the northern part of the state have not materialized. This experience is not unusual and resource-based industrialization, especially in developing countries, has rarely looked like producing the large-scale and lasting effects initiated by coal and iron in nineteenth-century Europe and North America (Auty, 1990).

Unrealistic expectations, often deliberately fostered by multinational companies anxious to secure mineral concessions on the most favourable terms, have contributed to these disappointments. There are many obstacles to the achievement of these expectations. Modern mining methods are highly capital-intensive by comparison with those prevailing in the nineteenth century. This limits not only the number, but also the type of job opportunities. Locally-recruited labour often lacks the skills required to operate complex machinery and management usually remains in the hands of imported, expatriate personnel. This in turn creates an enclave mentality in which mining communities remain isolated from the wider society of the country. This is not just a social phenomenon. It tends to generate levels of personal consumption among a relatively highly-paid élite which cannot be met by domestic production. The multiplier effects associated with the demands of the mining activities themselves are also often truncated because there are no established suppliers of equipment and specialized business services. The sophisticated nature of the technology makes it difficult for such suppliers to emerge within Third World countries. This situation is symptomatic of a technological dependence which contributes to the 'leakage' of multiplier effects outside the host country, usually to the home countries of the multinationals.

Many of the biggest disappointments have resulted from the failure of mineral processing and related downstream manufacturing to develop at or near the site of extraction. It is these activities which create the largest number of jobs and, frequently, the greatest profit. They are, therefore, highly desirable from a policy perspective. It is not only governments in developing countries which have been frustrated by the minimal extent of downstream processing. State or provincial authorities within developed economies have had similar experiences. The failure of iron ore mining to stimulate the manufacturing activities anticipated by the government of Western Australia was described earlier. In Canada, Alberta has discovered that its economic objectives do not always coincide with those of the private sector companies involved in the exploitation of its oil and gas resources. In particular, substantial volumes of gas suitable for petrochemical manufacture are being piped out of the province to Ontario and Quebec despite Alberta's desire to maximize the up-grading of

its resources within the boundaries of the province (Chapman, 1987). There are many reasons why downstream operations are often found well away from the site of resource extraction. The significance of technical changes associated with preliminary processing and advances in transportation has already been noted. These developments have improved the economics of market relative to raw material locations for manufacturing and have promoted the increasing geographical separation of the centres of mineral production and consumption. It is also worth noting that the search for resources has drawn mining and oil companies into increasingly remote and hostile environments. Although substantial settlements have been established in resource-frontier areas such as Siberia, common sense suggests that the obstacles to such developments are considerable and it is not difficult to understand why, with access to modern methods of bulk transportation, the processing of mineral resources is undertaken closer to the centres of demand. In the case of developing countries, this tendency is reinforced by the phenomenon of external control discussed earlier. The difficult relationship between host governments and foreign companies encourages the latter to reduce their exposure to the political risk of expropriation or nationalization by minimizing their investment in processing facilities at the mine.

Environmental impacts

Although the economic benefits of mining are not necessarily focused at the point of extraction, most of the environmental costs certainly are. Mining has always been recognized as one of the most environmentally damaging forms of economic activity. This characteristic is expressed in the title of Llewellyn's powerful fictional account of life in a Welsh coal-mining community – *How Green Was My Valley*. Many landscapes have been permanently scarred by mining. Such environmental costs or externalities are a feature of all forms of mining, but the economic preference for strip or open-pit methods of extraction for a growing range of minerals (see p. 175) is tending to increase rather than reduce the scale of environmental impacts.

Subsidence is one of the most distinctive impacts of mining and is responsible for damage to buildings and underground utilities as well as the disruption of drainage systems with adverse consequences for agriculture. All of these side-effects impose costs and inconvenience on others which are the subject of,

Box 8.2: Coal: abandoned pits and water pollution

The economic and social costs of pit closures for traditional mining communities may, at first sight, seem to be partially offset by the benefits derived from a reduction in the environmental impacts of mining. However, some of these impacts, such as derelict buildings or spoil heaps, may extend beyond the life of the mine which created them; others may be a direct consequence of closure. For example, mine-workings which become flooded when pumping ceases may release contaminated water to rivers and streams. Such water is typically acidic and often contains ferrous and other sulphides which produce an orange discoloration in receiving watercourses (see Robb, 1994).

The final deep-mine in the Durham coalfield closed in 1993. The National Rivers Authority (NRA), which is responsible for regulating river pollution in England and Wales, fears that minewater will contaminate the River Wear, leading to a need for further expenditure on treatment facilities. It may be argued that British Coal, which took the decision to close the mines, should pay for these facilities if the NRA's fears prove correct. However, the complexities of the hydrological processes involved will make it difficult to attribute responsibility and current UK legislation makes a successful prosecution of British Coal unlikely. The problem is illustrated by case of a farmer with land at St Helen Auckland. In 1979, a concrete plug on a disused pit shaft near his farm was blown off by the pressure of underground water. Since then, his field drainage ditches have begun overflowing with water, turning 10 acres of former useful land into a bog. The farmer believes that the water has risen through the many abandoned mineworkings under his land, but this is hard to prove and legal responsibility is even more difficult to establish.

often disputed, claims for compensation. It has been estimated that over 32 000 km^2 have been affected in the United States with a further 10 000 km^2 likely to be affected by the end of the century (Blunden, 1991: 90). The exploitation of coal is primarily responsible for mine-related subsidence, although it is also caused by other extractive activities. Land-surface subsidence of 2 to 4 feet (0.5–1 metre) is widespread in the Houston-Galveston area, for example, and exceeds 6 feet (2 metres) in some localities. This is due to a combination of factors including the extraction of groundwater, oil, gas, salt and sulphur. On the flat low-lying landscape of the Gulf Coast plain, the most serious consequence of subsidence is increased flood risk, especially in an area subject to hurricanes. This illustrates the indirect effects of mineral extraction and the difficulty of allocating responsibility (see Box 8.2). The collapse of a building due to subsidence immediately above a mine-working suggests a reasonably direct link between cause and effect; liability for increased flood damage resulting from a regional fall in the land-surface is such more difficult to establish.

Mining and mineral processing frequently result in water pollution. Contamination by heavy metals, such as copper, lead and zinc, may have especially dangerous effects because of their toxicity. Such materials may enter groundwater or surface streams as a result of runoff from waste-tips or discharges from processing facilities. Even in a country such as the United States, which is usually associated with rigorous environmental legislation and enforcement, modern mining may result in serious pollution. For example, massive leaks of water contaminated with cyanide and heavy metals into the headwaters of a tributary of the Rio Grande were partly responsible for the closure of a large gold mine at Summitville in Colorado in 1993, only eight years after it was re-opened. This example demonstrates the way that streams and rivers may extend the problem over a wide area and the well-publicized case of deaths due to the eating of fish contaminated by mercury in Japan emphasizes that the sea cannot be regarded as a waste-sink of infinite capacity. There are even fewer obstacles to the long-distance transportation of atmospheric pollutants and emissions from mineral smelters have been responsible for serious downwind vegetation damage similar to that associated with coal-burning power stations (Chapter 7).

The generation of waste material or 'spoil' is, perhaps, the biggest single environmental problem associated with mining. Large, unsightly tips have been left behind as a legacy of both deep- and open-pit mining. Several factors influence the extent of this problem. The method of extraction is important. Open-pit operations may, for example, require the removal and dumping of large quantities of 'over-burden' to secure access to underlying coal or other minerals. The ratio of waste to useful material varies widely between minerals. This is implicit in the differences in threshold concentration ratios required to justify extraction (Table 8.1). When the mineral content within an ore-bearing rock is very

Plate 8.4 The environmental impacts of mining and mineral-processing are often considerable as demonstrated by this cement plant at Exshaw, Alberta. (*Source:* K. Chapman)

low, as exemplified by the precious metals, large quantities must be removed to extract the target mineral. In these circumstances, the mass of waste material left after the extraction of the metallic content may be very little less than that of the rock mined in the first place, but it will occupy a much greater volume on the surface than it did when it was in the ground. Furthermore, the magnitude of this problem quite literally grows as lower-grade ores are used.

Mining waste is an externality in both a geographical and a temporal sense. It imposes an environmental 'cost' not only upon those living in the vicinity of the mine, but also on those who may live in or visit the area in the future, possibly long after extraction has ceased. This is evident in inherited mining landscapes such as the 'Cornish Alps' resulting from more than 200 years of china clay removal. The intergenerational issue acquires even greater significance when the legacy of environmental damage reflects activities undertaken by an external agency such as a foreign company or colonial power. For example, tin mining on the Jos Plateau in Nigeria finally ceased in the mid-1980s after a long history of commercial exploitation. It has left behind what the state government has described as a 'disaster area' consisting of more than $300 \, km^2$ of damaged or derelict land. Most of this damage occurred between the beginning of the century and the 1950s, when Nigeria was under British rule. This example should not be interpreted as confirmation of the particular evils of colonialism, but rather as an illustration of the way in which 'responsibility' for environmental damage may extend across both geographical and historical boundaries.

Responsibility and rehabilitation

The desirability of reclaiming mine-damaged land has been widely recognized by public authorities and numerous programmes have been established in the last 20 to 30 years, especially in former coal-mining areas in countries such as the United Kingdom and Germany. It is often impossible to return such landscapes to their former condition and efforts are directed towards encouraging alternative land uses. Open-pit mines, for example, create holes in the ground which may be converted from unsightly repositories for domestic and industrial rubbish to recreational resources used for water sports and other

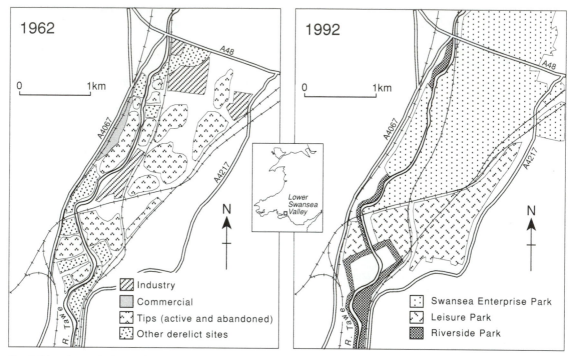

Fig. 8.5 Land uses in Lower Swansea Valley, 1962 and 1992. (*Source:* compiled from data supplied by City of Swansea Planning Department.)

purposes. The motivation for reclamation may be as much economic as environmental. Blighted landscapes discourage investment in new industries and this was an important consideration in the Lower Swansea Valley where environmental rehabilitation was regarded as a prerequisite for economic regeneration. Swansea was the world's principal centre of smelting and mineral processing in the late nineteenth century, accounting for more than three-quarters of refined copper production. Approximately 100 years of these operations created a nightmare, lifeless landscape resulting from atmospheric pollution and the dumping of toxic waste. As late as 1962, over 7 million tonnes of slag and furnace waste covered much of the valley floor, which was the most extensive contiguous area of industrial dereliction in the United Kingdom (Fig. 8.5). The last metal processing plant closed in 1981 and the appearance of the valley has been transformed as a result of a major redevelopment programme. The waste tips and abandoned factories have been replaced by the Swansea Enterprise Park, which incorporates a range of economic activities, especially services, and by a leisure park (Fig. 8.5).

The Lower Swansea Valley is an example of

rehabilitation undertaken long after the damage has been done with the costs borne not by those directly responsible, but by the succeeding generations with funding mainly derived from the public sector. This situation may be avoided in the future by requiring existing projects to meet standards designed to limit environmental impacts and, where feasible, to restore the landscape when mining ceases. The controversial nature of mining and related activities has already been emphasized and there is no doubt that they are subject to increasingly stringent regulations in many countries. Such regulation is not without its problems. The proposition that the polluter pays seems entirely reasonable, but the application of this principle imposes costs upon the mine operator that were previously imposed on others. These costs, therefore, enter into the decision to exploit a discovery. It has been suggested, for example, that the financial burdens imposed by environmental legislation and tax changes in Ontario during the 1970s, were responsible for a substantial reduction in mining investment (quoted in Blunden, 1985: 55). Similar arguments may become relevant at the international scale if mining corporations seek to avoid environmental opposition to their activities within the core economies. For example, Manners (1992: 129) notes that 'Already, Australian mineral and energy production are lower than would have been the case in the absence of environmental policies' and he argues that the deterrent effect of such policies, not only within Australia, but also in the United States and Canada, will be enhanced by the recent adoption of more welcoming attitudes to investment by foreign multinationals in developing countries. This argument is a reversal of the strategic concerns which have hitherto discouraged exploration and production outside the core economies (see above). The relative strength of the opposing forces of strategic interest and environmental concern will, no doubt, vary with prevailing economic and political circumstances and it is difficult to predict the outcome for the future geography of the mining industry.

Developing countries in particular may be faced with some very hard decisions. The economic attractions of a major mining project may be considerable, but their realization may depend heavily upon investment by foreign multinationals. The investors are, therefore, in a strong position to resist environmental controls which may be considered desirable by the host government. The history of the Ok Tedi gold and copper mine in Papua New Guinea illustrates the problem (Mikesell, 1992: 120–21). The government was very conscious of the potential environmental impacts of the project and preliminary negotiations paid considerable attention to these matters. Although agreement was reached with the mining consortium in 1980, operating experience has confirmed many of the original fears about the environmental impacts of the project. Ultimately, the dependence of Papua New Guinea upon the project for income to sustain its economic programmes has made it difficult to impose the environmental standards which would be required of a similar operation in a developed economy.

Although some of the most alarming predictions in *Limits to Growth* were focused on the availability of mineral resources, the prospect of physical scarcity seems more remote in the 1990s than it did in the 1970s. The availability of minerals, however, is still an issue. The very uneven distribution of many minerals, especially at the international scale, has ensured that questions of security of supply have been and remain of major concern. Many of the fundamental questions surrounding the relationship between developed and developing countries are evident in the political economy of mineral resources. The equitable distribution of the costs and benefits of mineral production is central to this relationship, which emphasizes the importance of essentially 'cultural' factors in understanding resource issues.

Further reading

Blunden, J. (1985) *Mineral resources and their management.* Longman, London.

Blunden, J. and Reddish, A. (1991) *Energy, resources and environment.* London: Open University, pp. 43–131.

Rees, J. (1990) *Natural resources: allocation, economics and policy* (2nd edn). London: Routledge, pp. 60–239.

CHAPTER 9

Water resources

Water is the most abundant environmental resource, covering more than 70 per cent of the Earth's surface. Most of this (more than 97 per cent) is salt water and this chapter is concerned only with the residual balance of fresh water (see Chapter 10 for a discussion of marine resources). In fact, most of this residual is not available for human use and estimates of the global water balance suggest that more than 75 per cent of it is 'locked up' in the polar ice caps. (McDonald and Kay, 1988). Overall, the rivers and freshwater lakes of the world account for an estimated 0.02 per cent of the water on the planet with groundwater and soil moisture contributing a further 0.58 per cent and the atmosphere an infinitesimal 0.001 per cent. Expressed in these terms the apparent abundance of fresh water begins to appear less impressive and only 0.6 per cent of the total volume of water on the planet is available in this form. This resource is uniquely versatile and is used for drinking, to sustain agriculture and industry, to serve as a means of transportation and waste disposal and as a focus for a wide range of recreational activities. This list is by no means exhaustive and the multiple uses of water not only place pressures upon available supplies, but also create serious conflicts of interest between, for example, the requirements of industry and those of amenity groups. Such conflicts often revolve around the inter-relationships between the two key attributes of water quantity and water quality. These attributes are considered separately in this chapter before turning to some of the management issues involved in coordinating the multiple uses of water.

Water supply and demand

Trends in water use

The operation of the hydrological cycle as a closed system ensures a theoretical global equilibrium which makes water an infinitely renewable resource, but this equilibrium can be disturbed by human intervention and it is customary to identify two measures of water 'use'. *Withdrawal* involves removing water from an underground or surface source and transporting it to a place of use. *Consumption* occurs when water that has been withdrawn is not immediately available for re-use. Consumption implies a reduction in water quantity, but it is important to appreciate that this is often closely related to a deterioration in water

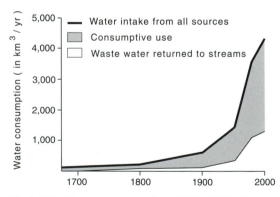

Fig. 9.1 World water consumption, 1650–2000. (*Source:* modified after L'vovich *et al.*, 1990.)

quality. Many 'uses' of water, such as cooling water for factories and power stations, ultimately result in return flows to rivers and lakes which are not significantly less than the original abstraction. However, the quality of this return flow is frequently poor and this effectively limits the opportunities for re-use by degrading the resource. A distinction may therefore be drawn between uses which impose few constraints upon other subsequent uses of the resource and those which deny such possibilities.

However measured, there is no doubt that demands upon water supplies are increasing. This is revealed not only in levels of consumption, but also in impacts upon water quality. Figure 9.1 emphasizes that the pressure on the global resource accelerated sharply during the twentieth century. This is due partly to population growth and partly to increases in per capita withdrawals. Overall, it is estimated that these withdrawals have increased four-fold over the last 300 years (L'vovich *et al.*, 1990). This reflects a close correlation between water use and levels of economic development that is evident in contemporary comparisons between developed and developing countries. Although there are important exceptions reflecting variations in water availability, per capita withdrawals are generally higher in the developed countries (Table 9.1).

Table 9.1: Annual water withdrawals by world region, 1987

Region/country	Withdrawals (Per capita m^3)
Africa	244
North and Central America (inc. US and Canada)	1 692
US	2 162[1]
Canada	1 752[2]
South America	476
Asia (inc. Japan)	526
Japan	923[3]
Europe	726
USSR	1 330[3]
Oceania (inc. Australia and New Zealand)	907
Australia	1 306[4]
New Zealand	379[3]
World	660

Source: World Resources Institute (1992)
Notes: [1]1985
[2]1986
[3]1980
[4]1975

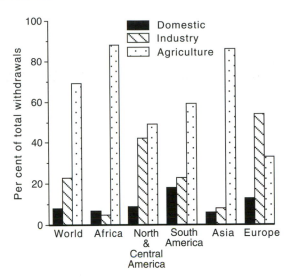

Fig. 9.2 Water consumption by sector, 1990. (*Source:* compiled from data in World Resources Institute, 1992.)

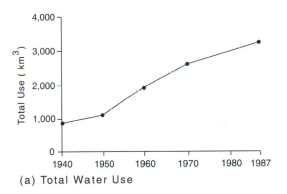

(a) Total Water Use

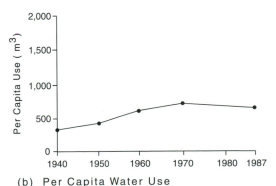

(b) Per Capita Water Use

Fig. 9.3 Total and per capita global water use, 1940–87. (*Source:* compiled from data in World Resources Institute, 1988 and 1992.)

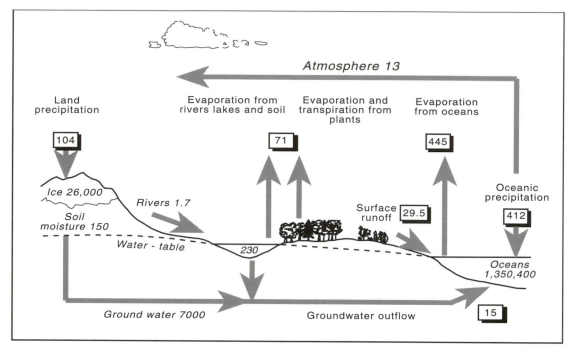

Fig. 9.4 The hydrological cycle: all volumes in thousands of cubic kilometres. (*Source:* Newson, 1992.)

The pattern as well as the level of water use varies between developed and developing countries. Generally speaking, agriculture is the principal use in the latter, while much larger proportions are diverted to industrial and domestic uses in the former (Fig. 9.2). These differences are directly related to the structure of economies and, by inference, to levels of economic development. Domestic water use is linked to advances in basic public health as well as less fundamental improvements in the quality of life associated with such appliances as washing machines and dishwashers. Industrial demands grow as manufacturing assumes greater importance within national economies and the biggest increases in water use are expected to occur in developing countries as a consequence of the inter-related processes of urbanization and industrialization. The acceleration of these processes in the developing world together with a trend towards a more conservation-orientated approach in many developed countries suggest a narrowing of the differentials in per capita water use evident in Table 9.1. Figure 9.3 suggests that the historic growth in per capita use may have stabilized, but the continuing rise in world population will inevitably lead to a similar rise in the absolute volume of withdrawals at the global scale.

Supplies and shortages

It has already been noted that the global water mass is incorporated within a closed self-regulating system by the fluxes and storages of the hydrological cycle (Fig. 9.4). Fresh water therefore lies at the perpetually renewable end of the resource continuum (Fig 1.4). The abundance implicit in such a renewable and infinite supply has not, however, prevented water shortage from becoming a major cause for concern. This paradox rests upon mismatches between supply and demand which reflect geographical and temporal realities. The fact that supply far exceeds demand in one location does not preclude a chronic deficit in another; similarly, water may be abundant in one year and scarce in the next. The existence of sophisticated water collection, storage and distribution systems is testament to the difficulties involved in adapting the hydrological cycle to human requirements – difficulties which, in certain circumstances, make it more appropriate to regard water as a conditionally, rather than a perpetually, renewable resource.

The amount of *stable runoff*, defined as the base flow from groundwater into rivers plus stable surface runoff added by water storage in lakes and reservoirs, determines the availability of water for human use.

This is in turn linked to the relationship between the input of precipitation and moisture returned to the atmosphere by evaporation. It is self-evident that there are wide geographical variations in water availability. Figure 9.5 indicates the global distribution of water surplus and water deficit regions as defined by the requirements of the prevailing vegetation cover. Water is in surplus when precipitation is large enough to meet the demands of the vegetation cover; it is in deficit when precipitation is lower than this potential demand. By this yardstick, Fig. 9.5 suggests that most of Africa, the Middle East and Australia are in deficit together with substantial parts of North America and Asia. Although Fig. 9.5 provides a valid general impression of water availability, it is important to note that it exaggerates the magnitude of the problem from a human perspective. What is significant in resource terms is the extent to which water shortage constrains human activities. Table 9.2 attempts to address this issue, but it begs the question of exactly what 'limited by drought' means. Indeed, the concept of drought is a difficult one and it can be defined both in climatic terms with reference to specific rainfall deviations or upon the basis of its impacts upon economy and society.

By considering water as a 'resource', it follows that shortages are defined with reference to demand as well as supply. Figure 9.6 indicates the per capita availability of annual internal renewable water resources by country (defined as freshwater runoff resulting from precipitation within the border of a country). This calculation uses population as a measure of potential demand which is, of course, simplistic given the previously noted link between levels of economic development and water consumption. Nevertheless, comparison of Figs 9.5 and 9.6 reveals some significant differences. If the arbitrary threshold of less than $1000 \, m^3$ and more than $10\,000 \, m^3$ per capita per year of internal renewable resource is adopted to define 'water-poor' and 'water-rich' countries respectively, Canada, with $109\,370 \, m^3$ in 1990, could be described as 'super-rich' despite the apparent deficiency over much of its territory indicated in Fig. 9.5. Conversely, several European countries including Belgium and Hungary are apparently 'water-poor'. The contrasts between Figs 9.5 and 9.6 reflect the difficulties involved in both defining and measuring water deficits, but they do at least emphasize the considerable spatial variations in the availability of the resource at the global scale.

Table 9.2: Areas of land limited by drought

Region	% of land limited by drought
North America	20
Central America	32
South America	17
Europe (mainly Spain)	8
Africa	44
South Asia	43
North and Central Asia	17
South East Asia	2
Australia	28

Source: Finkel (1982)

Geographical mismatches between supply and demand also exist at the national level. Despite its relatively small size, there are significant variations within the United Kingdom. Householders in south and south-east England are familiar with hosepipe bans, but such restrictions are less common in Wales and Scotland. These variations are minor by comparison with the situation in the United States where the average per capita water availability of $9\,940 \, m^3$ in 1990 concealed major differences between, for example, New England and the south-west. The problem has been aggravated by high rates of population and economic growth in the arid and semi-arid 'Sunbelt' states. Phoenix has been one of the fastest-growing cities in the United States since 1970 despite the fact that it is located in a region with a very low annual rainfall.

Average values for water availability are generally expressed relative to a time scale as well as a geographical unit. Thus an annual figure which suggests abundance may conceal seasonal shortages with significant effects upon, for example, the farming community. On a longer time scale, annual rainfall in arid and semi-arid areas is notoriously variable and human use of water is only sustainable if it acknowledges this characteristic. There are, for example, serious doubts about the sustainability of existing agricultural practices over much of the High Plains of the United States. (Box 9.1, Fig. 9.7).

Increasing supplies

The preceding section emphasizes that human perceptions of physical water scarcity are related to spatial variations and temporal irregularities in

Fig. 9.5 Areas of water surplus and water deficiency (Units are mm/y). (*Source:* Falkenmark, 1986.)
Note: please see text for explanation of 'surplus' and 'deficiency'.

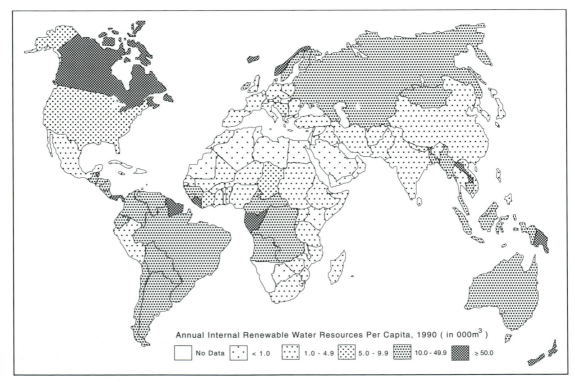

Fig. 9.6 Internal renewable water resources per capita by country, 1990. (*Source:* compiled from data in World Resources Institute, 1992.)

Box 9.1: Water 'mining' in the High Plains

Irrigation demands are, almost by definition, greatest where water supply is most scarce. In these circumstances, water may become a stock (non-renewable) rather than a flow (renewable) resource. This is illustrated by the exploitation of the Ogallala aquifer which underlies much of the High Plains of the United States (Fig. 9.7). This aquifer varies in thickness from less than 1 foot (0.3 m) to approximately 1300 feet (396 m) with by far the greatest volumes of water beneath Nebraska. Much of the water it contains has accumulated over very long periods of time and the current rate of replenishment is limited by an annual average rainfall of less than 500 mm over much of the High Plains.

Despite low and variable rainfall, the High Plains produce a substantial proportion of US agricultural output including sorghum, maize, wheat and feedlot beef. This is a startling achievement for an area that was regarded as so inhospitable by an 1820 expedition that it became known as the Great American Desert. The exploitation of the Ogallala aquifer has contributed to the change in the perception of the High Plains, although commercial agriculture was firmly established before the extensive use of groundwater. The 'Dust Bowl' of the 1930s encouraged a re-appraisal of irrigation by High Plains' farmers who had previously regarded it as supplementary to available rainfall. Agricultural production has been largely sustained by groundwater withdrawal since the 1950s and more than 90 per cent of the water pumped from the aquifer is used for agriculture. The rate of withdrawal has exceeded the rate of replenishment and a resource that appeared infinite in the 1950s began to look like a well running dry 20 years later. Major falls occurred in the water level (Fig. 9.7) and a government study in 1982 predicted that parts of the aquifer, especially in Texas, would become 'dry' by the year 2020. This message seems to have had an effect and the overall rate of decline

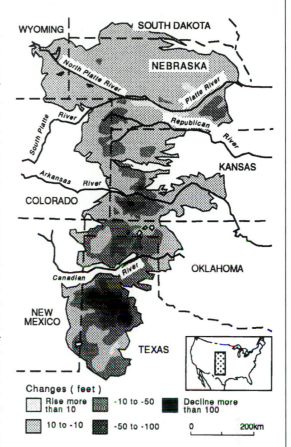

Fig. 9.7 Predevelopment to 1980 water-level changes in the High Plains Aquifer. (*Source:* Schwarz *et al.*, 1990.)

in the water table slowed down considerably in the 1980s as farmers have re-discovered dry-farming techniques and adopted more efficient irrigation practices.

supply. Attempts to overcome these problems provide some of the most spectacular examples of deliberate human interventions in the operation of natural environmental processes. As early as the sixth century BC, civilizations in Egypt and Mesopotamia developed sophisticated systems of water management which incorporated flood control as well as irrigation. Underground channels (qanats) were constructed in many parts of the Middle East, including modern Iran, in prehistoric times to transport water over significant distances from

springs to consumers. Generally speaking, the scale of human interventions in the hydrological cycle has increased over time.

Dams are the most visible manifestation of engineering solutions to supply problems and, in a wider context, may be regarded as symbols of the human domination of nature (see Chapter 1). It is estimated that there were more than 36 500 dams world-wide exceeding 15 m in height in 1986 with a further 1026 under construction (World Resources Institute, 1992: 330). The United States accounted for

15 per cent of the total. Its share 50 years earlier was certainly much greater and activities in the United States provided much of the impetus to a sharp increase in dam construction during the twentieth century (Fig. 9.8) (Beaumont, 1989). Indeed, a combination of physical and cultural circumstances ensured that the American south-west became something of a laboratory for the application of science and technology to water-supply problems (Graf, 1992). Numerous developing countries have since tested the products of this laboratory, acquiring since the 1960s massive dams such as the Aswan on the Nile in Egypt and the Akosombo on the Volta River in Ghana. Many of these projects have multiple objectives including flood control and HEP generation (see p. 200), but they have also been designed to boost the supply of water for human use and dams have played a major part in the expansion of the world's irrigated area (see Chapter 4).

Although big dams such as Aswan and Akosombo have major impacts upon individual river basins, consideration has been given to even more ambitious schemes to divert river flows and transfer water between basins. A controversial Soviet scheme would have involved diverting the flow of rivers such as the Ob and Pechora, which drain into the Arctic Ocean, southwards into the semi-arid steppes of Central Asia (Micklin, 1984). Conceived as the 'project of the century' by the Communist Party's Central Committee in the context of a prevailing philosophy which stressed the human domination of nature, this scheme had already been abandoned before the political fragmentation of the Soviet Union. Its rejection was partly due to the astronomical costs of a project which, in its most comprehensive form, would not have been completed until the middle of the twenty-first century. It was also reported that the decline in the level of the Caspian Sea, which had been a factor in the original formulation of the scheme, was reversed in the early 1980s (World Resources Institute, 1988: 138). Many also expressed concern about the uncertain environmental effects of the scheme upon, for example, salinity conditions in the Arctic Ocean. Similar economic and, especially, environmental reservations make it unlikely that a proposal for continental-scale water transfers in North America will ever be implemented. The North American Water and Power Alliance (NAWAPA), first mooted in 1964, envisaged the movement of water from rivers in Alaska and northern Canada to areas of need in central Canada, the south-western United States and north-

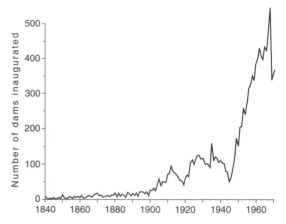

Fig. 9.8 World dam construction, 1840–1970. (*Source:* Beaumont, 1978.)

ern Mexico. Quite apart from the political sensitivities involved in the export of Canadian water to the United States, prevailing environmental attitudes have produced a reaction against this type of mega-project and a preference for less intrusive and drastic methods of matching water supply and demand. This is reflected in the withdrawal in 1993 of advisory support by US government agencies for a Chinese plan to build the world's largest dam on the Chang Jiang (Yangze River).

Conserving supplies

More efficient use of water is an obvious response to supply shortage. This may be achieved by reducing losses in the delivery and application of water or by limiting the growth of demand. Recycling is another aspect of water conservation which provides a link between the supply and demand sides. The opportunities for recycling may be expected to be greater than for most other resources as a result of the versatility of water which may be used for a succession of different purposes with less demanding quality requirements. There is evidence of a shift in emphasis from initial attempts to boost supplies to a later concern with demand management and greater efficiency of use as water resource development proceeds. This shift is often associated with a growing recognition of the true costs of providing secure water supplies and attempts to reflect these costs in pricing mechanisms which may be applied to agricultural, industrial and domestic consumers.

Viewed in global terms, the greatest potential for reducing water losses and using water more efficiently

Plate 9.1 The Bonneville dam in Washington is one of many established on the river systems of the western United States in the 1930s and 1940s. To some extent, these dams provided a model for similar large-scale engineering solutions to water supply problems in other parts of the world. (*Source:* A. Mather)

lies in agriculture, which accounts for almost 70 per cent of total world use (Fig. 9.2). Water is 'lost' to agriculture in two main ways: first in the form of surface runoff which finds its way to the oceans via rivers and streams; second as evaporation from water and land surfaces.

Large dams permit the capture of surface runoff and its collection in artificial reservoirs. However, relatively cheap, low technology measures may significantly reduce runoff losses. Small earth dams are a traditional method of water capture and storage in China. They retain water which is used for fishponds and the generation of community-based HEP as well as for agriculture. Small dams are not unique to China, but Chinese experience is encouraging other countries such as India to reconsider the potential benefits of such structures for rural communities. Earth dams represent a sophisticated technology by comparison with some other methods of water harvesting. Channelling runoff into natural depressions is an ancient skill still practised in mountainous areas of the Middle East and in parts of India. Another approach involves the construction of low earth or rock walls across hill-slopes. This retains a portion of the runoff behind the barrier

creating a micro catchment suitable for cultivation. The technique, which is being promoted in development projects in the Sahel in Mali and Burkina Faso, has the additional benefit of reducing fertilizer loss and soil erosion.

Much irrigation water is 'lost' by evaporation. This is partly due to over-watering and partly related to the methods of delivering water to the plants. The growing importance of irrigation has encouraged efforts to improve the efficiency of the technology. Although there are difficulties in defining and measuring water efficiency, it is generally accepted that there are wide variations between the different methods of irrigation. Flooding of water over the soil surface and then allowing it to reach plant roots by infiltration is by far the most common method. It is not difficult to appreciate that in a hot climate major losses will occur by evaporation, especially if the infiltration rate is slow as a result of over-watering. Evaporation losses also occur with aerial methods of irrigation involving sprays and sprinklers, although it is easier to control the rate of application and avoid over-watering. Trickle irrigation is even more efficient in this respect because it is characterized by a continuous but low rate of water application at or near the plant roots. Unfortunately, the more efficient methods of irrigation are also the most capital-intensive and they are, therefore, more common in the developed than in the developing countries.

Industrial uses of water become more important with advancing levels of economic development and they exceed agricultural withdrawals in Europe (Fig. 9.2). Recycling of cooling and process water has certainly contributed to the apparent savings in industry and such practices have become common in most developed countries. By 1975, US industry had, for example, achieved a water recycling rate of 2.2 which means that, on average, a unit of intake was 'used' more than twice before being discharged (World Resources Institute, 1986: 131). The principle of recycling may be adopted on a much larger scale than the individual industrial plant. Tokio has established a Water Recycling Centre which takes wastewater from sewage treatment facilities, chlorinates it and pumps it to high rise buildings for use in flushing toilets. This follows the earlier example of the Dan Wastewater Reclamation Project built to treat sewage from Tel Aviv. Israel, as a 'water-deficient' country (Fig. 9.6), has made recycling an important part of its national water policy. It re-used 35 per cent of its wastewater in 1986, mainly for irrigation, and plans to increase this to 80 per cent by the year 2000 (Shuval, 1987). It is estimated that recycling will effectively increase Israel's renewable water supply by 25 per cent.

Water pricing

Incentives to conserve water, especially in agricultural and domestic uses, are often conspicuously lacking and many of the inefficiencies of water use in irrigation are institutional rather than technical in origin. One of the most fundamental relates to the pricing of water. It is a paradox that a resource essential to life has been widely regarded as almost a 'free good'. For example, where irrigation is based on supplies derived from large-scale water storage and distribution systems, it is usually chronically underpriced in relation to the long-run costs of provision. This has been true of the western United States where much irrigation water is effectively subsidized by the federal government. A similar situation prevails in Australia where it is argued that '...wasteful applications of water [to pasture, fodder and wheat crops] give unfair economic advantages to irrigators over dryland farmers' (Bell, 1988: 364). Table 9.3

suggests that several developing countries are following the same path and it is clear that charges to private farmers for water from public irrigation facilities often do not generate sufficient revenue to cover operating and maintenance, let alone capital costs. It is, of course, reasonable to argue that the wider benefits, justify the capital investment especially when irrigation is linked to social and economic objectives in comprehensive plans (see below). Nevertheless, it cannot be sensible to set the price of water at a level which encourages profligacy. Indeed, under-funding of public irrigation systems often results in water losses as a result of poor operation and maintenance. Vested interests make it difficult to raise charges for irrigation water to levels corresponding to the marginal cost of provision, but evidence from the United States, China and Mexico confirms the intuitively obvious proposition that higher prices encourage more efficient use.

Water pricing is not only relevant to irrigation practices. Domestic consumers typically pay very low prices for the water from their taps. It may be argued that this is entirely appropriate within a civilized society in which provision of such an essential commodity should not be constrained by ability to pay. On the other hand, the failure to acknowledge the real cost of supplying water can result in patterns of use that are, ultimately, not sustainable. The swimming pools, lawns and golf-courses, which are standard features of desirable property developments in parts of the American south-west with desert climates, such as Coachella Valley in California, are not consistent with the underlying environmental realities. The city of Tucson in Arizona recognized this in the mid-1970s when significant increases in water charges were introduced in an attempt to encourage more careful use. Although initially unpopular, this was successful in reducing demand and it illustrates changing attitudes towards the resource. Beaumont (1989: 123) notes that 'Up to the early 1970s it was viewed as a resource which was too important to be allotted in strict monetary terms and was therefore allocated by regulations and rates drawn up by the [city] Water Department.' The sharp increase in charges was an acknowledgement of the 'real' value of water.

Industrial users are most familiar with the concept of a direct link between levels of consumption and the 'cost' of water. There is a striking difference between trends in metered and unmetered water supplies in the United Kingdom. The former have declined slowly since the mid-1970s; the latter have grown steadily.

Table 9.3 Irrigation service fees paid by farmers compared to public irrigation systems costs in selected Asian countries[1]

Country	Actual revenue from service fees (dollars per hectare)	Operating and maintenance costs (dollars per hectare)	Revenue as a proportion of operating and maintenance costs (%)	Total capital and operating costs		Revenue as a proportion of total costs	
				Moderate estimate (dollars per hectare)	High estimate (dollars per hectare)	Moderate estimate (%)	High estimate (%)
Indonesia	25.90	33.00	78.5	191	387	13.6	6.7
South Korea	192.00	210.00	91.4	1 057	1 523	18.2	12.6
Nepal	9.10	16.00	56.9	126	207	7.2	4.4
Philippines	16.85	14.00	120.4	75	166	22.5	10.2
Thailand	8.31	30.00	27.7	151	272	5.5	3.1
Bangladesh	3.75	21.00	17.9	375	–	1.0	–

Source: World Resources Institute (1988), p.206
Note: [1] Estimates obtained by converting local currency values at official exchange rates prevailing in June 1985

Generally speaking, this corresponds to the distinction between industrial and domestic/commercial users. Domestic water metering is restricted to a few parts of the United Kingdom such as the Isle of Wight and households are subject to a fixed annual water rate regardless of the level of use. Industrial users are metered and charges are based on the volume recorded. The evidence suggests that domestic water use has grown steadily despite an almost static population, but industrial use has declined. This decline may partly reflect changes in the structure of the economy as heavy industries, which use large quantities of water, have become less important. Nevertheless, it is reasonable to assume that it is also related to the influence of water metering. Indeed, the UK experience is not unusual and similarly contrasting trends in domestic and industrial water use related to charging policies are evident in other countries. It is worth noting that the installation of water metering at the point of consumption tends to reduce water demand whether or not it is directly linked to price increases (Bauman and Dworkin, 1978). This suggests that psychological pressures which encourage individual consumers to think about the consequences of their actions may be used to reinforce economic incentives designed to promote more efficient water use.

Water quality

There is a direct connection between water quantity and water quality. Comprehensive recycling schemes such as the previously described Dan Wastewater

Reclamation Project in Israel rest upon the concept of a hierarchy of water uses linked to differing quality requirements. These requirements are higher for drinking water and lower for agricultural and industrial uses. It follows that the widespread pollution of a country's water sources is not only undesirable on aesthetic grounds, but also seriously limits its supply options.

Trends in water quality

The chemical composition of water in the environment varies widely as a result of 'natural' solutes related to, for example, the presence of mineral salts. Nevertheless, it is human impacts which have attracted the greatest attention. Despite the lack of data, it is reasonable to assume that these impacts have increased over time (Schwarz *et al.*, 1990) as a result of urban growth, industrial development and agricultural activities. Progressive deterioration in water quality may, therefore, be regarded as an inevitable consequence of economic growth. This conclusion is too pessimistic, however, and recent experience within developed countries suggests that inherited water quality problems may be reversed and future pollution reduced by the adoption of better water-using practices. The introduction of such practices is related to levels of economic development, however, and whatever improvements have been achieved in the core economies have been offset by a deterioration in many developing countries, especially in the vicinity of the major cities.

It is ironic that water is not only essential to life, but is also heavily implicated in the transmission of many fatal diseases. Several, including typhoid and cholera, are related to pathogens excreted in the faeces of a carrier. There is, therefore, a circular link between the prevalence of such diseases and the discharge of untreated sewage. This link was first identified in the middle of the last century following a series of cholera epidemics in London. It was established that the incidence of the disease varied with the source of drinking water. The appalling condition of the River Thames, which was used as an open sewer, was certainly a factor and the main impetus for controlling water pollution was, and remains, the protection of public health. Significant improvements have been achieved in most developed countries in recent years. The proportion of the population served by waste-water treatment plants in OECD (broadly synonymous with 'developed') countries increased from 34 per cent in 1970 to 57 per cent in 1985 (World Resources Institute, 1992: 167), although there are wide variations ranging from 9 per cent in Portugal in 1985 to 94 per cent in Sweden.

Water pollution and industry

Industrial expansion is, of course, closely linked to urban growth and industrialists have traditionally regarded the environment not only as a source of raw materials, but also as a waste-sink for discharging unwanted by-products. Watercourses have proved very convenient 'sinks' and the dead rivers which blighted the landscapes of many centres of the Industrial Revolution were an indictment of the pursuit of profit without regard for environmental responsibilities. Discharges from factories tend to have different characteristics and cause different problems from untreated sewage. Although some have directly affected human health, it has been adverse impacts upon natural ecosystems with both economic and aesthetic consequences which have stimulated policy measures to restrict industrial pollution. These measures have achieved significant successes in some rivers and lakes which have been badly affected by industrial discharges. Water quality in the Rhine has improved sharply as a result of an international clean-up programme (see pp. 201–2). The Great Lakes, around which a substantial proportion of the population and industrial capacity of the United States and Canada is located, have also benefited from a similar programme initiated in 1972 (Ashworth, 1986; Jackson, 1993).

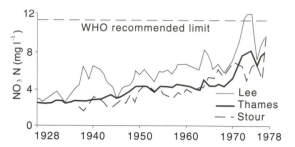

Fig. 9.9 Nitrate concentration in selected UK rivers, 1928–78. (*Source:* Royal Society, 1983.)

Water pollution and agriculture

Water pollution from agricultural activities has attracted increasing attention within developed countries where the sector generally accounts for a very small proportion of employment and national economic output. This paradox is related to the intensive nature of modern commercial agriculture which has achieved dramatic increases in productivity, partly as a result of the use of synthetic fertilizers and pesticides (see Chapter 4). Nutrients, especially nitrates, from both animal wastes and artificial fertilizers are the principal agricultural pollutants, but the growing use of pesticides world-wide is an additional cause for concern in view of the uncertainties surrounding the health effects of trace amounts in drinking water. It is generally accepted that nitrates are a real health hazard and they are implicated in gastric cancers, fetal malformations and blood poisoning. Nitrate concentrations in many European rivers have been steadily rising for more than 30 years and have reached dangerous levels in some cases. Figure 9.9 indicates long-term trends in selected British rivers relative to the World Health Organization's recommended limit of 11.3 mg/litre. Growing concern over the implications of these trends which have been observed in many rivers has led to a significant reduction in the rate of increase in the use of synthetic fertilizers in the developed countries (Fig. 9.10).

The nitrate problem demonstrates the interrelationship between water quality and human use of the land. For example, 12 Nitrate Sensitive Areas (NSAs) were established in the United Kingdom in 1990, mainly in areas of intensive arable farming. Farmers in these NSAs qualify for financial assistance to encourage the adoption of measures to reduce the application of fertilizers and animal manure. These measures include the conversion of

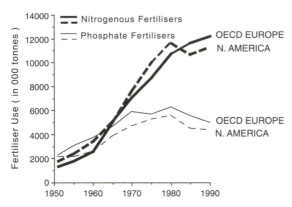

Fig. 9.10 Fertilizer use, 1950–90. (*Source:* compiled from data in FAO annual.)

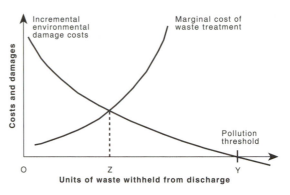

O - Z   Optimal level of waste reduction

Y   Point at which level of waste disposal is low enough to cause no environmental damage

Fig. 9.11 Optimal level of pollution abatement. (*Source:* Rees, 1985.)

arable to unfertilized and ungrazed grassland, and the growing of cover crops rather than the maintenance of bare land which is more liable to leaching. Such efforts to control water pollution are based upon a recognition of the need to integrate land use and pollution control policies.

Costs and benefits of pollution control

Efforts to reduce water pollution in developed countries have been stimulated by a growing realiza-tion that the deteriorating quality of rivers and lakes imposes both economic and social costs. However, the control of pollution itself imposes costs. Water treatment facilities, for example, are expensive and the marginal costs of pollution control rise sharply as the ultimate theoretical goal of zero discharge is approached. Figure 9.11 identifies an optimal level of pollution at the point of intersection between the

Plate 9.2 Intensive methods of food production, such as cattle feed-lots, are often responsible for the pollution of rivers and groundwater. (*Source:* K. Chapman)

rising marginal cost of treatment and the falling curve of incremental environmental damage costs. The latter suggests that the benefit to the environment of a unit reduction of pollution discharge is subject to diminishing returns as the first unit reduction produces the greatest gain and the last produces the smallest gain. Indeed, Fig. 9.11 suggests that a point is reached at Y (pollution threshold) when further reductions produce no benefit because the discharge falls below the assimilative capacity of the receiving environment. From an economic perspective, the optimal level of pollution control occurs at Z; removal of the units of waste represented by OZ yields benefits in improved environmental quality which exceed the costs of control; the costs are greater than the benefits from further reductions ZY.

It is one thing to acknowledge the concept of an optimal level of pollution, quite another to translate it into policy practice (see Rees, 1990: 281–6 and 297–302). There are great difficulties in measuring environmental damage costs. Converting the recreational and aesthetic 'losses' imposed by a polluted river into monetary terms presents major problems (see Mitchell, 1989: 199–241), but these are secondary to the issue of defining 'environmental damage costs'. For example, the closure of a polluting factory may result in a dramatic improvement in the quality of a particular river, but this benefit may be achieved at the cost of the jobs of its former employees. These social and economic costs may also be regarded as 'environmental damage costs' broadly conceived. Despite all these difficulties, the concept illustrated by Fig. 9.12 is an important one because it draws attention to the trade-offs involved in pollution control. These trade-offs, which invariably involve political rather than scientific judgements, must be made whatever mechanisms are adopted to control pollution.

Approaches to pollution control

A distinction is often made between point and non-point sources of water pollution. The former, which include discharges from factories and sewage treatment plants, occur at specific locations; the latter incorporate all the diffuse sources that are associated with human activity on the land such as runoff into surface waters and seepage into groundwater derived from cultivated ground, pig farms and miscellaneous urban land uses. A glance at the litter and other 'deposits' lying around on the average city street emphasizes the problem. The overall impact of non-point sources on water quality is very difficult to assess, but is probably greater than that of point sources. This is certainly true of developing countries where so much sewage and industrial waste remains untreated. The distinction is an important one from a practical point of view because point sources are much easier to identify and most pollution control policies have, therefore, focused upon them.

One approach to pollution control, generally favoured by economists, is to charge a fee or tax for any discharges to the environment. The theory of *pollution pricing* is an extension of the analysis presented in Fig. 9.11. This implies that a rational producer required to meet the environmental damage costs caused by his or her activities would invest in control facilities to eliminate OZ units of waste, but would pay the damage costs caused by ZY units. Essentially, this approach requires the polluter to bear the costs of the externalities imposed by his or her actions. Despite the intuitive appeal of such a system, it seems to excite academic economists more than it does policy-makers. Three European countries, Germany, France and the Netherlands, have incorporated effluent charges into their water management policies. These require polluters to pay a fee linked to the quality and quantity of their discharges into watercourses. The fee is paid to agencies responsible for water quality management, but its level is the outcome of political bargaining and not the result of objective analysis. Rees (1992: 389) notes that 'Only in Germany have pollution charges been introduced which contain a significant economic incentive for dischargers to reduce their waste water loads.' Given the difficulties involved in calibrating 'environmental damage costs', it is not surprising that the practice of pollution pricing bears little relationship to the theory.

State regulation is the preferred mechanism of pollution control in most countries. This involves setting standards of performance which are defined, monitored and enforced by government agencies, such as the Environmental Protection Agency (EPA) in the United States and the National Rivers Authority (NRA) in England and Wales, backed when necessary, by the courts. Although the instruments of such bureaucratic regulation vary widely, it is possible to identify two basic approaches to water pollution control from point sources – *quality objectives and effluent standards*. The former are defined with reference to the assimilative capacity of the environment; the latter are driven by the technology of pollution control. Quality objectives reflect the fact that the impact of a given unit of pollution will normally be greater on a small river than on a large one. They also rest on the proposition that higher water quality objectives should be set for a river which is used as a source of drinking water than for a heavily industrialized estuary. Thus policies based on quality objectives implicitly accept that the optimal level of pollution may be greater than zero. Effluent standards are set for specific categories of point source such as chemical plants or oil refineries. They are defined with reference to the technology of pollution control. In the United States, for example, the EPA has produced masses of technical documents specifying the effluent standards which may be achieved by using 'Best Practicable Technology' (BPT) and 'Best Available Technology' (BAT). The assumption is that similar factories will achieve common standards regardless of the characteristics of the water bodies into which they discharge. Although quality objectives and effluent standards are not mutually exclusive and many countries incorporate elements of both approaches in their pollution control policies, there are significant differences in the balance between them.

Quality objectives have been the guiding principle of water pollution-control policy in the United Kingdom ever since the mid-nineteenth century. This has been associated with a consensual rather than a confrontational approach towards the achievement of these objectives. The conditions attaching to discharge permits issued to individual point sources are based on negotiation between the pollution control agency and the discharger. Breach of these 'consent conditions' is illegal, but relatively few prosecutions have been brought and UK policy has been based upon what might be termed purification by persuasion. This approach has achieved some success and national surveys of the quality of rivers and lakes in the United Kingdom since 1958 showed a steady overall improvement to the mid-1980s. Nevertheless, the UK approach is not without its critics. Prior to the formation of the NRA in 1990, the regional water authorities established in 1974 were responsible both for enforcing pollution control and for overseeing the operation of the country's sewage works, which are by far the most important single category of point source. In addition to this conflict of interest, it has been suggested that the consensual approach encourages too close a relationship between polluters and regulators (Hawkins, 1984).

Politics of pollution control

Establishing pollution-control requirements by negotiation draws attention to the politics of pollution control. The outcome of such negotiation may create a situation in which two identical manufacturing plants are obliged to meet different standards. This may be justified by differences in the characteristics of the water-bodies into which they discharge, but such differences may offer a real competitive advantage to the plant facing lower standards and it is easy to see how environmental considerations may be subordinated to economic interests. The European Commission, for example, wishes to establish a standards-based approach to avoid such 'distortions' to competition in the enlarged market (Haigh, 1989). Similar considerations encouraged a major change in US water pollution control policy in 1972 when legislation effectively shifted the balance of responsibility for water pollution control from state to federal level and proposed that the previous practice of writing permits on a case-by-case basis, depending upon a presumed relationship between discharge and water quality, be replaced by the imposition of specific effluent standards on particular categories of point source.

Technology-based standards narrow the areas of uncertainty, but they are more precise in theory than they are in practice. The conditions imposed in US waste-discharge permits still reflect the outcome of lengthy discussions between the issuing agency and the applicant company. Personalities may be important and the terms of a discharge permit may vary depending upon its author, with certain senior staff within the same agency adopting a more lenient interpretation than others of what constitutes the appropriate 'standard'. Further complications are introduced by the existence of tensions between federal authority, in the form of the EPA, and the state agencies which deal with individual companies on a daily basis. The vigour with which policies are applied not only varies between individuals and agencies, but also over time. It is generally accepted that the Reagan administrations during the 1980s were less committed to environmental issues than that of President Carter which entered office in 1976.

The political influences upon the control of water pollution in the apparently scientific technology-based system of the United States emphasize the subjective considerations which affect all environmental policies. In practice, the optimal level of pollution control is established on the basis of value-judgement rather than objective analysis. This is evident in the different standards applying in developed and developing countries. Considerable interest has been generated in the possibility that multinational companies may deliberately locate dirty or dangerous industrial activities in developing countries with less demanding environmental and occupational health regulations. The available evidence suggests that the extent of such 'industrial flight' to international 'pollution havens' has been limited (Leonard, 1988). Nevertheless, there is no doubt that significant differences in environmental standards exist at the international scale which ensure that the externalities associated with industrial activity are often more acutely felt in the developing world. It is worth noting, however, that such externalities may reflect a judgement by policy-makers in developing countries that the optimal level of pollution is higher than might be considered appropriate in a developed country. The assimilative capacity of the environment may be greater in a relatively undeveloped area or it may be felt that faster rates of economic growth will be achieved with less restrictive pollution control regulations. In these circumstances, the additional economic and social benefits may be judged to more than offset the environmental costs. Such a trade-off may appear short-sighted from the relative comfort and prosperity of the developed world, but it is important to appreciate the resentment generated in developing countries by the applications of external value-judgements.

Water management

Regulating water use

Ensuring the efficient, equitable and, especially important in the context of rapidly increasing demand, sustainable use of water is one of the most formidable challenges of contemporary resource management. The critical importance of water, together with its multiplicity of uses, raises considerable potential for conflict and emphasizes the importance of creating administrative structures capable of resolving such conflict. The importance of such essentially cultural systems was demonstrated by the early hydraulic civilizations in Mesopotamia and Egypt which manipulated the waters of the Tigris, Euphrates and Nile to sustain settled agricul-

ture. In particular, these civilizations recognized the need to establish systems of control which reconciled individual rights of water access with the wider interests of the community. The nature of water resources ensures that the reconciliation of individual and community concerns is very important. Uncontrolled exploitation is especially damaging for a resource in which the actions of upstream users discharging pollutants may, for example, affect the position of downstream users abstracting drinking water on the same river. This type of situation has led to the establishment of legal frameworks defining the rights and responsibilities of water users. Newson (1992: 5), for example, refers to the Code of Hammurabi based on Sumerian and Babylonian water law. Such 'codes' acknowledge the fact that open-access regimes rapidly become inappropriate as pressure upon freshwater resources increases (Chapter 2). Nevertheless, regulation of abstractions from and discharges to river systems is very difficult because of their dispersed, linear form and their interdependencies with land use in their catchment.

The importance of water management is directly related to the pressure on the resource. Industrial growth and urban development in Europe and North America during the Industrial Revolution encouraged the development of more sophisticated systems of control and similar trends are evident in developing countries today. Urbanization promotes high levels of demand in specific locations and also generates substantial pressure on water quality. England and Wales have had longer to come to terms with these problems than most countries and may, therefore, be regarded as providing a suitable case study of institutional response (see Fernie and Pitkethly, 1985: 278–83; Parker and Sewell, 1988).

Water management in England and Wales

Before the eighteenth century, water management in England and Wales was dealt with at the local level using stream-based by-laws. By the end of the century there was a growing concern with providing supplies to meet the needs of the rapidly growing towns and cities. The urgency of this problem was emphasized by outbreaks of disease, especially cholera which arrived in the United Kingdom in 1831. As the connection between public health and water supply was understood, greater efforts were devoted to securing access to uncontaminated sources. This was regarded as a municipal responsibility and parliamentary legislation facilitated the creation of supply and sewage systems under the control of local authorities. This was reasonable while supplies could be obtained locally, but as demands increased the larger cities were forced to look further afield, leading to a chaotic pattern of water-resource development. The local administration of water supply survived until 1945 when legislation transferred certain responsibilities to central government. The overall objective was to promote a more rational use of water resources by encouraging collaborative schemes and the amalgamation of smaller authorities.

Further progress towards national water-resource planning on the basis of larger administrative units was made in the Water Resources Act 1963 which established a system of River Authorities throughout England and Wales. These replaced a larger number of River Boards in carrying out pollution control. More significantly, they were explicitly charged with the task of preparing plans to reconcile discharges to and abstractions from rivers. In undertaking this task it became apparent that a major obstacle to the more effective use of rivers in England and Wales was the inadequacy of many sewage treatment plants which were still owned and operated by local authorities. Many had neither the financial nor technical resources to improve the performance of these plants. The Water Act 1973 finally removed sewage treatment from municipal control and established a system of ten regional water authorities. The boundaries were matched to those of the major river basins rather than previously established administrative units. The status of these authorities changed with privatization in 1990 when they became utilities responsible, together with a large number of private water companies, for supply, sewerage, sewage treatment and disposal. Water resource planning and pollution control were transferred to the new NRA.

Certain general themes relevant to the management of water resources emerge from this review of experience in England and Wales. First, a progressive increase in the geographical scale of organization culminating in the attempt to match administrative boundaries to the fundamental hydrological unit of large river basins. Second, a closely related shift, motivated by the need to adopt a more comprehensive view as pressure on the resource increases, in the locus of decision-making from local to regional and, in some cases, national levels. Third, a recognition of the interdependencies between water quantity and water quality and of the need to coordinate the planning of these different aspects of the resource within multi-functional agencies. (The separation of functions between the privatized water authorities and the NRA in 1990 may be regarded as a backward step from this goal of integrated water management.)

River basin planning

The trend towards catchment-based administrative units in England and Wales has been matched by similar trends in other countries such as India. River basins have also been used as the framework for comprehensive planning schemes which attempt to situate water management within a much wider economic and social context designed to promote an integrated approach to development. The origins of river basin planning are usually traced to the Tennessee Valley Authority (TVA) which was created in 1933 to oversee an area equivalent to 80 per cent of England and Wales. This institution was charged by Franklin Roosevelt (quoted in Lilienthal, 1953: 52) with '... the broadest duty of planning for the proper use, conservation and development of the natural resources of the Tennessee River drainage basin and its adjoining territory for the general social and economic welfare of the Nation'. Considerable emphasis was placed on the construction of dams to generate HEP and to prevent flooding, but the TVA also stimulated agricultural development by pioneering work on land reclamation and soil conservation and by encouraging innovation in farming practices. The TVA has been subject to numerous retrospective assessments and its achievements in improving economic and social conditions in a relatively backward region of the United States have been offset in the eyes of critics by its preoccupation with large dams and its alleged neglect of the environmental consequences of its actions (cf. Owen, 1973; Chandler, 1984). Although the balance of its success and failure is disputed, the TVA has been very influential in promoting the concept of river basin planning. Figure 9.12 emphasizes that such planning involves much more than water resources management and there is no single model for river basin development. The objective is typically expressed in terms of maximizing welfare benefits to the population of the river basin, but there are variations in the specific tasks associated with the achievement of this goal. Recreation and conservation are often important in developed countries, but the supply of HEP for industry and the control of water for agriculture are usually the priorities in developing countries.

Several African countries, including Kenya,

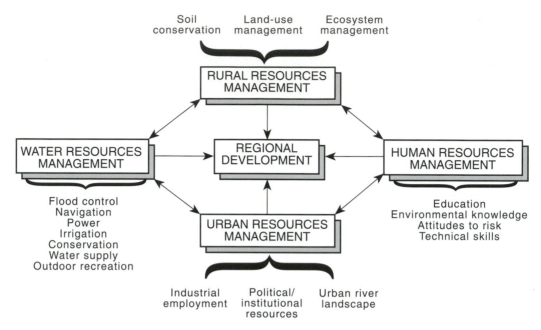

Soil
conservation   Land-use
management   Ecosystem
management

RURAL RESOURCES
MANAGEMENT

WATER RESOURCES
MANAGEMENT

REGIONAL
DEVELOPMENT

HUMAN RESOURCES
MANAGEMENT

Flood control
Navigation
Power
Irrigation
Conservation
Water supply
Outdoor recreation

URBAN RESOURCES
MANAGEMENT

Education
Environmental knowledge
Attitudes to risk
Technical skills

Industrial
employment   Political/
institutional
resources   Urban river
landscape

Fig. 9.12 Integrated river basin management. (*Source:* Newson, 1992.)

Plate 9.3 The controlled conservation and exploitation of available water supplies is central to integrated river basin planning. Sophisticated water management systems in the Milk River basin have been important in the development of irrigated agriculture in southern Alberta. (*Source:* K. Chapman)

Nigeria, Ethiopia and Senegal, have established river basin development agencies (Fig. 9.13). The extension of irrigated agriculture has been an important goal in all of them, but there is often a significant gap between objectives and achievements. A common theme in assessments of river basin planning in Africa is the tendency for over-optimistic projections (see Adams, 1985, 1991; Rowntree, 1990). This is a result of political factors as well as of technical forecasting errors. Large schemes have been used by politicians to enhance their own prestige. For example, the Akosombo Dam on the Volta River was promoted by President Nkrumah as an instrument for the economic transformation of the newly-independent Ghana in the early 1960s. It is doubtful that he believed his own rhetoric, but he found it politically advantageous to present this dazzling prospect to the population. Responsibility for encouraging unreasonable expectation does not lie exclusively with politicians. In reviewing the outcome of large-scale irrigation schemes in Nigeria, Adams (1991) suggests that over-optimism was endemic in the scientists and engineers responsible for their conception and implementation. This was most apparent in the shortfalls in the area of irrigated land created and in the associated crop yields. If the benefits of these schemes were overestimated, the costs were equally under-estimated. Both capital and operating costs have often proved higher than anticipated, while negative side-effects have been discounted. The environmental consequences of large dams have, for example, been neglected, together with the social and economic costs associated with the resettlement of those displaced by the creation of massive artificial lakes.

The essence of river basin planning is *integrated* development reflecting the interdependencies between land, water and people within a catchment. These interdependencies are implicit in the connections between the various activities identified in Fig. 9.12. The creation of an administrative structure with the technical expertise to understand these interdependencies and the power to manipulate them is essential to effective river basin planning – a conclusion which emphasizes the cultural dimension of resource management (Box 9.2).

International water management

Water management is generally easier for an island state such as the United Kingdom relative to a landlocked one such as Hungary. Indeed, effective water management becomes very difficult when the hydro-

> **Box 9.2: River basin planning in Kenya**
>
> The Tana and Athi Rivers Development Authority (TARDA) is responsible for an area containing more than 60 per cent of Kenya's population including the capital, Nairobi (Fig. 9.13). The Tana River Development Authority was created in 1974 and its remit was extended to incorporate the Athi River Basin in 1982. The need for water resource allocation was the immediate reason for establishing a river basin authority, but its functions included '...all matters affecting the development of the area' (quoted by Rowntree, 1990: 33). In practice, this comprehensive role has not been fulfilled and TARDA has concentrated on the technical problems of reconciling the water needs of HEP schemes in the middle and irrigated agriculture in the lower reaches of the Tana River. Rowntree identifies several factors in the failure of TARDA to meet its wider objectives. The relative lack of indigenous expertise and finance has led to dependence on external consultants and funding agencies committed to individual projects rather than the overall concept. The large area (132 700 km^2) and diverse ethnic composition of the Tana and Athi Basins has created the potential for internal tensions based on perceptions of conflicting tribal interests. In these circumstances, the opportunities for corruption are considerable and politicians have been accused of diverting funds to benefit their particular constituencies. Suggestions that patronage rather than competence has determined appointments to key positions have further weakened the credibility of TARDA. The demands which river basin planning places on administrative and political structures are implicit in the complex relationships between the different agencies involved in regional development in Kenya (Fig. 9.13). These relationships find expression in lines of authority and political accountability as well as flows of information and funding. It is important to appreciate that the Kenya situation is not unusual and similar problems have been identified in other African countries (see Winid, 1981; Adams, 1985, 1991).

logical unit straddles international boundaries. The probability of such fragmentation obviously increases with the area of the river basin, but is also inversely related to country-size. The watershed area of the Danube, for example, is less than 14 per cent of that of the Amazon, but it extends across a larger number

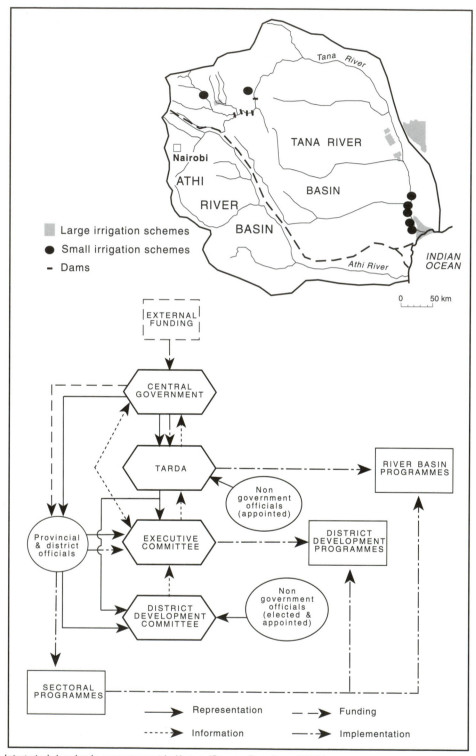

Fig. 9.13 Integrated river basin management in Kenya. (*Source:* Rowntree, 1990.)

Table 9.4: Rivers or lakes with five or more nations forming part of the basin

River	Number of nations	Watershed area (km^2)
Danube	12	817 000
Niger	10	2 200 000
Nile	9	3 030 700
Zaire	9	3 720 000
Rhine	8	168 757
Zambezi	8	1 419 960
Amazon	7	5 870 000
Mekong	6	786 000
Lake Chad	6	1 910 000
Volta	6	379 000
Ganges–Brahmaputra	5	1 600 400
Elbe	5	144 500
La Plata	5	3 200 000

Source: Gleick (1987)

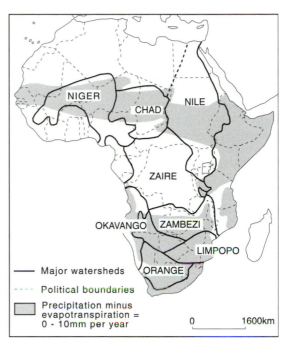

Fig. 9.14 Watersheds and political boundaries in Africa.

of countries because of the multiplicity of relatively small political units in Eastern Europe (Table 9.4). The importance of international relations in water management is emphasized when it is appreciated that 214 river or lake basins, accommodating 40 per cent of the world's population and covering more than 50 per cent of the Earth's land area, are shared by two or more countries (World Resources Institute, 1992: 171). Africa has a greater number of international river basins (i.e. involving two or more countries) than any other continent and 12 of these are shared by at least four countries. This situation is partly due to the fact that several of the world's largest rivers, including the Nile and Zaire (Table 9.4), drain the continent. Political boundaries, often inherited from the nineteenth-century colonial scramble for Africa, are superimposed on these river basins without regard to 'natural' frontiers defined in physical or cultural terms. The implications of this mismatch between basic hydrological and political units are all the more important because of widespread water deficits in many parts of the continent (Fig. 9.14) (Falkenmark, 1986).

Viewed in an historical perspective, there has been a clear trend towards greater cooperation in the management of international river basins as individual states have come to appreciate the potentially disruptive consequences of independent action. The so-called 'Helsinki rules' express this trend towards limited territorial sovereignty or equitable utilization based on the premise that a state may use shared freshwater resources to the extent that its uses do not interfere with the reasonable uses of another state. International organizations, especially the United Nations, have promoted this approach. A UN conference in 1977, for example, called for an international code of conduct regarding the rights of upstream and downstream countries and a 1991 document prepared for the UN Conference on Environment and Development restated the need for cooperation (World Resources Institute, 1992: 172). In practice, cooperation depends upon the attitudes of the parties involved and international law in this area remains weak. The greatest progress, as measured by the number of international treaties, has been made in Europe and North America. Several agreements relating to Africa, Asia and Latin America have been signed, but, in many cases, these have not prevented continued wrangling between signatories. The extent of international cooperation in water resource management is strongly influenced by diplomatic considerations as the recent history of cooperation and conflict related to the waters of the Rhine and Jordan respectively demonstrate.

The Rhine: international cooperation

The River Rhine has long held an important place in the economic life of Western Europe as reflected in an early navigation agreement involving five riparian countries in 1868. The orientation of the river basin relative to existing political frontiers poses particular problems for the Netherlands. Several upstream countries, especially France and Germany, have traditionally discharged heavy pollution loads into the river. This imposes cleaning costs upon the Dutch who extract drinking water from the Rhine. Not surprisingly, they have played an active role in promoting cooperative arrangements for the reduction of pollution. Discussions began in the 1950s and the International Commission for the Protection of the Rhine Against Pollution was established in 1963. This agreement was extended and strengthened with the signing of the Convention for the Protection of the Rhine Against Chemical Pollution in 1976. The Rhine Action Programme initiated in 1986 aims to produce drinkable water from the Rhine, reduce sediment loads and create conditions favourable to the return of indigenous aquatic life. These efforts have produced significant improvements in water quality since the early 1970s (Kiss, 1985). The existence of the European Union, which has provided both a predisposition towards, and an institutional framework for cooperation, has facilitated these achievements.

The Jordan: international conflict

The enhancement of water quality has been the principal motivation to international cooperation not only in the case of the Rhine, but also in other drainage basins such as the Great Lakes and the Danube (see Teclaf and Teclaf, 1985; Linnerooth, 1990). Water quantity or availability is usually the major concern in developing countries. Indeed, water may become a 'strategic' resource by the same criteria of economic importance and security of supply that are used to define 'strategic' minerals (Chapter 8). The potential vulnerability to supply disruption of a downstream country which derives much of its water from sources beyond its frontiers is self-evident. This possibility may be exploited by hostile neighbours to undermine its economic position in a manner which is the antithesis of the general trend towards cooperation in the management of international river basins. Several factors make the risk of a 'water war' especially great in the Middle East (see Starr, 1991; Kliot, 1993). First, whatever criteria are adopted, it is conspicuously 'water poor' (Figs 9.5 and 9.6). Second, population and economic growth, especially in the oil-rich states of the Gulf, is increasing demand. Third, there are very few major rivers and those that do exist cross international frontiers, creating opportunities

for conflicts of interest between neighbouring states. For example, the South East Anatolia Development Project, which was initiated by Turkey in 1983, will eventually involve the construction of 13 dams on the Tigris and Euphrates, provoking fears in Syria and Iraq that their own plans further downstream will be compromised. Syria and Iraq themselves came close to war in 1975 when Syria reduced the flow of the Euphrates to fill its Ath-Thawrah dam (Starr, 1991). The fact that two Arab states should even consider this possibility emphasizes that the additional complication of Israeli demands upon the scarce water supplies of the Middle East further increases the risks of conflict. The confrontation has focused on the waters of the River Jordan.

Figure 9.15 identifies the main features of the Jordan Basin and Fig. 9.16 indicates the inputs to and outputs from the system. The headwaters of the River Jordan are formed by three streams, the Hasbani, Dan and Banias. The Hasbani rises in Lebanon, while the Dan and Banias lie entirely within the disputed, but Israeli-occupied Golan Heights. Further inputs are provided by springs and streams flowing into Lake Tiberias and, especially, by the Yarmuk River which has its source in Syria and passes through Jordan to enter the River Jordan below Lake Tiberias. Approximately 3 per cent of the basin lies within the pre-1967 boundaries of Israel, but it is estimated that Israel (including the West Bank) currently uses 40 per cent of the available supplies (Joffé, 1993). Although the Jordan Basin is relatively small by international standards (11 500 km^2), the superimposition of the hydrological pattern upon the political one has created a complex situation in a region where all of the riparian states face water shortages (Wolf and Ross, 1992).

Viewed from an Israeli perspective, secure water supplies have been a primary concern ever since the creation of the state in 1948. It began construction work in 1953 on a scheme to divert water from the Jordan above Lake Tiberias to serve the proposed National Water Carrier. Objections from Syria, supported by the United Nations, resulted in the relocation of the extraction point to Eshod Kinrot on the shores of the lake. The National Water Carrier was established primarily to meet the needs of irrigated agriculture and it is clear that competition between the riparian states is a zero-sum game in which gains by one will inevitably result in losses by others. For example, Israel objected to a Syrian and Jordanian proposal in 1953 to construct the Maquarin dam across the Yarmuk on the grounds

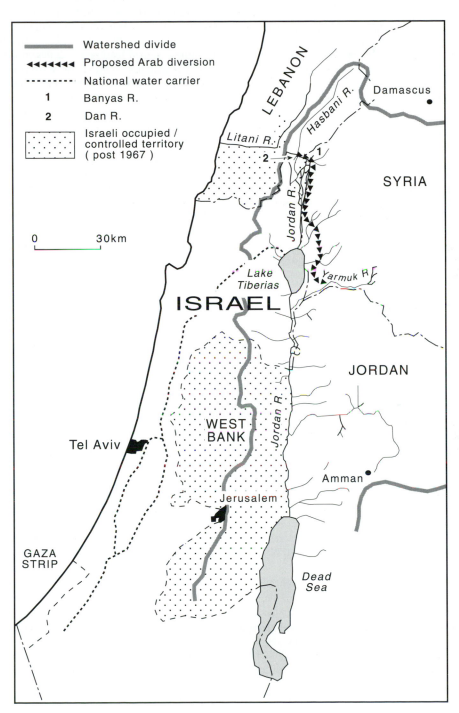

Fig. 9.15 The Jordan River Basin.

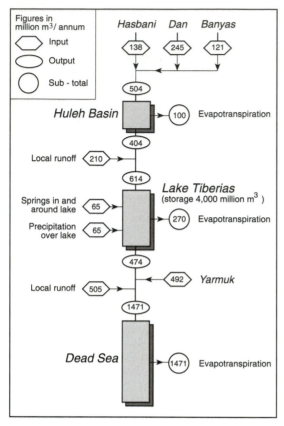

Fig. 9.16 Water balance in the Jordan River Basin. (*Source:* Beaumont, 1989.)

that it would reduce the flow into the Jordan. The most explicit threat to use control of water supplies as a weapon was the Arab Headwaters Diversion Plan in 1964. This consisted of several options of which the preferred one was the diversion of the Hasbani and Banias to the Yarmuk (Fig. 9.15). Israel estimated that this would reduce the intake to the National Water Carrier by 50 per cent. Work on the scheme began in 1965, but was disrupted by Israeli military strikes which continued until the 1967 war. This conflict resulted in significant territorial expansion by Israel. Many factors motivated Israeli occupation of the Golan Heights, southern Lebanon and the West Bank, but there is no doubt that these actions have strengthened the country's water supply position. It now controls the headwaters of the Jordan system and has re-opened the possibility, first raised in 1953, of diverting the Litani River in Lebanon (Amery, 1993). Furthermore, occupation of the West Bank has added substantial groundwater from underlying

aquifers and Israel draws more than 40 per cent of its water requirements from the West Bank (Lonergan, 1993: 209). Whatever the outcome of efforts to secure peace in the Middle East, the water issue remains a central feature of the negotiations. Indeed, a serious attempt to move away from the essentially competitive approach to water resource development in the Jordan Basin which has characterized the last 50 years would represent one of the most hopeful signs of a genuine agreement.

Water resource issues exemplify many of the more general issues of environmental resources. Fresh water in rivers, lakes or groundwater stores is a tiny proportion of the water content of the Earth: in other words the practical resource is only a tiny proportion of the theoretical resource or the resource base. Like many environmental resources, fresh water is widely regarded as an inheritance or as an entitlement of every human: not everyone can readily see it just as a commodity to be sold for profit. Equally, however, it is often wasted if it is treated as a free good or if it is undervalued in monetary terms.

Consumption usually rises with standard of living. Basic metabolic requirements of drinking water are minute compared with other forms of domestic consumption, and almost negligible compared with industrial and agricultural uses. Water *needs*, in the biological sense, are insignificant relative to the *wants* of modern Western lifestyles. In this sense, increasing demand arising from population growth is a matter of little consequence, and absolute shortages at the global scale can scarcely be envisaged. In some parts of the world, however, humankind has faced a chronic shortage of water for thousands of years. As human wants increase, more and more areas face the prospect of water deficits. As scarcity bites, management regimes come under increasing pressures, at both local and international scales.

Like some other environmental resources (e.g. land, wild animals), water is valued for its aesthetic as well as for its physical properties. The age-old human fascination with water is increasingly manifested in its use as a recreational environment and as a means of enhancing amenity, whether in private gardens or civic squares. As humans become richer and increasingly value water for these aesthetic reasons, water-management regimes need to be modified if conflict is to be avoided. Multi-purpose management may be technically possible (as, for example, in the case of the forest resource), but the necessary institutional adaptations may prove more difficult to achieve.

Further reading

Agnew, C. and Anderson, E. (1992) *Water resources in the arid realm*. London: Routledge.

Beaumont, P. (1989) *Environmental management and development in drylands*. London: Routledge.

McDonald, A.T. and Kay, D. (1988) *Water resources: issues and strategies*. Harlow: Longman.

Newson, M. (1992) *Land, water and development*. London: Routledge.

Thomas, C. and Howlett, D. (eds) (1993) *Resource politics: freshwater and regional relations*. Buckingham: Open University.

CHAPTER 10

Ocean resources

Human use of the Earth's environmental resources has focused on the land masses, but 71 per cent of the surface of the planet is covered by water. The world's oceans, together with the seabed and underlying rocks, are sources of each of the three main groups of environmental resources identified in Chapter 1. They yield raw materials used as inputs to the economic system; they contribute to the global life-support system; and they provide environmental services such as serving as a waste-sink for pollutants associated with human activities, especially at the coastal juxtaposition of land and sea. It is not feasible to attempt here a comprehensive review of all of these different contributions to the quality of human life and the objective of this chapter is to illustrate some general issues of resource use with reference to the marine environment.

Ownership of the oceans

Territoriality has long been and remains a basic human instinct. It is expressed in the establishment of more or less clearly defined boundaries at a variety of geographical scales. Such boundaries are taken for granted on the Earth's land surface, but they are less evident on the seas. Indeed the vast extent of the world's oceans has made concepts of ownership and control seem inappropriate; it has also promoted a sense of the infinite abundance of their resources. This situation is changing, however, and is reflected in the evolving political geography of the oceans (see Prescott, 1975; Glassner, 1990).

Sovereignty and maritime boundaries

Some degree of territorial control over adjacent waters has been exercised by coastal states at least as far back as the classical civilizations of the Mediterranean Basin. Strategic and political considerations were paramount in such extensions of territorial sovereignty from land to coastal waters. Repelling invaders was, for example, more difficult after a landing and a territorial sea was regarded as the assertion of a first line of defence. This was implicit in the association of the long-established and widely-accepted 3 nautical mile (nm) (i.e. 5.55 km) territorial limit with the range of shore-mounted cannons. However, such cannons also afforded protection for native from alien fishermen and, therefore, established a connection between land-based political authority and the ownership of marine resources.

While diplomatic considerations promoted the almost universal acceptance of the principle of territorial waters paralleling the coastlines of nation states at distances of 3, 4 or 12 nm for almost 500 years prior to the mid-twentieth century, such considerations resulted in a very different view of the seas beyond these boundaries. Following the invocation of papal authority by Spain and Portugal in the late fifteenth century to justify their claims to all known and unknown oceanic waters to the west of Europe and Africa, the principle of *mare liberum* or freedom of the seas became established in the seventeenth century. Freedom of navigation was the primary concern and was of particular benefit to the great maritime empires established first by the Dutch and then by the British. Both of these countries required such freedom to challenge the trade monopolies of the Spanish and Portuguese with the Spice Islands (i.e. Indonesian archipelago). Access to fishery resources was a secondary consideration to these powers, although conflicts between English and Dutch fishermen in the relatively enclosed waters of the North Sea prompted a British lawyer, John

Selden, to publish a treatise advocating the extension of territorial limits to create a closed sea (*mare clausum*) where the economic interests of a state were threatened. In the event, English (and later British) economic interests were best served by the concept of *mare liberum* as advocated by the Dutch jurist Hugo Grotius in 1608. Ultimately, freedom of navigation was regarded as fundamental to the operations of a trading empire and it seemed inconceivable in the seventeenth century that marine resources could be expropriated by any one country or could be exhausted by over-exploitation.

Resources and maritime boundaries

Marine fishing has traditionally been based upon local catching for local markets and this remains the pattern in much of the Third World. However, commercial operations involving long-distance movements between home ports and fishing grounds have existed for over 2000 years (Coull, 1993a). For example, the Basques were exploiting cod and whales in the Bay of Biscay as early as the twelfth century and they subsequently extended their operations throughout the north-east Atlantic as far north as Norway. Nevertheless, such activities were the exception rather than the rule prior to the introduction of steam engine propulsion in fishing craft in the 1880s and the power hauling of fishing gear at about the same time. The initial application of these new technologies was concentrated in Western Europe and associated particularly with operations out of British ports. Further advances occurred with the introduction of the marine diesel in the inter-war years and improved techniques of fish preservation, notably the development of large factory-freezer trawlers after the Second World War. These technical changes not only increased catching power, but, more significantly, they greatly extended the geographical range of commercial fishing operations. By the early 1960s, Soviet and Japanese fleets in particular were operating on a global scale in waters as far apart as the North Atlantic and South Pacific.

These extensions of the fishing frontier drew attention to the growing pressure on the resource. Conflicts between coastal fishermen and the international nomads in the form of distant-water fleets provided visible evidence of this pressure. It was, perhaps, ironic that British trawlermen who, in many respects, pioneered the development of long-range industrial fishing from ports such as Grimsby and Hull should find themselves the victims of the first really serious attempt by one country to claim and, more especially, to assert its claim for ownership of fisheries well beyond the traditionally restricted extent of territorial waters. The cod wars brought direct confrontations between British trawlers, sometimes escorted by naval vessels, and Icelandic gunboats seeking to enforce a claim to exclusive fishing rights within a 12 nm zone declared in 1958 and later a 50 nm zone announced in 1972. Despite considerable resistance from the United Kingdom, these claims were eventually conceded and the resolution of this dispute in favour of Iceland '... became a trigger mechanism in the acceptance of general limits extensions in the North Atlantic area, and indeed beyond' (Coull, 1993a: 159).

More or less coinciding with the zenith of distant-water fishing in the 1950s and 1960s, another source of pressure upon marine resources emerged to challenge the concept of open access. This was related to the growth of interest in the prospects of offshore areas as sources of oil and natural gas. The principle of producing oil from under the sea had been established as early as the 1890s when wells were drilled from piers extending into the Santa Barbara Channel from the coast of California. These were physically connected to the land and it was not until 1947 that the first well was drilled out of sight of the shore at Ship Shoal in the Gulf of Mexico. By the 1950s, the oil industry was pushing beyond the boundaries of established territorial waters and this raised obvious questions regarding the ownership not only of the living resources of the sea, but also of the mineral and energy resources believed to lie on or beneath the seabed. The resolution of these questions has witnessed a progressive shift away from the essentially open-access resource regime which has characterized the exploitation of the resources of all but the immediate coastal margins of the world's seas and oceans for most of human history.

Enclosure of the marine commons

Coull (1975: 103), draws a parallel between the evolution of the field systems of Britain and Europe and the 'enclosure' of the marine commons, noting that the organization of resource use becomes more defined in a situation of pressure. An important element in the clarification of these rules has been a progressive extension of the offshore areas subject to national control by coastal states (Fig. 10.1). This geographical extension has been accompanied by an expansion in the scope of government regulation

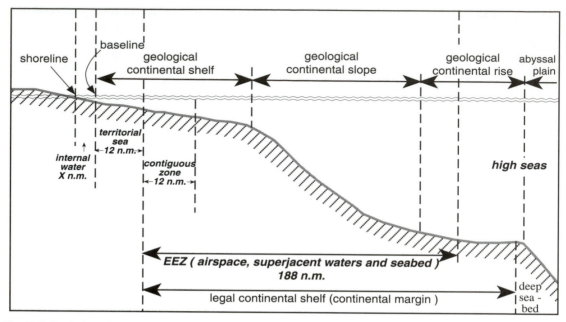

Fig. 10.1 Physical and legal maritime boundaries.

affecting the exploitation of marine resources. It has also drawn attention to significant conflicts of interest between countries based upon accidents of geography and upon differences in their technological capabilities.

Territorial sea

The origins of and motives for the establishment of territorial seas have already been reviewed and, as late as 1950, the majority of claims were for limits not exceeding 4 nm. This situation reflected both the historical influences and the geographical circumstances of the major European countries with their maritime empires and their immediate experience of the relatively enclosed waters of the Mediterranean, Baltic and North Seas. However, the collapse of these empires was accompanied by the creation of a large number of independent states. Securing control over ocean resources adjacent to their coastlines was of more pressing concern to these states than a liberal view of the freedom of the seas. The pressure in the international community to establish wider territorial seas was reflected in a shift from a modal value of 3 nm in 1950 to 12 nm by 1972 (Prescott, 1975: 68). Several states, including Peru, Chile and Ecuador,

had unilaterally declared 200 nm territorial seas by the mid-1950s and such actions emphasized the need for international agreement. Following protracted preliminary discussions, the first UN Law of the Sea Conference (UNCLOS I) was convened in 1958. In fact, it was only after the Convention produced by a third conference (UNCLOS III) in 1982 that a limit of 12 nm, paralleled on its outer margin by a *contiguous zone* of similar width, was clearly established as the international norm if not the universal rule. The significance of the contiguous zone is mainly related to the domestic law enforcement of the coastal state over such matters as customs and immigration.

The continental shelf

The issue of sovereignty in the broadest sense has been central to the establishment of a wider territorial sea. Thus customs and immigration affairs as well as matters of legal jurisdiction are linked to the boundaries defined by the territorial sea and the contiguous zone. By contrast, resource issues have driven international agreements which have witnessed the partition of vast areas of the continental shelves between coastal states. Unlike the boundaries of the territorial sea and the contiguous zone which exist

Plate 10.1 The growth of the offshore oil industry was a major factor stimulating the development of international agreements governing rights of access to the continental shelves. Intensive development of major offshore oil provinces such as the North Sea could only take place after territorial divisions between adjacent coastal states had been established. (*Source:* K. Chapman)

only as lines on a map, the edge of the continental shelf is a physical reality. Figure 10.1 emphasizes that the coastal margin is paralleled by a zone of relatively shallow water lying above the continental shelf. This zone varies in width from as little as 20 km to as much as 1500 km and its outer edge is defined by the sharp break of the continental slope. Much of the known resource wealth of the oceans and certainly their most accessible resources are concentrated in the water above or the rocks below the continental shelves. A growing awareness of this situation, stimulated especially by the prospect of oil and gas discoveries, resulted in an attempt at UNCLOS I to define the rights of national access to these resources.

The UN Convention on the Continental Shelf was ratified and become international law in 1964. The

definition of the continental shelf adopted by the Convention was flexible. Thus a water depth of 200 m was regarded as the outer limit, but it was acknowledged that this may be extended in accordance with advances in technological capabilities to exploit resources in deeper water. Despite this ambiguity, the Convention had important practical consequences. In particular, it was a prerequisite for the expansion of the offshore oil and gas industry not only because it granted sovereign rights to coastal states over resources lying on and beneath the seabed of the continental shelf, but also because it specified a method for partitioning the shelf where the configuration of the coastline leads to competing claims in enclosed basins such as the Mediterranean. In the North Sea, for example, the acceptance of the so-called median line principle by most of the interested parties made it easier to establish the territorial framework within which the search for oil and gas could take place (Chapman, 1976).

An important feature of the Convention on the Continental Shelf was a clear separation of the seabed and sub-seabed from the overlying waters. Thus its terms did not affect the position of the resources living within the sea. This distinction made it easier to reach agreement because decisions were made before the location and extent of any mineral resources were known. Vested interests make it more difficult to define (or redefine) rights of access to proven resources as the history of attempts to regulate marine fisheries testifies.

Exclusive Economic Zones (EEZ)

Efforts to secure international agreement on exclusive fishing zones beyond territorial waters ended in failure at UNCLOS I (1958) and UNCLOS II (1960). However, this did not deter a succession of unilateral declarations following the earlier examples of most seaboard countries in Latin America. The experience of the first cod war and the growing assertiveness of newly-independent countries were reflected in these declarations which sought to establish preferential or monopoly rights within these zones for domestic fishermen. The *de facto* extension of national fishing zones was legitimized by UNCLOS III which formalized the concept of the EEZ. This grants coastal states jurisdiction over all resources within the EEZ. The outer limit of the EEZ is arbitrarily set at 200 nm except where boundaries established with reference to opposing coastlines overlap. In circumstances where the continental shelf extends beyond 200 nm, UNCLOS III permits a state to claim the resources on or below the seabed to a distance of 350 nm.

More than 70 states were claiming and enforcing an EEZ of 200 nm by the beginning of the 1990s and many others had established exclusive fishing zones of equivalent extent. The details of the situation are confused, but the general trend is clear. The last 40 years or so have seen a fundamental change in the political geography of the oceans driven by efforts to regulate access to their resources. It is estimated that if all coastal and island states choose to establish a 200 nm EEZ, approximately 36 per cent of the world's oceans would be placed under national jurisdiction leading to the total 'enclosure' of vast resource-rich areas such as the North Atlantic and North Pacific.

Protecting the marine commons

The progressive 'enclosure' of the marine commons has been motivated by the desire of coastal states to appropriate the resources of the oceans. The delimitation of the boundaries which reflect this trend has, in many cases, been the outcome of processes of conflict, negotiation and compromise. However, cooperation is more appropriate in protecting the marine environment from the adverse effects of pollution. Such pollution reflects the role of the world's oceans as a waste-sink for various by-products of human activities. These by-products are both accidental, as demonstrated by events such as the *Exxon Valdez* disaster in 1989, and deliberate, as exemplified by the dumping of toxic wastes. There is evidence that the assimilative capacity of the marine environment is limited (GESAMP, 1990). This capacity varies from place to place as a result of the interaction of such factors as water temperature, tidal-flushing and wave energy. The imposition of uniform standards or controls which disregard such geographical variations is, therefore, difficult to justify. Nevertheless, there is growing concern about the problem of marine pollution despite a conspicuous lack of scientific evidence directly linking pollution to the deterioration of fish stocks. Experimental work in controlled environments emphasizes that pollutants adversely affect the health of fish, but it is very difficult to isolate these effects and to identify threshold levels in the open sea.

The London Dumping Convention of 1972 was the first international effort to control marine pollution. The signatories are committed to cease all dumping of industrial wastes at sea by 1995. Although such activities are very visible sources of marine pollution, they are estimated to be responsible for only 10 per cent of the total, slightly less than the 12 per cent contribution of maritime transport (i.e. tanker spills, etc.) (GESAMP, 1990: 108). The long-term effects of pollutants originating from land-based activities such as discharges from industrial facilities as well as contaminated runoff (see Chapter 9) are potentially much more damaging. They are already serious in many enclosed seas with intensive economic development around their coastal margins and relatively weak natural flushing mechanisms. The Baltic and Mediterranean Seas, for example, share these characteristics and the condition of both is cause for concern.

Marine pollution is another example of the concept of environmental externalities. The difficulties involved in controlling such externalities have already been emphasized with reference to atmospheric pollution (Chapter 7). Nevertheless, the need for international cooperation to control marine pollution, especially from land-based sources, has been recognized in various conventions and plans associated with the UN Regional Seas Programme. Agreements extending to land-based pollution had been reached for six maritime regions by 1992 (i.e. the North Sea, the Baltic Sea, the north-east Atlantic, the Mediterranean, the Gulf and the south-east Pacific). The effectiveness of such agreements is often constrained by the differing circumstances and priorities of the signatories. For example, progress in implementing a 1992 agreement to spend £12 million improving the quality of the Baltic is being hampered by financial problems. Countries such as Sweden and Denmark are better able to meet the costs than relatively impoverished states such as Poland and Latvia. Despite these difficulties, the existence of such agreements emphasizes that, ultimately, protecting the marine commons is a shared responsibility despite the apparently conflicting trend towards the extension of national 'ownership' over the world's seas and oceans.

Fisheries resources

Ocean resources have been synonymous with fish for most of human history and many distinctive societies have been built upon the exploitation of marine fisheries. The importance of fish as a food resource varies widely depending upon geographical and cultural factors. The proximity of the resource

Plate 10.2 Conflicts between different users are becoming more common as pressure upon marine resources grows. Pollution, related especially to the activities of the oil and petrochemical industries, has adversely affected the livelihood of these shrimpers in Galveston Bay and the Gulf of Mexico. (*Source:* K. Chapman)

encourages its exploitation in coastal communities, although religious beliefs have sometimes restricted the level of consumption. Viewed at the global scale, fish contributed almost 16 per cent of the total average intake of animal protein in 1988 (FAO, 1992:162). It is generally more significant in developing countries and accounted for over 60 per cent of the 1988 animal protein (i.e. 'meat') intake in, for example, Ghana and North Korea. Japan (50.6 per cent) is the only developed country in which fish occupies such an important place in the national diet (cf. 6.6 per cent in North America and 9.7 per cent in Western Europe). Nevertheless, with a substantial proportion of the world's population heavily dependent upon marine fisheries, it is clearly important to understand the dynamics of what is, theoretically, a perpetually renewable resource.

Spatial distribution

Fish are highly mobile and some species, notably salmon and tuna, can migrate enormous distances during their life cycle. Notwithstanding the biological limits of different species which define their environmental ranges, such mobility might be expected to encourage a relatively even distribution of fish resources throughout the world's oceans. In fact, fish are abundant in some areas and very scarce in others. The zooplankton community, including protozoa and larval fishes, forms the link between the primary production of plant material (i.e. phytoplankton) and the commercially exploited species. It is, therefore, an indicator of their distribution. Figure 10.2 emphasizes that the biological productivity of the oceans varies from one location to another with some areas qualifying as the marine equivalent of deserts. The North Atlantic and North Pacific lie at the other end of the spectrum together with most waters overlying the continental shelves. Many factors contribute to the pattern in Fig. 10.2 including the interaction of bottom topography, salinity, water temperature, sunlight and currents. The latter are especially important in contributing to upwelling. This occurs when surface currents driven by offshore winds have the effect of drawing deep, nutrient-rich water to the surface. It is associated mainly with the western edge of continents where it is a major influence upon the productivity of the fisheries of countries such as Peru and Chile.

The geography of the world's fish catch reflects the importance of the environmental variables implicit in the distribution of zooplankton. The UN Food and

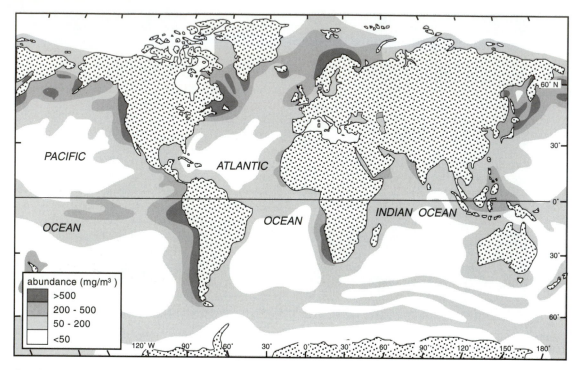

Fig. 10.2 Biological production of world oceans, expressed as the distribution of zooplankton in the upper 100 m layer. (*Source:* Times Books, 1983.)

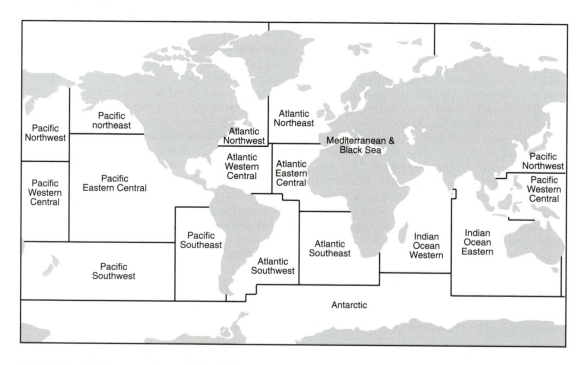

Fig. 10.3 World fishing areas. (*Source:* FAO, 1992.)

Table 10.1: Nominal world marine catch by major fishing areas, 1990

Area	Catch (000 tonnes)	Catch (% of total)
Atlantic, north-west	3 221	3.9
Atlantic, north-east	9 183	11.1
Atlantic, western central	1 697	2.0
Atlantic, eastern central	4 098	4.9
Atlantic, south-west	2 029	2.4
Atlantic, south-east	1 530	1.8
Mediterranean and Black Sea	1 489	1.8
Indian Ocean, western	3 376	4.1
Indian Ocean, eastern	2 828	3.4
Pacific, north-west	25 688	31.0
Pacific, north-east	3 428	4.1
Pacific, western central	7 311	8.8
Pacific, eastern central	1 519	1.8
Pacific, south-west	1 031	1.2
Pacific, south-east	13 945	17.2
Antarctic	428	0.5
TOTAL	82 801	100.0

Source: FAO (1992).

Agricultural Organization (FAO) makes annual estimates of the catch within broadly defined areas (Fig. 10.3). Table 10.1 shows the distribution of the total marine catch between these areas in 1990. The overwhelming dominance of the Pacific is clear. The degree to which the catching effort is concentrated in specific locations is understated by Table 10.1 because of the very large size of the areas to which the FAO statistics relate. Indeed it is estimated that 99 per cent of the total catch is obtained from within 200 nm of the coastline.

The uneven distribution of fisheries resources within the world's oceans has several implications. First, it emphasizes that simple extrapolations of the resource base cannot be made by translating experience in the traditional fishing grounds to other areas. Second, it implies that the opportunities for pursuing the spatial-expansion response to resource scarcity (see Chapter 3) are constrained despite the apparently limitless extent of the oceans. Third, it has the effect of creating pressure points which promote conflicts over access to the resource. These conflicts involve not only adjacent coastal states, but also the deep-sea fishing fleets of other countries. Although the freedom of these fleets has been restricted by the 'enclosure' movement described earlier, six countries (i.e. former USSR, Japan, Spain, Poland, South Korea and Taiwan) accounted for

almost 90 per cent of the total non-local catch in 1989 (FAO, 1992: 143). The continued pressure of such fleets upon what is clearly a localized resource makes it important to monitor trends in production.

Trends in production

Estimates of long-term trends in the world fish catch are necessarily speculative. There is, however, little doubt that the rate of exploitation has accelerated sharply during the twentieth century. Coull (1993a: 107) quotes a Russian source suggesting a figure of between 1.5 and 2 million tonnes in 1850, rising to 4 million tonnes in 1900 and 9.5 million tonnes in 1913. This had reached 22 million tonnes by 1938. Viewed in the context of these historical trends, the two decades beginning in 1950 witnessed exceptional rates of growth averaging approximately 6 per cent per year (Fig. 10.4). The corresponding percentage for the 20 years between 1971 and 1990 was only 2.3 (FAO, 1992: 132). Indeed, the catch actually fell in 1990 and 1991. The full significance of these most recently published figures is difficult to judge, but 'There is little reason to believe that the global catch can continue to expand' (FAO, 1992: 134) and an annual production of 100–120 million tonnes (including inland fisheries) using conventional fishing methods is widely regarded as a ceiling.

Global trends obscure differences between the world's major fishing grounds. The principal fishing areas of the Pacific (the north-west and south-east) continued to increase production during the 1980s, although the fisheries of Peru and Chile were, to some extent, sustained by catching lower-quality species for reduction to fish meal. By contrast, trends in the North Atlantic, which has the longest history of intensive commercial exploitation, have been downward since the mid-1970s (Fig. 10.4). The implications of such exploitation are readily appreciated when attention is focused upon particular species and fishing grounds.

The cod has been a target of commercial fishermen in the North Atlantic since medieval times, but pressure on the resource first became a cause for serious concern in the 1960s. Despite the greater control afforded by the expanding territorial claims by such states as Canada, Iceland and Norway adjacent to the fishing grounds, the total catch of Atlantic cod has declined steadily since 1970 (Fig. 10.5). The Peruvian anchovy has had a much shorter history of commercial exploitation, but it provides one of the clearest illustrations of the consequences of

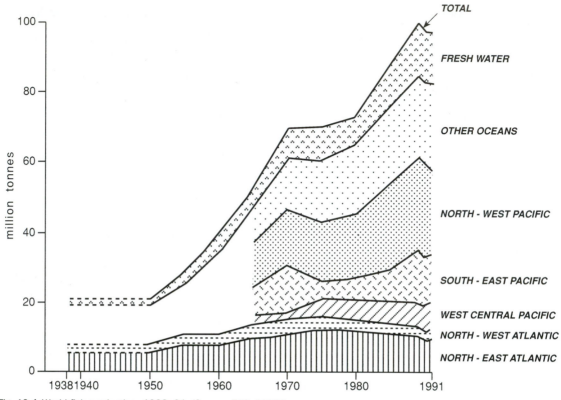

Fig. 10.4 World fish production, 1938–91. (*Source:* FAO, 1992.)

over-fishing. The anchovy became attractive in the late 1950s for conversion to fish meal for use as an animal feed in the developed countries, especially the United States. A fleet of more than 1400 boats was operating off the coast of Peru by the late 1960s and Chimbote become the world's leading fishing port as measured by landed tonnage. Doubts about the sustainability of the fishery were expressed at a relatively early stage in the 'boom' and progressively more rigorous attempts at regulation were introduced from 1965. Despite these efforts, the catch fell drastically in 1972 and has never recovered to its earlier peak (Fig. 10.5).

Maximum sustainable yield (MSY) in fisheries

The examples of the Atlantic cod and the Peruvian anchovy highlight the need for effective management of fisheries resources. The basic objective of such management is to ensure that present levels of exploitation are consistent with the replacement of fish stocks to ensure the long-term sustainability of the resource. The concept of maximum sustainable yield (MSY) is an important reference point in efforts towards achieving this objective. For any renewable resource, there is a maximum rate of growth in stock or population which may be achieved under prevailing environmental conditions. This MSY is closely related to the notion of carrying capacity in which population is limited by available food supplies (Fig. 2.4). It has already been emphasized in Chapter 2 that the optimal level of harvesting for a renewable resource is difficult to define because the physical or biological optimum is different from the economic one (Fig. 10.6) (see Gordon, 1954; Lawson, 1984). The potential yield is represented by a curve which rises steeply close to the origin because, in the case of fish, competition for food is reduced by the removal of some of the stock. This allows the rate of growth in the remaining population to increase until the MSY is reached. The enhanced biological productivity of the population is not maintained indefinitely, however, and the trend is reversed as the remaining stocks are progressively reduced. Eventually, a point is reached

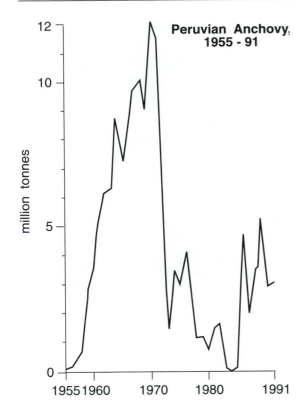

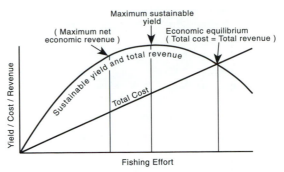

Fig. 10.6 Maximum sustainable yield in a fishery. (*Source:* modified after Rees, 1985.)

beyond which further efforts to increase the catch will be counter-productive because the population will become too depleted to reproduce itself.

The diagonal line in Fig. 10.6 represents the effort or costs involved in catching fish. It assumes that the costs increase in direct proportion to the number of fishing trips (Table 2.2) and intersects the yield curve at a point beyond the MSY. If the yield curve is also viewed as a measure of the total revenues earned by fishing activity, the intersection of the diagonal and the curve defines the economic equilibrium at which costs and revenues are equal. It is only when this point is reached that individual fishermen in an open-access fishery will receive clear signals that further effort does not make economic sense because costs will exceed returns. However, this signal is received too late from a biological point of view because it occurs only when the effort has pushed the catch beyond the MSY. Indeed, the economic optimum occurs at a much lower level of effort when the vertical distance between the cost/effort line and the revenue/yield curve is greatest (Fig. 10.6). In practice, the individualistic tradition of the fishing industry ensures that it is often impossible to identify when this economic optimum is reached. Furthermore, this same tradition makes it difficult to restrain the fishing effort below the point at which costs and revenues are equal despite the fact that this lies to the right of the biological MSY. The challenge of fisheries management is to control this endemic tendency towards over-fishing.

Fisheries management

Regulating access to the fishing grounds is one way of reducing pressure on the resource. Individual countries may limit fishing in spawning and nursery

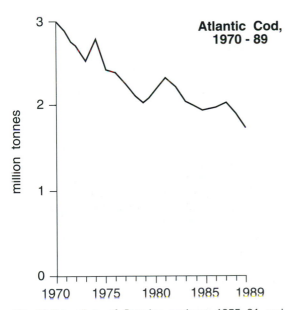

Fig. 10.5 Landings of Peruvian anchovy, 1955–91 and Atlantic cod, 1970–89. (*Sources:* Coull, 1993a; FAO, 1992.)

Plate 10.3 Advances in the technology and organization of commercial fishing have sharply increased catching-power. (a) A fish-carrier of the former Soviet fleet discharging at Riga, Latvia. (*Source:* J.R. Coull)

Plate 10.3 (b) A large tuna purse-netter unloading its catch by conveyor belt at San Pedro, California. (*Source:* J.R. Coull)

grounds in acknowledgement of the need to protect the biological future of the resource. The extended territorial jurisdiction of individual countries over the oceans has enhanced the scope for such measures, but fish do not recognize international agreements and their movements have no regard for legal boundaries. Many stocks straddle these boundaries and, in these circumstances, effective regulation of access depends upon cooperation. This is not always easy to achieve. In the north-west Atlantic for example, Canadian efforts to limit pressure upon stocks of cod within its 200 nm limit have resulted in disputes with the fishermen of other countries, notably Spain and Portugal, exploiting the resource outside this limit (Sullivan, 1989).

Advances in the technology of finding and catching fish have contributed significantly to the increasing pressure upon stocks. It is virtually impossible to reverse the direction of technical change, but national governments and international agencies have adopted various regulatory measures to limit the impact of enhanced catching capacity. Larger nets and more powerful hauling gear have increased the potential for the indiscriminate capture of species. Mesh size regulations, intended to prevent the taking of juvenile and under-size fish, are one response to this problem. They are often supported by stipulating limits upon the minimum size of fish which may be landed. Licensing of vessels is based on the premiss that catching capacity, as expressed by the number of boats and fishermen operating within a fishery, should match the available resource base. The imposition of quotas goes a step further by shifting the focus from indirect controls upon the methods of harvesting the resource to direct controls upon the size of the catch itself. The basic approach is to set annual (or seasonal) catch limits for a fishery on the basis of scientific assessments of its MSY. The specification of a total allowable catch (TAC) has, for example, formed the basis of the EU's Common Fisheries Policy (CFP) since the mid-1970s with the TAC divided into national quotas allocated to vessels registered in each member state.

The chequered history of the CFP draws attention to some of the difficulties of managing fisheries resources (Symes, 1991). Bitter disputes have arisen surrounding the CFP, especially as it affects activities in the North Sea, not only *between* member states, but also involving government and fishing interests *within* these states. The fact that such disputes have occurred within a union based upon an assumption of mutual interest emphasizes the difficulties involved in fisheries management. The mobile and invisible nature of the resource is itself a problem. Basic data are hard to obtain, leaving ample scope for honourable differences of scientific opinion in establishing the resource base. Even when the data are not disputed, interpretation is frequently difficult. It is, therefore, not surprising that estimates of MSY, which are prerequisites for sensible

management strategies, often display a very wide margin of error (Box 10.1). Such difficulties in the formulation of policies are compounded by problems of implementation. Surveillance and enforcement are problematic given that each fishing vessel is a mobile hunter/gatherer. The high costs of such operations place developing countries at a particular disadvantage in policing their waters, especially when challenged by the activities of distant-water fleets. Attempts to regulate fisheries inevitably result in winners and losers and the latter will, naturally, seek to resist or limit such regulation. The trawlermen of Hull and Grimsby were the principal losers following the cod wars and the United Kingdom resisted the extensions of Icelandic territorial waters. This example demonstrates that policies to reduce over-fishing and ensure the sustainability of the resource may have very adverse effects upon particular communities. Faced with effective opposition, difficult long-term decisions may be postponed in favour of political expediency. Overall, the proliferation of management measures during the last quarter century does not obscure the fact that '...the situation of global fisheries resources continues to give cause for serious concern' (Coull, 1993a: 176) (Box 10.2).

Offshore oil and gas resources

Drilling rigs and production platforms in many offshore areas provide a very visible reminder that fish are not the only resources recovered from beneath the waves of the world's seas and oceans. The first attempts to obtain oil from under the sea were associated with extensions of known onshore fields in which the geological structures happened to straddle the coastline. Despite early ventures in locations as diverse as the coast of California and, from the 1920s, Lake Maracaibo in Venezuela, the real impetus to the offshore oil industry was provided by activities in the Gulf of Mexico (D.W. Davis, 1991). Following an initial discovery in 1947, these activities received a major boost with the first large-scale allocation of leases by the US federal government in 1954. These permitted oil companies to search for and to produce oil and gas from beneath specific tracts of the seabed. This allocation paved the way for the creation of an international offshore oil and gas industry which, by virtue of its origins, has been dominated by US companies.

Box 10.1 The North Sea haddock: a case study in fisheries management

Two basic approaches are used to assess fish stocks. The first involves the controlled sampling of landings to monitor the species composition, size and age-structure of the catch. The second is based on evidence gathered by direct measurement of, for example, larval stocks by research vessels in nursery grounds. Such activities in the North Sea have been responsible for the assembly of the longest data series on fish catches and fish stocks in the world. The circumstances for the effective management of the haddock and other commercial species are, therefore, especially favourable. Annual total allowable catches (TAC) for various species were introduced in 1975 in response to mounting concerns about the sustainability of the North Sea fisheries in view of growing pressure from modern fishing vessels.

Haddock is the most important demersal (i.e. bottom-dwelling) species in the North Sea. Figure 10.7 shows the TAC proposed for haddock in each year by a team of marine scientists, the TAC actually adopted following negotiations by representatives of the various governments, and the catch which was eventually reported. In several years the proposed TAC identifies a 'debatable range' which reflects differences in scientific opinion based on uncertainties regarding the state of the existing stock and its rate of recruitment as well as doubts about the length of time required to permit recovery

for exploitation. The significance of these margins of error extends beyond the North Sea because they suggest that the corresponding margins are likely to be very much wider in most other fisheries for which fewer data are available.

Considered scientific advice is not always accepted by policy-makers as Fig. 10.7 demonstrates. Pressure from the fishing lobby resulted in the TAC being set at a higher level than recommended by the scientists in the first two years of the system. No agreement was reached in 1978 and fishing continued without reference to a TAC. A closer correspondence has been achieved between the recommended and the actual TAC in subsequent years, but there remains a tendency for it to be inflated as a result of the arguments of fishermen's organizations claiming that they cannot afford to catch less. Overall, the effectiveness of the system is debatable. The actual catch has exceeded the TAC in approximately half of the years suggesting that enforcement is difficult. On the other hand, the failure to reach the TAC in the remaining years could be interpreted as a sign of pressure on the resource. The clearest evidence of such pressure, however, lies in the decline of the catch during the 1980s. Indeed, it is clear that by the start of the 1990s the stock had been reduced to the lowest level ever known, despite the efforts to operate a conservation policy.

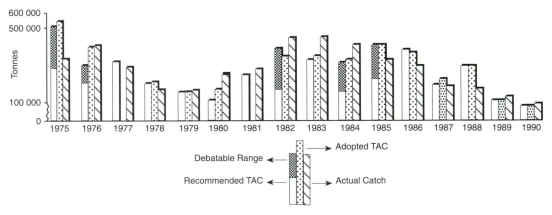

Fig. 10.7 Management of North Sea haddock stocks, 1975–90. (Source: Coull, 1993a.)

Extending the resource base

Oil and natural gas were perceived as resources well before the 'boom' in the Gulf of Mexico, but the opening-up of offshore areas as potential new sources

of supply was a very significant step. It required legal changes clarifying the 'ownership' of any discoveries (see above) and depended upon major advances in technology. Indeed, these essentially 'cultural' changes were a prerequisite to the exploitation of

Box 10.2 Aquaculture: from hunting to farming

Intensive fish farming is one approach to the problem of pressure on marine fisheries. It has been practised for at least 5000 years in China where fish taken from managed ponds continue to make a significant contribution to diet, especially in rural communities. The Chinese experience, together with growing concern over the adequacy of marine fish stocks and rising prices, has recently stimulated interest in fish farming. This is yet another example of the operation of the market-response model to resource scarcity as innovations in genetic engineering and environmental control are applied to the effective domestication of certain species of fish. Japan is, perhaps, the clearest example of this response as the activities of its distant-water fleets have been curtailed by the enclosure of the marine commons. In these circumstances, government encouragement has led to a sharp increase in Japanese aquaculture production since 1970.

Global production from fish farming is currently estimated at 15 million tonnes per year compared with a marine catch of 80–85 million tonnes.

Indeed, it is even more significant when expressed in terms of value. Over one-third of the harvest of commercial fisheries is destined for low-value fish meal, whereas fish farming concentrates upon high-value species such as carp and salmon. Parallels have been drawn between the impact upon food production of the so-called 'green revolution' associated with the introduction of high-yielding varieties of staple crops such as wheat and, especially, rice, and a possible 'blue revolution' based on fish farming (Coull, 1993b). This implies that the farming or husbandry of living resources in water could follow, with a time lag of several thousand years, the great transition from hunter/gathering to sedentary agriculture on land. The ultimate potential of aquaculture as a source of food is difficult to judge, although it is important to emphasize that current production is restricted to fresh water and, to a much lesser extent, shallow and sheltered coastal sea waters. There is no prospect that the fish resources of most of the world's seas and oceans will be 'farmed' rather than hunted.

offshore oil and gas which could almost be regarded as new resources in Zimmerman's terms (see Chapter 1) despite the fact that the geological circumstances of their occurrence are identical to those applying on land.

Recurrent fears about the adequacy of oil reserves, especially in the United States, were noted in Chapter 7. Growing imports and a realization that onshore production could not meet domestic demand were powerful stimuli to further exploration in the United States in the late 1940s. It had long been known that the prospects of finding oil and gas beneath the continental shelves were good and there was considerable interest in the geographical extension of the US resource base into adjacent offshore areas. This was reflected in a proclamation by President Truman in 1945 effectively claiming sovereignty over the natural resources of the subsoil and seabed of the continental shelf surrounding the United States.

Various estimates have been made of the extent of potentially oil-bearing sedimentary basins lying offshore, ranging from 30 to 50 per cent of the global total (see Earney, 1990; Shell Briefing Service, 1983). Estimates of the reserves contained within these basins are highly speculative and the uncertainties surrounding reserve estimates (see Chapter 7) are even greater for offshore environments because of the comparative lack of geological information. This is emphasized by two forecasts made in Shell publications. In 1983, it was suggested that '. . . in future, for every barrel of oil that will be found on land, two will be found offshore' (Shell Briefing Service, 1983: 1); by 1989, only 15 per cent of ultimately recoverable reserves of oil were predicted to lie offshore (Jennings, 1989: 5). Despite the differing judgements of the long-term future of offshore oil implicit in these forecasts, there is no doubt that it is a significant addition to the total resource base and it accounted for 26.8 per cent of total *proven* reserves in 1992 (Table 10.2).

The fear of physical scarcity has not been the only motivation driving the search for offshore sources of oil and gas. The role of geopolitical concerns in shaping events in the international oil industry was described in Chapter 7 and the development of offshore oil provinces such as the North Sea may be interpreted as an important element in efforts to diversify the sources of oil away from the Middle East. Table 10.3 indicates the results of these efforts as expressed by changes in the geography of offshore oil production. The absolute and relative decline of the earliest offshore producers, the United States and Venezuela, is evident. The growth in overall production has been accompanied by an increase in the number of producing countries from 19 in 1970 to 49

Table 10.2 Offshore oil and gas proven reserves, 1992

Area	Oil		Natural Gas	
	(million barrels)	%	(billion cubic feet)	%
Middle East	118 316	43.8	333 000	33.2
North America	6 390	2.4	52 540	5.2
Latin America	55 274	20.5	110 850	11.0
Europe	37 885	14.1	177 663	17.7
Asia/Pacific*a*	29 148	10.7	195 846	19.5
Africa	22 878	8.5	134 740	13.4
Total offshore	269 891	100.0	1 004 639	100.0
Total world	1 006 891		4 885 400	
Offshore as % of world total	26.8		20.6	

Sources: Compiled from data in *Offshore* and *BP Statistical Review of World Energy*
Note: *a*Including countries of former USSR.

by 1992. Nevertheless, relatively few account for a substantial proportion of the total and the distribution of offshore proven reserves is, like the onshore equivalent, heavily biased towards the Middle East (Table 10.2).

Defining the frontiers

The history of offshore oil and gas production illustrates the role of technology in 'creating' environmental resources (see Chapter 1). Even if the exact position and size of, for example, the giant Brent oilfield in the North Sea had been known in, say, 1950, it would have remained undeveloped. Prevailing technology would have been unable to cope with the problems posed by the combination of deep water and difficult wind and wave conditions associated with its location mid-way between the Shetland Islands and Norway. However, by the time of its discovery in 1971, the development of Brent was a formidable, but not impossible, technical challenge. This encapsulates the way in which advances in the technology of both offshore exploration and production have progressively extended the frontiers of commercial exploitation in two general directions – into deeper water and more extreme climatic conditions (Mangone, 1983).

Most of the cumulative production of offshore oil has been derived from near-shore fields in relatively shallow water. As late as 1949, more than 50 years after the first offshore wells were drilled from piers off

the coast of southern California, there was only one mobile drilling rig in existence. This was able to operate in waters up to 6 m deep. The pace of technical change accelerated after 1950 as different approaches to marine drilling were introduced (Box 10.3). Converting a discovery into a producing field creates a further set of problems. Whereas exploration drilling is a temporary activity, production requires investment in facilities to be used throughout the life of a field which may exceed 30 years. These facilities usually take the form of massive steel or concrete structures placed on the seabed and topped by platforms above the water surface. More recently, advances in the design of remotely-controlled submersible vehicles and deep-diving techniques have seen the introduction of sub-sea production facilities which reduce or remove the need for conventional platforms. By the early 1990s, the initial, hesitant steps of the oil industry into the marine environment which began more than 100 years earlier had led to production from fields in water depths approaching 350 m, well beyond the original 200 m legal definition of the continental shelf. Indeed new fields at more than 850 m will be commissioned in 1995 and 1996.

The drive into deeper water has been associated with a related move into areas experiencing more difficult surface conditions. Average and extreme wind velocities and wave heights in the North Sea, for example, are significantly greater than in the Gulf of Mexico. High-latitude environments pose particular problems, but many of the promising offshore prospects are in Arctic and sub-Arctic regions such as the Barents and Beaufort Seas. In the latter case, it is the continuing US search for new sources of domestic oil off the coast of Alaska which is responsible for the high levels of activity. Indeed this motivation has stimulated the development of technology to cope with such environments in much the same way that the US oil industry was drawn into the new offshore challenge of the Gulf of Mexico 40 years earlier. Ice is the principal difficulty associated with offshore exploration and production in high latitudes. Structures must be capable of withstanding the enormous pressures generated by ice in all its various forms (i.e. pack ice, ice floes, icebergs and land-fast ice) and numerous innovative techniques have been developed as a result of experience in the Beaufort Sea. It is also necessary to provide suitable living conditions for oil workers and it is self-evident that the cost of operations in such environments is very high.

It is important to qualify any impression that the exploitation of offshore oil and gas has involved an

Table 10.3: World offshore oil production 1972, 1982 and 1992

Area/Country	1972		1982		1992	
	(000 b/d)[a]	%	(000 b/d)	%	(000 b/d)	%
Middle East	3 209	33.9	4 102	30.4	4 141	22.8
inc. Saudi Arabia	(1 491)	(15.8)	(2 392)	(17.7)	(1 580)	(8.6)
North America	1 665	17.6	1 110	8.2	719	3.9
inc. US	(1 665)	(17.6)	(1 110)	(8.2)	(709)	(3.9)
Latin America	3 111	33.0	3 035	22.4	3 333	18.2
inc. Venezuela	(2 886)	(30.5)	(1 026)	(7.6)	(858)	(4.7)
Mexico	(38)	(0.4)	(1 639)	(12.1)	(1 720)	(9.4)
Europe	43	0.4	2 687	19.8	3 640	19.9
inc. UK	(–)	(–)	(2 060)	(15.2)	(1 550)	(8.5)
Norway	(32)	(0.3)	(534)	(3.9)	(1 800)	(9.8)
Asia/Pacific	595	6.3	1 616	11.9	3 454	18.9
inc. Indonesia	(70)	(0.7)	(536)	(3.9)	(1 150)	(6.3)
Africa	598	6.3	815	6.0	2 363	12.9
inc. Nigeria	(409)	(4.3)	(371)	(2.7)	(1 230)	(6.7)
USSR[b]	236	2.5	177	1.3	621	3.4
Total offshore	9 457	100.0	13 542	100.0	18 271	100.0
Total world	57 710		57 090		64 920	
Offshore as % of world total		16.4		23.7		28.1

Sources: see Table 10.2.

Note: [a]Barrels per day

[b]Data for 1992 include Azerbaijan, Russian Federation, Turkmenistan and Ukraine

inexorable advance of the frontiers of the feasible. Certainly, the history of the oil industry illustrates the importance of spatial expansion as a response to resource scarcity, but the limits are not determined exclusively by technology. The costs of applying technology may exceed the perceived value of the resource. The volatility of oil prices during the last 20 years or so was noted in Chapter 7 and it has already been emphasized that the uncertainties for oil consumers created by the price increases in 1973 and 1978/79 encouraged the search for new sources of oil outside the Middle East. These uncertainties effectively increased the value of oil from the North Sea and other offshore areas. However, oil prices have tended to decline since the early 1980s (Fig. 7.10). This decline has had both immediate and longer-term impacts upon the offshore oil industry. As with other mineral resources, levels of exploration activity are extremely sensitive to changes in price and the slump to $12 per barrel in 1986, for example, prompted the 'mothballing' of many drilling rigs which could not find work. Over the longer term, the fact that oil demand has increased much less rapidly than anticipated 25 years ago, together with the

weakness of prices since the beginning of the 1980s, have meant that offshore production has increased much less rapidly than once seemed likely. Some observers suggested that it would account for as much as 50 per cent of total world oil production by the end of the century but the 28.1 per cent share in 1992 (Table 10.3) suggests that the proportion ultimately achieved will be much lower than this forecast.

Marine mineral resources

Fears of the impending exhaustion of various mineral resources have been shown to be exaggerated (see Chapter 8), but the strategic importance of certain minerals ensures that diversification of sources of supply is desirable for many importing countries, especially where a small number of producing countries account for the bulk of output (Table 8.3). In these circumstances, there is considerable interest in the possibility of recovering minerals from the seabed. However, the fact that the 'ownership' of such minerals is not established beyond the limits of

Box 10.3 Finding oil in deeper water

Figure 10.8 relates the development of under-water drilling techniques to increasing water depth. The earliest offshore wells were drilled from piers connected to the shore and were used to tap seaward extensions of established onshore fields. Barges were important in drilling the first shallow-water wells in the Gulf of Mexico. They contained compartments which were flooded until the vessel sat on the seabed. The technique was limited to depths below 12 m because the superstructure had to remain above water for drilling operations to be conducted. The next development was the self-elevating barge. This had tall legs on which the drilling platform could be jacked up to sit well clear of the water. These craft, which evolved from floating docks, were succeeded by the first purpose-built offshore drilling platforms in the early 1950s. The legs of these rigs can be moved up and down relative to the platform deck. With the legs jacked up, the rig can be towed into position where they are extended downwards to support the platform on the seabed. The early jack-up rigs were restricted to relatively shallow water, but their current operating limit is approximately 140 m. Semi-submersible rigs were introduced in the 1960s

and they had become the most widely used drilling platforms by the 1970s. Unlike submersible barges and jack-up rigs, they do not usually sit on the seabed and are therefore able to operate in very much deeper waters (although they are capable of functioning as bottom-supported platforms in shallow water). The vast majority of semi-submersible rigs are moored on a drilling location by a system of multiple anchors. The limitations of this method constrain the maximum operational water depth to approximately 650 m. However, this constraint had been removed for a few semi-submersible rigs fitted with 'dynamic-positioning' by the 1990s. Dynamic positioning involves computer-controlled thrusters which automatically compensate for sea-movement to maintain the rig on-station. This technology was originally developed for use in drill-ships which represent the ultimate in mobility and water depth capability. They are able to drill exploration wells in water depths exceeding 3500 m. This is well beyond the range at which oil can be commercially produced under existing circumstances and discoveries at such extreme depths must be regarded as potential future resources.

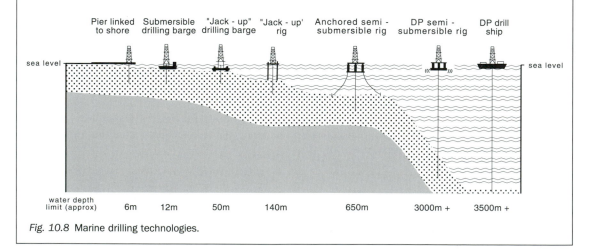

Fig. 10.8 Marine drilling technologies.

national EEZs creates major difficulties in determining the proper distribution of the economic benefits arising from their future exploitation.

Occurrence and economics

The recovery of minerals lying on or beneath the seabed is not new. For example, deep-mine workings

of coal and iron ore straddling the coastline have been and continue to be exploited in various locations. Sand and gravel for the construction industry are dredged from waters up to 35 m deep and alluvial deposits of numerous precious metals, including gold and platinum are known to exist below the high water mark. These have been mined in Alaska and deposits of tin have been exploited off the coast of Indonesia on a commercial scale since the 1960s. All of these examples share the common characteristics of occurring in shallow territorial waters relatively close to the shore. Any controversies usually relate to the environmental impact of such mining activities, while questions of ownership and access are not normally in dispute. Such questions are, however, central to the debate surrounding the prospects for the exploitation of minerals from the deep oceans well beyond the 200 nm reach of national EEZs.

Deep ocean minerals are divided into two categories which reflect differences in the circumstances of their formation. Ferromanganese nodules and crusts are formed by various processes linked to marine sedimentation and are, therefore, widely distributed on the ocean floors. Polymetallic sulphides are associated with hydro-thermal activity at the junctions of the oceanic plates and have a more restricted distribution. Knowledge of these minerals has been gained as a by-product of scientific research seeking to improve our understanding of the Earth's geological structure. Most of this research has occurred in the last 30 years or so, although the existence of marine nodules was known in the middle of the nineteenth century. The range of minerals contained in nodules and in the vicinity of active plate boundaries is considerable and includes manganese, copper, nickel and cobalt (Earney, 1990). Reserve estimates are highly speculative, but deep-ocean sources of copper and nickel, for example, are probably equivalent to current land-based reserves.

Despite the abundance of many strategic minerals beneath the deep oceans, these deposits are unlikely to be exploited in the foreseeable future. Feasibility studies have suggested that solutions can be found to the engineering problems posed by deep-sea mining, but the costs are prohibitive. A dramatic and sustained reversal of the secular downward trend in the real price of minerals (Fig. 8.4) will be necessary to offer any prospect of a commercial return on the massive capital investments that would be required. Paradoxically, improved knowledge of oceanic geology and metal genesis contributes, by analogy, to a better understanding of continental geology and mineral distributions. This in turn may inform the land-based exploration effort, further delaying the advent of deep-sea mining. Even more important than these economic and technical obstacles, however, are some fundamental political questions related to the distribution of any benefits generated by future deep-sea mining.

Ownership and exploitation

The progressive 'enclosure' of the oceans may have facilitated the management of marine resources by regulating access, but this process has worked to the benefit of some and to the detriment of others. Landlocked states in particular have argued that it represents the effective appropriation of resources that, at least in theory, were previously available to all. This view was expressed as early as 1967 in an influential speech by Dr Arved Pardo, Malta's Permanent Representative to the United Nations. He called upon the UN General Assembly to reserve all non-territorial resources as a 'common heritage of mankind' and to establish some form of international agency to act as trustee of this heritage. His initiative was a major stimulus to the convening of UNCLOS III (see p. 208). Although this Convention partially rejected Pardo's concerns about the morality of extending national jurisdictions further into the oceans by legitimizing the earlier unilateral actions of states claiming territorial control over resources to a distance of 200 nm, it nevertheless embraced the concept of a 'common heritage' beyond the EEZ by proposing the creation of an International Seabed Authority (ISA) to oversee any future exploitation of seabed mineral resources.

The proposed geographical responsibilities of ISA cover approximately 60 per cent of the seabed (i.e. the Area). UNCLOS III provided considerable detail on the structure of the new agency. It was envisaged that all formal signatories of UNCLOS III would be represented within the organization which would, therefore, bear some similarities to the United Nations itself. As well as administrative (i.e. a Secretariat) and executive (i.e. an Assembly and a Council) branches, the ISA would have an operational arm (i.e. the Enterprise) to conduct mining activities. A central feature of the proposed regime was the so-called 'parallel system'. This requires any organization wishing to mine within the Area to select two, not necessarily contiguous, sites judged to be of similar economic potential. The ISA chooses one of these sites to be reserved for its own activities; the other site may then be exploited by the applicant. This system is an attempt to ensure that any economic benefits of deep-sea mining are spread more widely within the international community than would otherwise occur under an open-access (i.e. free-for-all) regime. Implicit in this is a distinction between the economic interests of developing and developed countries, although this dichotomy is itself a gross over-simplification of a very complex situation.

Generally speaking, the creation of the ISA is supported by the developing countries because they recognize that the richer states are, by virtue of their greater financial and technical capabilities, in a much stronger position to exploit seabed mineral resources. They fear that the economic benefits from such operations will, therefore, be appropriated by the pioneer operators without regard to the common heritage concept of deep-sea resources (Leipziger and Mudge, 1976). In some cases, the concern has been heightened by a sense of economic injustice deriving from dissatisfaction with the revenues generated by conventional mining activities controlled by multi-nationals based in the developed countries (see Chapter 8). Another problem for countries heavily dependent upon mineral exports such as Zambia (see Table 8.6) is the possible effect of unwelcome competition upon already depressed prices.

The United States is the most vigorous opponent of the creation of the ISA and, like the United Kingdom and Germany, is a non-signatory of UNCLOS III. The arguments advanced by the United States for rejecting the proposals for regulating deep-sea mining are mirror-images of those deployed by most developing countries. As emphasized in Chapter 8, export dependence on minerals has its counterpart in import dependence. From a US perspective, the seabed offers the welcome prospect of an alternative source of supply for strategic minerals (Anderson, 1991). The ISA is also perceived as a form of collectivization of the seabed and is anathema to the capitalist ethic traditionally associated with the United States. Furthermore, it is argued that *any* exploitation of seabed minerals will be dependent upon expertise held by the developed countries, especially the United States, and that this reality is not acknowledged in the proposed arrangements which will devolve decision-making powers. The ISA will also act as a medium for the transfer of technology, a role which is resented by many mining companies which see no reason to surrender their commercial secrets.

The conflict over the exploitation of seabed minerals is unresolved. It may be regarded as the latest phase in a wider debate surrounding the regulation of access to and the distribution of benefits from the resources of the oceans. These questions are as old as human history in the context of land-based resources; their relevance to the marine environment is more recent. Disputes over marine fishing and navigation rights have flared up many times in the past, but much greater attention has been devoted to marine resource issues within the last 30 or 40 years. This reflects the increasing magnitude of human intervention and, in particular, the extension of the frontier of intensive marine resource exploitation away from the coastal zone. A parallel may be drawn between current trends affecting the world's oceans and the pioneer settlement of continental interiors. Colonial expansion in the Americas, for example, witnessed an outward spread from population centres initially established on or near the coast. Pressures on the world's seas and oceans are, similarly, greatest near to the coast, but are inexorably spreading outwards, as exemplified by the possibilities of deep-sea mining. These trends may be regarded as a positive demonstration of human ingenuity in widening the range of useful materials derived from the alien marine environment beyond the living resources which have traditionally been the focus of attention. On the other hand, there is evidence that the pressure on these latter resources is approaching or, in some cases, has exceeded, the limits of sustainability. Such evidence highlights the need to devise more effective systems for the management of ocean resources. This need will become more urgent as humankind pays greater attention to the resource possibilities presented by the vast areas of the planet covered by water.

Further reading

Coull, J.R. (1993) *World fisheries resources*. London: Routledge.

Earney, F.C.F. (1990) *Marine mineral resources*. London: Routledge.

FAO (1992) Marine fisheries and the law of the sea: a decade of change. In FAO, *The state of food and agriculture*. Rome: FAO, pp. 129–194.

CHAPTER 11

Environmental wealth and limits to growth

Throughout history, humankind has sought wealth and has feared famine. A land flowing with milk and honey has been the dream of many and the reality of few – at least until recently. Paradoxically, the modern age has witnessed unprecedented growth in both population and prosperity. More people are now better fed than at any time in history. Nevertheless, the essential insecurity of the human race is reflected in the chronic *angst* that is displayed about the continuing adequacy of environmental resources, including basic *needs* of food and shelter as well as *wants* of energy, minerals and other resources. And it is widely assumed that a rich resource endowment, at the national level, will lead to national prosperity – an assumption that has led to many wars and seizures of territory.

In this chapter, the relationship between national wealth and environmental resources is first considered. Attention is then turned to the question of limits to growth imposed by environmental resources and their availability.

Environmental resources and economic development

Intuitively, we expect that a land rich in environmental resources will enjoy a higher level of economic development and a greater level of prosperity than one that is poorly endowed. In detail, we may argue about the appropriate definition and measures of prosperity or wealth, and we may also debate the best measures of resource endowment. But in general we expect that a large country with fertile, well-watered land, extensive forests, and rich deposits of fossil fuels and minerals will enjoy a higher level of prosperity than one which is less richly endowed with these

resources, or which carries a larger population among which the resources are shared. The most cursory glance at the global scene today, however, is sufficient to convince us that the relationship between resource endowment and economic development is far from simple. The small, densely populated and mountainous land of Japan, for example, enjoys a level of prosperity (as measured by conventional indicators such as GNP per capita) far higher than that of the large and lightly peopled former Soviet Union. It also has a higher per capita income than Australia, although by most measures (e.g. land area per capita, minerals) Australia is more richly endowed. Similarly, Singapore and Hong Kong have achieved higher levels of economic development than, for example, Zaire or Zambia. Yet the former are by any standard poor in environmental resources, while the latter are rich. At the very least, these examples show that economic development does not depend on environmental resources alone.

Adam Smith and the *Wealth of Nations*

A milestone in the development of thought about the relationship between resources and economic development was the publication in 1776 of Adam Smith's *Inquiry into the Nature and Causes of the Wealth of Nations*. National wealth, in Smith's view, depended on how the labour force was deployed. At one level, he clearly recognized the fundamental significance of resources: 'Men', he concluded, '…naturally multiply in proportion to the means of their subsistence' (1776, 1970 edn: 250). And in the section in the *Wealth of Nations* entitled 'Of the Different Progress of Opulence in Different Nations', he referred to the key role of land resources in making possible the growth of cities, which in turn made possible the growth of specialized manufacturing and of interna-

Plate 11.1 Adam Smith. (*Source:* Illustration courtesy of the Mary Evans Picture Library.)

tional trade: 'It is the surplus produce of the country only ... that constitutes the subsistence of the town' (p. 480). He recognized that land rent varied with both location and fertility. Environmental resources, in short, were not ignored by Smith, as is sometimes assumed, but were rather accorded an important role. That role, however, was indirect, and *enabled* rather than *caused* economic growth . He noted that 'the most opulent nations ... generally excel all their neighbours in agriculture as well as in manufactures'; but went on to observe that 'they are commonly more distinguished by their superiority in the latter than in the former' (p. 110). The distinguishing features were less related to direct resource utilization (of land by agriculture) than to industrial activity. The crucial factors were the skill with which labour was applied and the relative proportions of those engaged in useful labour and of those not so employed. In essence the key issue was the division of labour, which could be achieved more fully in manufactures than in agriculture or in other forms of direct resource use. The key to national wealth therefore lay in the labour force rather than in environmental resources. Or as expressed in the more recent words of Max Singer

(1989: 23), 'Economic growth comes from people learning to work better'.

In Smith's day, the empirical evidence seemed to support his argument. In England, a higher level of economic development had been achieved than in France, although France had, by many measures, a richer endowment of natural resources. Indeed other commentators have argued that resource scarcity in England was the stimulus for the technical advances that heralded the industrial age (Chapter 3). And while Smith was aware of the theoretical role of the environment as a limiting factor (for example on population growth), in practical terms no limits were encountered. The world economy was expanding, and imports from overseas could be substituted for supplies of scarce resources from home. It is not surprising that he should see social factors rather than environmental factors as the key variables underlying national wealth.

The *real* wealth of nations?

The view that labour – and not land – was the primary source of value and wealth was not unique to Adam Smith. The other great inspirer of modern political philosophy, Karl Marx, held a similar view, reached on a very different basis and from a very different direction. In recent decades, however, a basic questioning has been directed at such views, and alternative outlooks have been developed in which very much greater emphasis is placed on resource endowment. For example in S.R. Eyre's *Real Wealth of Nations* (1978), it is noted that the population had increased ten-fold since Smith's day. Conditions in Britain, and indeed in the rest of the world, were now so different that it was argued that the old conventional wisdom no longer applied. Eyre proceeded to rank the countries of the world on the basis of their biological productivity and mineral resources related to size of population. This ranking is in stark contrast to that resulting from more conventional rankings of wealth (in terms of GNP per capita). At the top are countries such as French Guiana, Gabon, Surinam and Congo – lightly peopled countries with high rates of biological productivity (in terms of increments of biomass per unit area and time) and in some cases rich mineral resources. At the very bottom are densely populated island states such as Singapore, and in very lowly positions are European countries such as Germany, the Netherlands, the United Kingdom and Switzerland, as well as Japan (Fig. 11.1). While Eyre did not

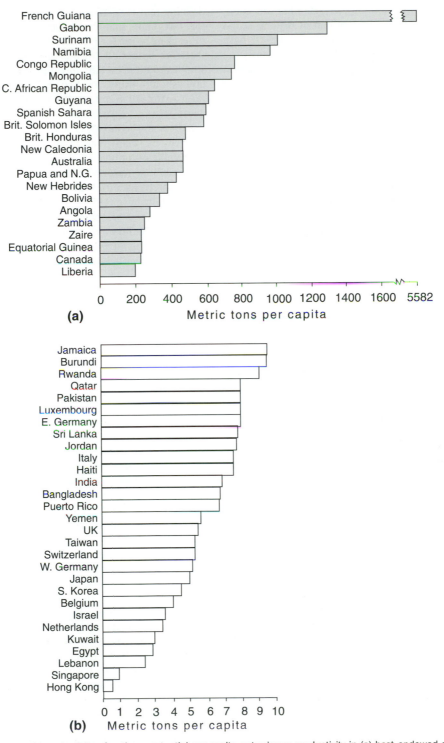

(a) Metric tons per capita

(b) Metric tons per capita

Fig. 11.1 The organic productivity of nations: potential per capita net primary productivity in (a) best endowed and (b) most poorly endowed countries. Note different scale in (a) and (b). (*Source*: modified after Eyre, 1978.)

argue that the population of, say, Zambia was about to enjoy the material wealth of that of Britain, he did highlight the fact that the centres of wealth have changed throughout history. More significantly, he also contended that primary-producer nations (i.e. in general those richly endowed with natural resources) were occupying increasingly strong positions in the modern world. His conclusion was that environmental resources were growing in significance in relation to national well-being, and that however significant the division of labour had been in this respect in the past, its role was fading in an age of large and rapidly growing population. At around the same time, Chisholm (1980) also concluded that the significance of resources in relation to the wealth of nations was growing. (The significance of the trends of the 1970s, which provided the backgrounds for both Eyre and Chisholm, is discussed later in this chapter.)

Eyre's thesis gives rise to many problems. At one level, human wants extend far beyond biological productivity and mineral resources. The extent to which humans value biomass and ore *per se* is strictly limited. Wealth and well-being imply the availability of a much wider range of environmental resources than these crude indicators can represent. At another level, the logical basis of the thesis is that of environmental determinism – the view that human activity is determined by the nature of the physical environment. The scope for human ingenuity to 'add value' to a limited resource endowment is at best strictly limited. Furthermore, as is discussed later in this chapter, the evidence that environmental resources are becoming scarcer and that they will soon impose limits on growth is by no means conclusive. While Eyre may have provided a useful service in questioning the relative roles of labour and land in relation to national well-being, to claim that the 'real' wealth of nations lies in their organic productivity and mineral resources is at the very least an over-statement of the case. It remains, however, to attempt to clarify the role and contribution of environmental resources to economic development in a manner which avoids sweeping generalizations and grand assumptions, and which pays closer attention to empirical evidence over the last few decades.

Environmental resources and stages of economic development

Perhaps the explanation for the continuing uncertainty about this role lies ultimately in the difficulty of defining and measuring environmental resources on the one hand and economic development on the other. The scope for rigorous quantitative analysis is therefore limited. A further problem is the unit of study: should it be the individual country, and if so, how should trade, trading blocs and empires be regarded? Like many other fundamental but thorny questions, the relationship between environmental resources and economic development has been largely ignored by many researchers, in favour of less intractable questions.

One commentator who has considered the relationship between resources and economic development is Norton Ginsburg (1957). He concludes that to assist in economic development, resources need not lie within a country but they must be accessible. Accessibility in turn implies transport and imports, both of which require capital. One means of accumulating such capital is through exploiting resources within a country, and he quotes examples such as Saudi Arabia and Venezuela. The role of resource endowment is therefore greatest, he argues, in the earlier stages of economic development, when it can act as a source of initial capital. When a country becomes more highly developed, that role fades. Other workers have referred to the role of rich resource endowments in facilitating the rapid early growth of the United States and Australia, and to their declining role subsequently over the long term. In short, the relationship changes through time and with level of economic development. If this is so, it is therefore not surprising that there should be no close correlation between national wealth and resource endowment in the world as a whole. (In later work, Ginsburg and colleagues (1986) found that the highest correlation between per capita GNP and resource variables was with 'non-renewable energy potential', or per capita availability of fossil fuels ($r = 0.624$). The fact that correlation coefficients were lower for other resource variables is a reminder that much may depend on the type of resources, and not just on their *quantity*.)

General conclusions similar to those of Ginsburg's earlier work were reached by Chenery (1965). For the world as a whole, he found that income levels were not strongly correlated with resource endowments, although these endowments appeared to be relevant in explaining differences within groups of countries having similar historical and cultural backgrounds. In his view, natural resources become less important as determinants of level of income as income levels rise. In low-income countries, rich resources may be of

Table 11.1 Economic growth rates in mineral-rich and other developing countries

	Hard-mineral exporters		Oil exporters		Other middle-income countries		Other low-income countries	
	1960–71	1971–83	1960–71	1971–83	1960–71	1971–83	1960–71	1971–83
Mean annual growth of per capita GDP (%)	2.5	−1.0	2.9	1.9	3.7	2.0	1.3	0.7
Number of countries	10	10	10	10	29	29	20	20

Source: compiled from Auty (1993).

great significance in terms of exports and foreign investment. Resource products often account for much larger shares of the exports of such countries than of more highly developed lands. (Food and other resource exports may have allowed some countries in the past to accumulate sufficient capital to allow development to 'take off'. Today, many low-income countries face serious difficulties in finding markets for food in the developed world, because of the protectionist policies pursued in many developed countries and because of the subsidized food exported by such lands (Chapter 4).)

More recently, the conventional wisdom that a rich resource endowment may be beneficial to countries at low and mid-income levels of development has been questioned, notably by Auty (1993). He concludes that the performance of mineral-rich developing countries since the 1960s has been worse – not better – than that of less well-endowed countries (Table 11.1). From this he has developed the 'resource-curse' thesis. In essence, his argument is that a rapid inflow of mineral revenues into an economy results in a strengthening of the exchange rate. This can make agriculture and manufacturing uncompetitive internationally. At the same time the 'enclave' tendency of much mining activity means that few linkages are formed and few benefits accrue to the domestic economy (Chapter 8).

If the relationship between resource endowment and economic development at the national scale is complex, the same is also true at the regional level. As Chapter 8 indicates, areas such as coalfields have frequently given rise to major urban–industrial complexes, which in turn had their own impact on resources such as water and agricultural land. Rapid growth around resource-rich areas has been less common in recent decades. Costs of long-distance bulk transport have fallen, and many areas of exploitation of both oil and minerals have been in remote and inhospitable areas. In many instances it has been easier to ship resource products to existing urban and industrial centres than to develop major new centres in the areas of production. At the regional level, therefore, the role of resource endowment in relation to economic development may be weaker at present than it has been in the past.

Most attempts to investigate the relationship between resources and economic development have focused on fossil fuels, minerals, and other physical resources. In comparison, the role of amenity resources has been largely neglected. By the second half of the twentieth century, however, that role has assumed growing importance, especially at the local and regional levels. At the local level, amenity resources can be enhanced by landscaping and other environmental improvements. Regionally, areas with favourable climates or attractive mountain or coastal settings have often, in recent decades, been associated with rapid growth. In the United States, for example, first California and Florida and then other parts of the 'Sun Belt' of the south-west have witnessed more rapid rates of growth in population and economy than most other parts of the country, although other areas may be better endowed with land and other physical resources. Transport improvements have meant a new freedom for economic activities to locate in environmentally attractive areas, although such areas may be poor in terms of physical resources. As Ullman (1954) concludes, 'For the first time in the world's history pleasant living conditions – amenities – instead of more narrowly defined economic advantages are becoming the sparks that generate significant population increase' (p. 119) and '...the climate of California and Florida takes its place as a population magnet along with the coal of Pittsburgh and the soil of Iowa' (p.131).

With urban growth in such areas, the age-old influence of town on surrounding countryside comes

Plate 11.2 The Mediterranean coast: favourable climate and other amenity resources, combined with cheap air transport from northern Europe, have provided the setting for rapid economic growth based on tourism. (*Source:* K. Chapman.)

into play (Box 3.1). New demands for local resources arise, and increasing stress may be placed on physical resources. Areas that are rich in amenity resources are not necessarily well endowed with material or physical resources. One obvious instance is the inverse relationship between a dry climate and the availability of water resources (Chapter 9). Another, at the wider scale, is that of mountains, now almost universally perceived as positive in terms of amenity, yet frequently poor in land and other material resources. These examples epitomize the growing general value attached to amenity resources *vis à vis* physical resources (Chapter 2). On parts of the Mediterranean coast, for example, amenity resources have provided the setting for rapid economic growth based on tourism, in the same way as the coalfields of Western Europe became the setting for rapid industrial-based growth in the nineteenth century. (And environmental impoverishment has resulted in some areas, as it did on the old industrial coalfields.) In other words, amenity resources can have an influence on patterns and rates of economic development that is similar to that of energy and mineral resources in the past.

Thus the influence of resource endowment on economic development may depend both on the type of resource and the stage of development, at both the national and regional levels. It is possible that the influence has changed over the long term, not least as a result of changes in the real cost of bulk transport. For example Maddison (1964) suggests that superior natural resources gave the United States a great advantage over Europe during the late nineteenth and early twentieth centuries. Since then, however, Europe has gained easier access to resources (and especially to energy resources) as a result of improvements in transport. He goes on to conclude that the significance of natural resources as an influence on the rate and pattern of economic development has therefore lessened overall, as well as in individual countries.

In detail the nature of the relationship between resource endowment and economic development is complex. No clear, straightforward relationship between resource endowment and economic development is apparent at the global scale. Both resource-rich and resource-poor countries rank among the richest and poorest in terms of GNP per capita. The relationship may depend on the stage of development and on the type of resources. Overall, however, the influence of resource endowment (at least in the traditional interpretation) is probably weakening.

This conclusion is based not only on Maddison's suggestion, but also on other types of evidence.

The dwindling economic significance of resource endowment?

Beyond the 'Era of Materials'?

In Chapter 7, attention is focused on the fact that the amount of energy consumed per unit growth in GNP has fallen in many developed countries. At early stages of development, large amounts of energy are consumed per unit output, but these amounts decrease as development proceeds. The relationship changes because of a combination of factors: increasing efficiency of use; changing industrial structure (as the emphasis shifts from heavy, energy-intensive industries to a wider range of diverse industries); and changing economic structure, as service industries grow more rapidly than manufacturing industries. The availability of large amounts of energy – and hence of energy resources – may therefore not have a determining effect on economic development.

Similar trends have been detected in the use of various basic materials in the developed world over the last two decades. For example in the United States the consumption of steel per unit of GNP (kg per 1000 (1958) dollars) had by the 1980s fallen to only 40 per cent of the peak it reached in 1920 (Larson *et al.*, 1986). Consumption per capita had also ceased to grow. Similar, if less dramatic, trends have been detected in several other bulk materials: consumption has stabilized or declined while economic growth has continued. On the basis of their examination of these trends, Larson *et al.*, (1986) have concluded that developed-world countries have reached a turning point. They are leaving the 'Era of Materials', which has spanned the period since the Industrial Revolution, and are now entering a new era in which the level of use of bulk resource products will no longer be an important indicator of economic progress. Larson and his colleagues conclude that four causes underlie this trend. One is the substitution of one material for another, which has slowed the growth of demand for individual materials. Another is design changes and increases in the efficiency with which materials are used. A third is that the markets which grew rapidly during the Era of Materials are now approaching saturation, for example fertilizers in the developed world (Chapter 4). Finally, new markets tend to involve products (e.g. electronic goods) that have

relatively small contents of materials. In short, the old relationship between resources and economic growth may have changed or be changing.

Scarcity and Growth

Eyre, Chisholm and numerous other commentators during the 1970s and early 1980s considered that the relationship between resources and economic development was changing: they perceived that resources were becoming scarcer. Intuitively, we would expect that scarcity would increase as resources are utilized. As Chapter 1 indicates, however, the picture is complicated by the fact that different kinds of scarcities can occur. The voluminous literature that has been generated on the subject of resource scarcity is witness to its complicated and controversial nature.

One of the major contributions to the debate was that of Barnett and Morse (1963). Their influential volume entitled *Scarcity and Growth* had as its primary concern the economic concept of increasing resource scarcity, and had as its background the fears expressed by successive American presidents in the early post-war period that their country was facing resource scarcity. The President's Materials Policy Commission (the Paley Commission) concluded in 1952 that over the next 25 years there would not be general exhaustion of resources, but it expressed concern about price trends and about some minerals. The research foundation Resources for the Future was established as a result, and it was under its auspices that Barnett and Morse undertook their work. They quickly concluded that the notion of absolute scarcity was untenable since technological progress means many resource definitions change over time, i.e. they ruled out Malthusian scarcity (Chapter 1). They first considered their 'strong hypothesis' – that the scarcity of resources, as measured by their real cost, increases over time. They concluded that this hypothesis was not supported by the evidence, except perhaps in the forest sector (Table 11.2(a)). In general the unit costs of resource products in the United States had shown a decline from the late nineteenth century onwards. They next examined their 'weak hypothesis' of increasing scarcity, which suggested that costs of resource products did not fall as much as those of non-extractive products. Again, this hypothesis was not supported by the time-series data on costs, except in the case of forest products. Relative costs of minerals were found to have fallen, while those of agricultural goods had remained constant (Barnett and Morse, 1963). They concluded that for the United States

Table 11.2 Trends in (a) real unit costs and in (b) real unit costs relative to the non-extractive sector of the economy (1929 = 100)

	Total extractive	Agriculture	Minerals	Forestry
(a)				
1870–1900	134	132	210	59
1919	122	114	164	106
1957	60	61	47	90
(b)				
1870–1900	99	97	154	37
1919	103	97	139	84
1957	87	89	68	130

Source: Barnett and Morse (1963).

during the period between the Civil War and 1957, these hypotheses of increasing economic scarcity were not in general supported. (Forest resources, which had played, such an important role in the first Conservation Movement (Chapter 3) were perhaps a special case.) The main reasons, in their view, were that economically more plentiful resources were being substituted for less plentiful ones; increased domestic discoveries and imports of minerals; and advances in knowledge and technology.

Some 15 years after *Scarcity and Growth* was published, Barnett. (1979) 'revisited' the work, extending his analyses of US statistics to 1970 and examining data for other countries. He concluded that the trend of declining real costs had continued. Support for neither the strong nor the weak hypothesis was found among US data. For other countries, no support was found for the strong hypothesis, and there was only limited evidence in favour of the weak hypothesis. A few years later, Hall and Hall (1984) attempted to update Barnett's work, and concluded that increasing scarcity was now displayed by energy as well as by forest products. In the case of the former, they concluded that the OPEC embargo was only partly responsible, as they considered that an underlying trend of increasing economic scarcity was already evident before it was imposed.

The influential general conclusions of Barnett and Morse have not gone unchallenged. Smith (1978), for example, concludes on the basis of analysis of comparable data extended to 1973 that there were different elements present in the price trends at different periods and that judgements on whether long-term trends of resource scarcity existed could not be made on these data alone. Even if this criticism

boils down to one of time scale, Smith's other major point remains: the resources examined by Barnett and Morse were all products for which markets existed. Their analyses say nothing about non-market resources, and hence general conclusions about resource scarcity cannot be reached. (Barnett and Morse did refer to environmental quality and deterioration in quality of life as areas of concern, but regarded them as separate from marketed resource products.) Various other criticisms have been raised by other commentators. For example Hall and Hall (1984) pointed out that resource-price data are affected by government intervention in markets. The data have also failed to reflect damage caused by pollution and other means on common-property resources such as rivers and the atmosphere. In other words, Barnett and Morse's data understate the real costs of resources, and hence confident conclusions about trends cannot be reached from their analyses.

The debate over *Scarcity and Growth* epitomizes the wider controversy that surrounds the question of whether resources are becoming scarcer or more abundant. This controversy has continued, off and on, ever since the days of Malthus. During the 1960s it re-emerged strongly, and in the 1970s in particular it reached a new peak of intensity. Ironically, at around the same time as Barnett and Morse were concluding that there was no evidence that resource products were becoming scarcer (at least in the United States), some other commentators in the developed world were warning that environmental limits to growth were looming. Such views were both a product of and a stimulus for the burgeoning Environmental Movement (Chapter 3), and they were fuelled by the accelerating rates of population growth and of consumption of resource products that characterized the third quarter of the twentieth century.

Resource shortages and limits to growth

Scarcity rather than abundance has been the lot of humankind throughout history. It is ironic that predictions of resource shortages should have become more explicit during the last two centuries, when humankind (or at least that part of it living in the developed world) has enjoyed unprecedented abundance.

Numerous examples could be quoted of specific

concerns being expressed in the decades following Malthus. Some are limited to particular issues and sectors, while others extend to more basic questions of lifestyle and food supply. In the mid-nineteenth century, for example, Jevons (1866) considered that the limited availability of coal would prevent Britain from maintaining its rate of industrial growth. British growth trends may have been affected by numerous factors since then, but a shortage of coal does not rank significantly among them. Similarly, at the end of the the century Crookes concluded that the finite nature of nitrate deposits would impose a limit on the growth of food production (Chapter 4). It is easy to dismiss such predictions, based as they are on the assumptions that resources can be defined in absolute terms as simple stocks of materials, with the possibility of technological change being disregarded. But such assumptions die hard, and similar kinds of predictions have continued to pour forth in recent decades. Two of the most notable of these are *Blueprint for Survival* and *Limits to Growth*. Another publication of the same genre – *Global 2000* – was based on a study commissioned by President Jimmy Carter in the later stages of the Environmental Movement of the 1970s (Chapter 3).

Blueprint for Survival

In 1972 *The Ecologist* magazine published *A Blueprint for Survival*. Later published in book form, the *Blueprint* was a lengthy statement from a group of eminent British scientists (including S.R. Eyre), acting on their perception of contemporary trends in the global use of resources and related environment problems. They considered that if contemporary trends were allowed to continue, 'the breakdown of society and the irreversible disruption of the life-support systems on this planet, possibly by the end of the century, certainly within the lifetimes of our children, are inevitable' (Goldsmith *et al.*, 1972: 9). They believed that the governments of the day were refusing to face facts or attempting to play down the seriousness of the situation, as they saw it, and effective correctional measures were not being undertaken. Essentially what was being advocated was a 'new philosophy of life', and a programme for bringing about the radical societal change necessary to ensure the survival of the environment.

One of the contemporary trends on which *Blueprint* was based was that of metals. The authors concluded that of 16 major metals, the reserves of ten would be exhausted within 100 years if contemporary rates of

extraction continued, and those of all but two would be worked out if the exponential growth rates in operation since 1960 were to continue. Like many gloomy predictions of earlier periods, this one has failed to materialize (Chapter 8). Consumption growth rates of the 1960s have not been sustained, and new discoveries of ores have been made. The simple projection of contemporary trends has proved once more to be fragile basis on which to make forecasts.

Like many other predictions, *Blueprint* advocated radical social change. The goal was to stabilize population and demand for resource products, and the optimal way of achieving this goal, in the view of the authors, was through the creation of a 'satisfactory social environment' (p. 62). The aim of the social engineering required to create such an environment was to establish decentralized, self-sufficient communities, in which people worked near their homes and assumed responsibility for running services such as schools and hospitals.

While there is no doubt that *Blueprint* made a strong impact at the time of its publication, its longer-term effect is difficult to assess. At one level it is easy to conclude that its authors were simply wrong. Their predictions about the fate of metal resources, for example, now seem as ill-founded as those of (for example) Jevons and Crookes. On the other hand, its publication at the very least encouraged more radical questioning of societal structures, goals and lifestyles. And it is of interest in conveying the general message that purely technical solutions, such as the seeking of increased yields of crops and better prospecting methods for minerals, would not avert future problems. The fact that such conclusions were reached by (mainly biological) scientists lent further strength to the message.

Limits to Growth

The authors of *Blueprint* were strongly influenced by the computer modelling pioneered by Forrester (1971) at the Massachusetts Institute of Technology and developed further by Meadows *et al.* (1972). This work, *Limits to Growth (LTG)*, became an international best-seller (a status rarely achieved by the results of computer modelling), achieving world-wide sales of millions of copies. Five major global trends were examined: accelerating industrialization, rapid population growth, widespread malnutrition, depletion of non-renewable resources, and a deteriorating environment. A key theme was that of exponential

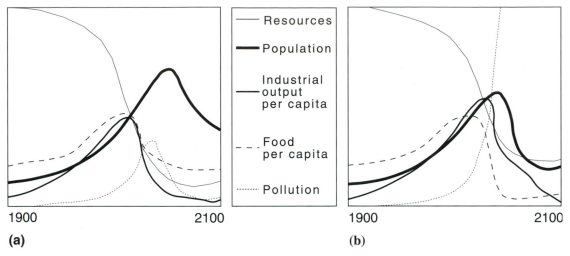

Fig. 11.2 Limits to growth: world model (a) standard run (b) natural resource reserves doubled. (*Source:* Meadows *et al.*, 1972.)

growth and its invidious effects. The availability of powerful computers allowed formal mathematical models of these trends and their interactions to be constructed and run in a way that had previously been impossible. In the 'standard' world model, run on the assumption of the continuation of contemporary trends, industrial production grew exponentially (like food production and population), but was eventually retarded by a rapidly diminishing resource base. Because of time lags and delays in the system, population and pollution continued to grow for some time after the peak of industrialization. Population growth, however, was finally halted by declining availability of food. In short, a scenario of overshoot and collapse was drawn (Fig. 11.2(a)), with collapse occurring in the mid-twenty-first century. The immediate cause was depletion of non-renewable resources, but a similar result occurred even if the amount of available resources were doubled (Fig. 11.2(b)). In this case industrialization could proceed to a higher level, but if this happened then pollution increased. As a consequence of increased pollution and declining availability of food, death rates increase, giving rise to rapid population decline. In other words, the eventual result was the same although the causal mechanisms were different. 'Overshoot and collapse', however, were not seen as inevitable. If growth rates stabilized, they could be averted. Like *Blueprint*, *LTG* saw an urgent need for changes to population growth rates and to rates and patterns of economic growth.

Much criticism was levelled at the exercise: the use

of over-simplified world-average figures for the selected variables; the highly aggregated scale of the model; the over-simplified set of interactions identified between the key variables. These and other criticisms are conveniently reviewed by McCutcheon (1979), among numerous other commentators (including, notably, Cole *et al.*, 1973). Economists were among the most scathing critics. Some of their rejoinders were curious amalgams of vitriolic dismissal and basic questioning. For example Beckerman (1972) regarded *LTG* as 'such a brazen, impudent piece of nonsense that nobody could possibly take it seriously' (p. 327) before going on to ask 'why it should matter all that much whether we do run out of some raw material' (p. 337). The main grounds for criticism were the scant reference to economic theory and to empirical data. The flavour of this type of criticism is conveyed by the title of a paper by Nordhaus (1973) 'World dynamics: measurement without data'. The fact remains, however, that *LTG* made a huge impact in its own right, and stimulated the construction of many other more sophisticated and less highly aggregated models. Recently, the *LTG* team has repeated the exercise, with similar results (Box 11.1).

Global 2000

In 1977, after the effects of the OPEC oil embargo and images of famine in the Sahel had reinforced the message of *LTG*, President Jimmy Carter ordered a major study of projected trends in global population,

---

Box 11.1 Beyond the limits

In 1992 a 'sequel to the international bestseller *The Limits to Growth*' was published. Entitled *Beyond the Limits*, it was written by three of the four authors of the 1972 volume. They repeated in 1991 the computer modelling that formed the basis of the earlier book. In essence, the conclusions were similar. In the earlier volume it was concluded that the physical limits to human use or materials and energy were some decades ahead. In the update, it was concluded that 'many resource and pollution flows had grown beyond their sustainable limits' (Meadows *et al.*, 1992: xiv). Unless the flows of materials and energy are reduced, uncontrolled decline will occur in future decades in per capita food production, energy use and industrial production.

In the update, a combination of limits operates. Just after the year 2000 in the computer simulation, pollution begins to have a serious effect on land fertility, and food production begins to fall after 2015. Also the combination of population growth and growth in industry means that as many non-renewable resources are consumed in the first two decades of the twenty-first century as in the whole of the twentieth century. More capital and energy are required to utilize remaining resources. The need to shift more investment into agriculture and non-renewable resources in order to maintain levels of production means that industrial capital plant declines. 'Industrialized' agriculture suffers as a consequence. Eventually population also begins to decline in the second half of the century, as a result of decline in food supply and medical services. As in the 1972 version, they then repeated the computer run with a doubling of the natural-resource endowment assumed for the first scenario. Industrial growth continues for an extra 20 years, but again more pollution ultimately leads to a decline in food production and hence to a rising death rate. In short the end result is the same as in the first scenario apart from a slightly later timing.

The authors are careful to emphasize that their scenarios are not predictions, and that there is still time for the outcomes of decline to be averted. They (like many others: see Chapter 12) urge that we begin to make the transition to sustainability, acknowledging that radical changes would be required in modern society if this were to be achieved.

---

resources and environment through to the end of the century. This study was to serve as the foundation for long-term government planning. The outcome of the trend projections on which the study was based was gloomy though not surprising: by 2000 the world would be more crowded, more polluted and less stable ecologically. Severe stresses would occur in the resource sectors of agricultural land, forests, water and biodiversity in particular. By 2100, the world population could be as much as 30 billion, close to estimates of maximum carrying capacity (Barney, 1980).

The authors of *Global 2000* were careful to emphasize that they were dealing with projections rather than predictions: their conclusions would come to pass only if contemporary policies continued. They noted that these policies were already beginning to change, but concluded that more rapid and radical change would be required if gloomy outcomes were to be avoided. In particular, they emphasized the need for strengthened international cooperation to deal with problems of population, resources and the environment. The United States as the world's largest economy could expect its policies to have a major influence on global trends, but it should also cooperate 'generously and justly' with other nations to seek solutions to global problems, and especially to those involving the global commons.

Like *Blueprint* and *LTG*, *Global 2000* made a big impact at the time of its publication. 'Official' confirmation was now available for the conclusions previously reached only by individuals and private bodies. In his farewell address to the nation, President Carter referred to its subject-matter as one of the three most important problems facing the American people (the others being arms control and human rights). The long-term impact, however, is another matter. A change of President did not help in this respect, and indeed the report is sometimes dismissed as having been 'commissioned by Jimmy Carter and ignored by Ronald Reagan'. Also like the early studies, it was essentially Malthusian in tone.

These three studies were at the same time products of the Environmental Movement and stimuli for it. Events of the 1970s seemed at the time to vindicate the pessimists and doom-mongers about resource futures, drawing attention as they did to a rapidly growing world population and rapidly escalating demand for resource products during the third quarter of the twentiethh century. The assumption was widespread that Spaceship Earth (Chapter 3) was a finite and limited storehouse of resources, as was the belief that trends of increasing demand for resources would soon meet a rigid ceiling of potential supply. By the 1980s, however, many of the concerns were fading. By then it had become apparent that the shortages of oil and food that had been experienced in some parts of the world during the 1970s had not necessarily been the portents of permanent scarcity. The upheavals that followed the oil price rises of the 1970s dampened economic growth, and marked the end of the era of rapid growth in consumption that had characterized many resource products (and especially oil and metals) between the Second World War and the early 1970s. Initially, it probably seemed that as soon as the economic recession was over, trends in resource consumption would soon revert to their former levels. It gradually became apparent, however, that longer-term changes were occurring. As Chapter 7 shows for energy, consumption growth rates have not yet reverted to their former levels. Fundamental changes appear to have occurred in the relationship between economic growth and growth in energy consumption. And as is suggested earlier in this chapter, per capita consumption of many bulk resource products has, in the developed world, levelled off or declined. Furthermore, even if we are not 'beyond the era of materials', there is more widespread awareness of the possibilities of resource substitution and of improvements in the efficiency of resource use.

This does not mean that concern with environmental problems has evaporated, but rather that it has undergone a change of focus. Today, concern is less about the adequacy of material resource products such as food, wood and metals, and more with issues such as biodiversity and in particular with the maintenance of global life-support systems. Issues relating to environmental 'waste-sink' functions (such as global warming and marine pollution) now attract as much attention as the adequacy of global food and energy supplies did during the 1970s (Box 11.1).

Population and technology

Publications such as *Blueprint* and *Limits to Growth* publicized the question of resource adequacy at a time of unprecedented population growth and of environ-

mental concern. The role of population growth in relation to resource adequacy and to general environmental problems became a controversial issue and the subject of heated debate. Like many debates on resource issues, this one was both polarized and confused, the problem of disentangling immediate and underlying causes of resource trends being especially intractable.

Paul Ehrlich highlighted the role of population, as he perceived it, in a succession of publications including *The Population Bomb* (1968). He considered that population growth resulted in greater per capita demand on energy and mineral resources as poorer and less accessible materials were utilized. In addition, population growth was associated with increasing costs of transport, communications and the maintenance of public institutions (Ehrlich and Holdren 1971). Eventually, in Ehrlich's view, the environmental impact of population growth would be pushed beyond the buffering capacities of natural systems.

A gloomy view of population trends, and the advocacy of stringent population-control policies, have been a persistent theme in discourse about environmental resources since the days of Malthus, but they became much more prominent and extreme in the early 1970s. Emotive phrases such as 'the population explosion' became a familiar ingredient of the debate, and provided the setting in which extreme views were propounded. For example in *Famine 1975* (Paddock and Paddock, 1967), a policy of triage was advocated as the basis for disbursing food aid. On the analogy of military medicine, 'casualties' have to be classified for priority treatment. Some will not survive even if treatment is provided, and therefore these cases should not receive treatment. On this basis the prospects for some countries were dismissed as so hopeless that food aid should be withheld. Ehrlich himself advocated that compulsion and coercion should be used, if necessary, to achieve population control. The repressive and draconian means to the end of 'Zero Population Growth' (ZPG) are obviously repugnant to many.

Furthermore, the motives of many of those espousing ZPG policies have been questioned. In particular, the self-interest epitomized as 'lifeboat ethics' has been identified as an underlying issue. The analogy of the lifeboat was first drawn explicitly by Hardin (1974): ten men are safely provisioned in a lifeboat in a sea full of drowning men crying for help. If they pull an eleventh man on board, there will be insufficient provisions for all, and all will perish. The correct course of action, according to this view, is to ignore the cries for help and to ensure the safety of those already on the lifeboat. This scenario was readily transposed to the global scale and to the relationship between the developed and developing worlds. In a world of fixed carrying capacity and rapid population growth, the provision of food aid was not 'helpful'. It is not altogether surprising that the same author should advocate coercion in population control: in his view the community should have the power to decide how many children should be borne (Hardin, 1972).

Ehrlich and a number of fellow Americans, including Hardin, Isaac Asimov and Paul Getty, proceeded to set up an organization known as 'The Environmental Fund'. In the mid-1970s it propounded the view that the provision of food aid violated the principle of natural carrying capacity of the Earth. Food aid would mean improved nutrition and hence increased fertility which would in turn exacerbate the problems. Such views are one manifestation of what Pepper (1984) terms 'ecofascism'. Whether the environment itself is the primary concern, or whether there are other motives, is debatable. In essence, the environment becomes the vehicle for extreme and authoritarian views, the effect of which is to perpetuate the domination of some (poor) humans by other (rich) humans.

Ehrlich is one of many commentators who perceive population growth as a major cause of environmental problems, and who consider that it explains a large part of the depletive human use of the the environment in the developing world in particular. Some of these commentators, however, go on to reject simple Malthusian explanations as insufficient or inadequate. For example, Southgate *et al.* (1990) point to the frequent response of extending agriculture on to fragile marginal lands. With better agricultural extension services and different regimes of land tenure, sustainable intensification of agriculture could take place on the better and less fragile areas. In their view, tenure regimes in many countries also discourage tree planting and encourage deforestation. Many agree that to blame population alone is untenable, as in some lands environmental problems are worse than might be expected from rates of population growth and in some countries they are less severe (e.g. Repetto and Holmes, 1983). In such cases factors such as the breakdown of traditional resource regimes, commercialization and inequality of access to land and other resources are also significant issues (Chapter 2). Indeed population growth may be a

convenient scapegoat: it can allow inept or corrupt governments to divert attention from their failure to introduce appropriate measures for resource management and conservation by blaming the prolific breeding behaviour of the poor (e.g. Bromley, 1990).

A particular focus of debate in the early 1970s was the relative roles of population and technology in environmental problems. The debate was polarized around Ehrlich and Barry Commoner, who believed that technology, rather than population, was the main culprit. He argued that technology was now producing synthetic products, which unlike natural materials, broke down into persistent and sometimes toxic residues. Later, a measure of agreement between the Ehrlich and Commoner schools was achieved in the form of the expression I = PAT, where I stands for environmental impact, P for population, A for affluence or consumption per capita, and T for the effect of technology (strictly speaking the expression should relate to *change* in these quantities). Two main features characterize this expression. The first is the obvious one that factors other than population growth are acknowledged to be significant. The other is the multiplicative effect: a growth of population combined with increasing per capita consumption and changing technology might be expected to have an enormous effect. Perhaps it should be stressed that while this expression may be valid over the short or medium term, it does not necessarily apply in the longer term. If it is true that environmental resources can influence the wealth of nations especially during early stages in their development, it may also be true that the environment is affected by development especially during its early stages. Previous reference has been made to how per capita consumption of energy and materials may decline as the more advanced stages of development are reached. It is also suggested that discharges of wastes into the environment increase rapidly as economies grow, but then stabilize or decline (Bernstam, 1991). Such declines can be due both to improved technologies and to conscious efforts to reduce emissions.

Population and carrying capacity

Kingsley Davis (1991) has highlighted two major issues that arise from the inter-relationship between population and resources. The issues are closely inter-related. The first is that any theory which links population and resources and which ignores or overlooks culture (in the widest sense, including the making of tools, the exchange of goods and the social organization of groups) is likely to be deficient. The second issue is a preoccupation with food and its adequacy in discourse on the population-resources relationship. As Davis suggests, nowhere in the world are people satisfied with having just enough to eat.

Central to the concerns of commentators such as Ehrlich and Hardin and of the authors of *Blueprint* and *Limits to Growth* is the view that biological notions of carrying capacity can be applied to humans. Since the days of Malthus, the view has been widespread that carrying capacity is fixed by the nature of the environment, and human population presses against the limit that it represents. Intuitively, such a view seems irrefutable. The Earth and its land area are finite, and humans are part of nature, just like other species for which the carrying capacities of the environment are limited.

But humans are not *just* part of nature, as are other species (Chapter 1). As Engels (1873) points out, 'The most that the animal can achieve is to *collect*; man *produces*' (original italics, quoted from Parsons, 1977: 140). With this fundamental qualitative difference, the extent to which notions of fixed resources and limited carrying capacities can be applied becomes questionable. As early as the 1840s, Engels observed that science was advancing at least as rapidly as population, and that food production increased with it. The validity of the Malthusian hypothesis was therefore questioned. Furthermore, Marx viewed the concept of over-population quite differently from the Malthusians. In his view it was determined by social conditions rather than by biological factors. In other words carrying capacity and over-population could not be defined in purely biological terms. Instead the destitution, starvation and misery stemming from what was regarded by some as over-population resulted from a rotten socio-economic system (i.e. capitalism), rather than from the operation of biological laws of population growth and carrying capacities.

While many of the biologists and other commentators associated with pessimistic views such as those expressed in *Blueprint for Survival* assume that biological laws of population and carrying capacity apply to humans just as to other species, many social scientists (in the broadest sense) take up quite different positions. Reference has already been made to economists such as Beckerman and Nordhaus, as well as to political philosophers such as Engels and Marx. It seems that disciplinary background is a strong influence on the stance adopted in the debate on population and carrying capacity. Many biologists adopt negative positions, deriving from their assumptions that 'natural' biological laws can be applied to humans. On the other hand many social scientists are less convinced of the relevance of such laws. Indeed some are of the view that population growth is a positive influence on resources – that resources are *created*, rather than depleted, as population increases.

The Ultimate Resource?

One of the best known advocates of this view is Julian Simon. He considers that raw materials and energy are becoming less scarce, and that population growth has long-term benefits. He claims that at least in the United States, where he lives, pollution is decreasing. He regards humans and their brain power as the 'ultimate resource', which he chose as the title of one of his books (Simon, 1981). Simon is impatient with the 'bad news' that characterizes the debate about resources, population and the environment. In one paper he asks why we hear so much 'false bad news' (Simon, 1980), and speculates that there may be several reasons. Among them may be the idea that bad news sells books and promotes funding for research scientists, while good news does neither. He also suggests that psychological factors may be involved, and that models involving exponential growth have the power to bewitch. In addition, he considers that some may publicize gloomy predictions in the hope that such warnings will stimulate governments and institutions into remedial action. Whatever the reasons, he concludes that the bad news is over-played and that more should be heard of what he regards as good news.

While population growth is perceived negatively by many, it is perceived positively by Simon; 'in the long run the effect of more people is *positive rather than negative* (1990: 2, original italics). He accepts that population growth may give rise to some short-term problems, but argues that the pressures on resources resulting from population growth are the driving force behind innovation and the creation of new resources. Underlying his position is the concept that the quantity of resources available on Earth is not fixed, but rather is changeable in response to human efforts. In this debate as in so many others in the field of resources, the outlook adopted depends on the basic concept. If a static view of resources is assumed,

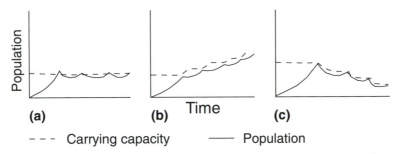

Fig. 11.3 Trends in population and carrying capacity: (a) constant carrying capacity; (b) carrying capacity rising in response to population pressure (e.g. technical innovation, new resource perceptions); (c) carrying capacity declining (e.g. as result of land degradation).

depletion is inevitable and a fixed carrying capacity must exist. If, on the other hand, a dynamic view is adopted (i.e. that 'resources are not, they become'), then population growth may be seen as one of the driving forces in the creation of resources.

In adopting this position, Simon resembles Ester Boserup, a Danish agricultural economist whose original work focused on the early development of agriculture. In her view, population pressure brought about the transition from shifting cultivation to increasingly intensive forms of agriculture. As has been indicated in Chapter 3, others have suggested that population pressure was also responsible for the transition from the hunter-gatherer stage to that of agriculture. In other words, the pressing of population against current resource limits resulted in the pushing back of these limits: necessity is the mother of invention. The challenge of feeding a growing population was required before the response of technical innovation was forthcoming. According to Boserup (1983), the 'green revolution' and modern birth-control techniques were developed as a result of research prompted by a fear of rapid population growth. More generally, she contends that the need to feed large populations led to technology transfers from one society to another and to the invention of new methods of production.

The contrasting outlooks of the neo-Malthusians and those of the Simon–Boserup school can be summarized in the simple diagrams that constitute Fig. 11.3. In Fig. 11.3(a), carrying capacity is constant, and population presses against an unyielding limit. In Fig. 11.3(b), carrying capacity is more elastic, and can yield in response to growing population. But in addition to the possibilities of a constant or increasing carrying capacity, there is a third possibility. If resources are exploited beyond the critical zone (Chapter 1) or if the environment is

otherwise degraded, the carrying capacity may decrease (Fig. 11.3(c)). Whether this downward trend may in turn stimulate new forms of resource management, and ultimately give rise to the creation of new resources, is debatable.

Rigid or flexible limits: the empirical evidence

The scattered and fragmentary evidence that exists suggests that the Malthusian and Boserupian hypotheses may both apply in some circumstances but that neither is inevitable nor universal. In considering this evidence, we should remember that carrying capacity is not a simple concept. At the basic level, it relates to biological needs rather than to wants, but few human societies are today content with basic subsistence. It must also be related to type of economy. The concept of carrying capacity in the Alps or Mediterranean mountains in the nineteenth century is much less meaningful in the late twentieth century. Today, the economies of parts of such areas are based on tourism rather than on subsistence agriculture. The ability or inability of a local population to feed itself directly from local land resources is irrelevant. It can simply purchase food from other parts of the world, using its income from tourism. Nevertheless, the concept of carrying capacity is one which continues to fascinate humankind, and many believe that it will apply at the global scale even if it is problematic at the local scale.

Numerous specific examples of receding limits could be quoted. For instance the introduction of a chemical process for synthesizing nitrate fertilizers from atmospheric nitrogen (Chapter 4) removed the previous limits of agricultural productivity imposed by the availability of mineral nitrate deposits. On a broader scale, the global population today is far larger than could have been supported under a hunter-gatherer system, and indeed also much greater than could have been maintained under the types of agriculture carried on in medieval Europe. Limits of both level and extent of production have been pushed far back, the former through science and technology and the latter through the growth of the world-economy and of international trade. To accept that the limits have receded, however, is not the same as to accept that population growth or population pressure has been the driving force. Big questions about cause and effect in scientific advance are notoriously difficult to answer, and it is perhaps not surprising that conclusive evidence is hard to find. Simon, however, has attempted to compare the rate of scientific discoveries with population levels in ancient Greece and Rome. He contends that an increase in population or in its rate of growth is associated with an increase in scientific activity, and a population decline with a decrease. He also asserts that in modern countries with similar standards of living, scientific output is proportional to population size. He quotes in support the cases of the United States and Sweden and their respective numbers of scientific writings and of patented processes (Simon, 1981).

Even if such evidence or contentions are found convincing and it is accepted that population pressure may under some circumstances be the driving force for scientific discoveries that lead to the pushing back of limits, it does not follow that population pressure invariably or inevitably has such results. Cases can also be quoted where populations have apparently overshot the carrying capacity of their environment, and collapsed or declined.

The case of Europe

Much debate has focused on medieval Europe, and on the extent to which it was over-populated. There is widespread agreement that the population grew rapidly from the ninth to the fourteenth century, and that population expansion was a driving force for the extensive colonization of forest lands, especially in Eastern Europe. Some historians considered that the population of France and other parts of Western Europe was pressing against the carrying capacity of the land, and that dramatic technical breakthroughs were not forthcoming. Some commentators therefore interpret the Black Death of 1348 and other plagues, famines and wars of the fourteenth century as Malthusian checks (e.g. Abel, 1980). (Some deterioration of climate may also have occurred, adding to the stresses.) But Boserup argues that innovation occurred in the form of the replacement of pastoralism and long-fallow agriculture by short-fallow systems. She also contends that people adapted to pressures by consuming more vegetable and less animal material during periods of stress (carrying capacity being higher for a population whose diet was based on vegetable material). What is clear is that after a lull in the fourteenth century population growth resumed, and presumably once more began to press against the limits of carrying capacity. In England, population quadrupled between 1550 and 1820. Since self-sufficiency in food was apparently maintained, agricultural production must have quadrupled. Wrigley (1986) concludes that food prices were under less pressure at the end of the period than at the beginning, and that the population was less vulnerable to poor harvests. In the Netherlands, on the other hand, the need to feed a rapidly growing population led to the development of sophisticated financial and commercial institutions, to the rapid expansion of production from new areas in Central and Eastern Europe, and to the growth of the world-economy (e.g. Bogucka, 1978). In short, population growth in Europe may have met with Malthusian checks, but it also led to a variety of responses. There is no single outcome, either in terms of population overshoot and collapse, or of scarcity-induced technical change.

In the wider world, environmental problems have also been blamed by some commentators for the decline or collapse of several ancient civilizations, including that in the Tigris–Euphrates lowlands in the Middle East and the Harappan civilization in the Indus valley (in the Indian sub-continent). Many commentators caution against the acceptance of simplistic conclusions, in which population decline is 'explained' simply in terms of environmental problems. Nevertheless, the sober conclusion reached in the review of ancient civilizations in the massive tome entitled *The Earth as Transformed by Human Action* (Turner et al., 1990) is that population declines were associated with radical environmental transformations. At a much smaller scale, the case of Easter Island is often quoted as an example of population and societal decline resulting from environmental problems. Indeed it is sometimes presented as a literal microcosm – that is of a small-scale model of what may happen to the Earth (see *Easter Island, Earth Island* (Bahn and Flenley, 1992)).

The case of Easter Island

Easter Island is one of the remotest spots on Earth. It is a small island, of less than 200 km^2, lying in the

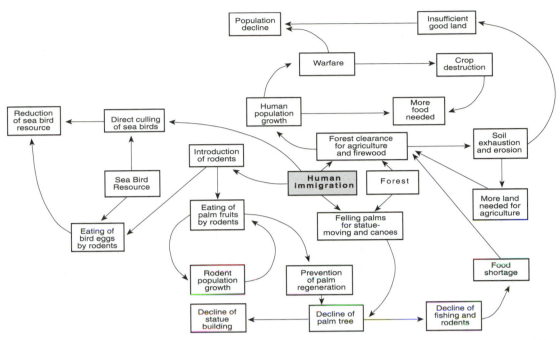

Fig. 11.4 Easter Island: immigration, environmental degradation and population decline. (*Source:* Bahn and Flenley, 1992.)

Pacific Ocean over 3500 km to the west of South America and more than 2000 km from the nearest inhabited land. When the first Europeans visited the island in 1722, the population of around 3000 lived in squalid huts and caves and engaged in almost perpetual feuds. Yet the island was characterized by hundreds of stone statues, averaging around 6 metres in height. How could such a miserable society have achieved the organization and technical skills required to carve and erect such statues? Clearly the Easter Island society of the eighteenth century was only a pale shadow of the society responsible for the statues.

The first settlers on Easter Island probably arrived during the fifth century AD. By 1550 the population had grown to around 7000, but then a collapse appears to have occurred. Many statues were abandoned half completed around the quarry from which the rock was taken. By the time of the first European visit, the population had fallen to less than half of its peak level – a collapse likened by some to that indicated by *Limits to Growth*. The island had become extensively deforested by the time of the population collapse, and part of the reason for deforestation is likely to have been connected to the erection of statues. Tree trunks are believed to have been used as rollers on which the heavy stones were

transported from quarry to erection site. The large number of statues could have involved huge numbers of trees and hence massive deforestation. By the sixteenth century, a shortage of wood was forcing a switch from timber houses to reed huts and caves, and prevented the construction of canoes. Rates of soil erosion increased, and crop yields declined. In the face of a food shortage the population turned to cannibalism. The inability to erect statues undermined belief systems and social organization, and conflicts over diminishing resources resulted in almost permanent warfare. In short, the result was societal disintegration and population decline (Figs 11.4 and 11.5).

The pathways of decline illustrated in Fig. 11.4 must of necessity be conjectural to some degree, and again cause and effect are difficult to disentangle. While there is little doubt that society decayed and population declined, the causes of these unhappy trends are more difficult to identify with certainty and some of the evidence can be little more than circumstantial. And even if it is accepted that depletion of environmental resources was the primary cause, the significance of this conclusion is debatable. Is it valid to project the conclusion from the microcosm of Easter Island to the Earth as a whole? The probability of a scientific or organiza-

Earth Island

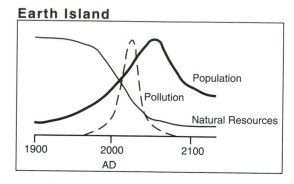

Easter Island

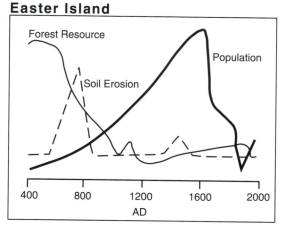

Fig. 11.5 Limits to growth on Earth Island and Easter Island. (*Source:* Bahn and Flenley, 1992.)

tional innovator emerging in a closed society consisting of a population of a few thousand on a small oceanic island is very much smaller than that in a global population of several billion. In other words, the likelihood of population pressure leading to improved resource management and the pushing back of environmental limits at the scale of Easter Island is much less than at the scale of Earth Island. On the other hand, it is interesting to note that it was 'conspicuous consumption' of resources, in the form of timbers used for transporting statues, that was the death-knell of Easter Island society – not consumption to meet the basic biological needs of humans for food and shelter. And warfare and societal decay, rather than simple starvation, seem to have been heavily involved in the decline.

Easter Island confirms, if confirmation be required, that societies and populations can decline rapidly, in a manner reminiscent of the collapse mode of *Limits to Growth*. Equally, however, many

examples can be quoted of cases where the model depicted in Fig. 11.3(b) applies – where carrying capacity rises as a result of innovations in resource management. On that model, the critical point is where the population presses against the supply of resource products under the prevailing system of management. Two main alternatives are possible: either the limits hold, and the population declines, or the limits yield and recede, allowing population growth to continue. At present it is impossible to isolate the conditions that determine which of these paths will be followed. Whether it will ever be possible to do so is doubtful, since both human and environmental conditions vary so much and since a virtually infinite range of possible combinations exists. Major innovations may be by their very nature unpredictable, and it is perhaps facile to expect that any particular combination of human and environmental conditions will necessarily produce them in any deterministic manner. Nevertheless, it can be speculated that small, isolated populations are less likely to give rise to the innovations required to push back the limits than are larger populations. It is certainly the case that some environments are more susceptible to environmental degradation than others. For example high mountains and semi-arid zones are especially fragile, and in at least some such areas the trend of carrying capacity is downwards (Fig. 11.3(c)).

Fragile environments

One specific case is that of the Mediterranean mountains. McNeill (1992) considers that carrying capacity has declined in these areas as a result of population overshoot accompanied by deforestation and soil erosion. He proceeds to note that the carrying capacity of these lands had previously risen, pointing in particular to the arrival of new crops such as maize and potatoes after the sixteenth century. The same area, therefore, can experience both increases and decreases in carrying capacity, but according to McNeill, marginal and fragile environments such as mountain areas feel population pressure more often than gentler environments, and the constraints on successful responses are greater. But in his view under-population can also be a threat. Below a certain level of population, there is insufficient labour to ensure that terraces and irrigation are maintained. Agriculture becomes less productive, and a spiral of decline develops. Community morale declines, and the result is mass emigration. Population undershoot, therefore, as well as population overshoot, can be damaging. Furthermore, if either 'overshoot' or 'undershoot' gives rise to rapid emigration, the brighter individuals, including potential innovators, are likely to be among the first to leave. In the case of the Mediterranean mountains (as in those of many other fragile areas) the picture is complicated by changing external relations. With the advent of the Industrial Revolution, the Mediterranean Basin began to play a new and dependent role in the wider European economy, as a supplier of resources and labour. Previous equilibria were disturbed. 'Market integration', for the Mediterranean mountains in the nineteenth century as for parts of the Andes and other frontier and mountain areas today, could be a trigger for destabilization. With it, resource production is increasingly geared to external demands. Local communities, facing worsening conditions, join in the process of over-exploitation (Coppola, 1978). Changing external relations, therefore, can seriously disrupt trends in population and carrying capacity.

Global or regional collapse?

The Earth's surface varies greatly in its resource potential (under a given set of cultural and technological conditions) and in its environmental fragility. Global collapse, along the lines of the *LTG* model, is therefore less likely than regional collapse. How the wider global community would react to the prospects of regional collapse in this age of increasing inter-connectedness is debatable. While it is true that food aid is now available on an unprecedented scale, a readiness to supply assistance during short-term emergencies does not necessarily mean a willingness to accept long-term responsibility. The long-established safety-valve for regions under stress – outwards migration – is increasingly circumscribed by international controls, and the potentials of new frontier regions are more limited than they were in (for example) the nineteenth century. Another possible response to resource pressures is annexation or seizure by force. Mass migration and pioneer extension are perhaps one form of resource 'war': conflicts with indigenous peoples have been widespread in the areas in which they have occurred (Chapter 3). Numerous examples of how scarcities of renewable resources have contributed to social instability and civil strife in areas such as Bangladesh, north-west and south Africa, the Philippines and Central America are

reported by Homer-Dixon *et al.* (1993). They conclude that significant causal links can be demonstrated between scarcities of renewable resources and violence, and that the incidence of such unrest is likely to grow. Water resources in particular are a potential flashpoint in areas such as the Middle East (Chapter 9). In many states in the developing world in particular, underlying tensions of religion or ethnicity can be exacerbated by environmental problems. For example the distribution of resources such as food and water plays an important role in the current civil war in Sudan (Thomas, 1992). What is usually seen primarily as an ethnic or religious conflict also has a significant ecological dimension. Ecological degradation is seen by some as the driving force behind the onslaught of the mainly northern Sudanese élite on the peoples of the south (e.g. Suliman, 1993).

An impoverished people deficient in basic food supplies is unlikely to have the capability of engaging in resource wars, even if their migration as environmental refugees can have serious impact. On the other hand, richer nations facing shortages of resource products such as oil and 'strategic' minerals have the capability and may have the will to do so. Perhaps conflicts of this type are a greater threat than shortages of the basic means of subsistence, and more likely to lead to catastrophic collapse. This is perhaps simply another way of saying that resource problems are more social than environmental in nature.

The nature of the debate about resource futures has changed radically over the last quarter-century. Global famine and imminent scarcities of major resource products are not feared as they were in the early 1970s. Surplus food is available in much of the developed world, and the prices of oil and many other resource products are at or near historic lows. Resources, it seems, are abundant rather than scarce. This does not mean that all concerns about environmental resources have evaporated. Other issues have emerged, including aesthetic and non-material values and the ability of the environment to absorb the waste products of modern living. But even if little concern is expressed about the adequacy of material resources at present, it is apparent that growth in population and resource consumption cannot continue forever: ultimately the world population and its resource demands must stabilize (or collapse). Equally clearly, many aspects of modern resource use are not sustainable, and if continued over long periods can only lead to the impoverishment of human life on Earth, as well as of non-human life. It is not surprising, therefore, that recent years have seen the emergence of the concept of sustainability in the management of environmental resources, and of the possibility that we are at present on the brink of a major transition – to sustainability – of which the significance will be as profound as that of the earlier Agricultural and Industrial Revolutions. This concept is the subject of the next and final chapter.

Further reading

Barnett, H.J. and Morse, C. (1963) *Scarcity and growth: the economics of natural resource availability*. Baltimore: Johns Hopkins Press for Resources for the Future.

Barney, G.O. (ed.) (1980) *The Global 2000 report to the President of the US*. New York: Pergamon.

Meadows, D.H., Meadows, D.L., Randers, J. and Behrens, W.W. (1972) *Limits to growth*. London and Sydney: Pan.

Meadows, D.H., Meadows, D.L. and Randers, J. (1992) *Beyond the limits*. London: Earthscan.

Simon, J.L. (1981) *The ultimate resource*. Oxford: Martin Robertson.

CHAPTER 12

Transition to sustainability?

As Chapter 11 shows, the 1970s were a decade of unprecedented concern about the adequacy of environmental resources. Impending scarcities and threats to continued economic growth were the common themes of the 'doomsday' environmental literature that characterized the decade. It soon became apparent, however, that some scarcities had been misidentified. They were essentially geopolitical rather than physical in nature, and it also soon transpired that the high growth rates in the consumption of some resources during the third quarter of the twentieth century were unlikely to be maintained. Above all, the narrow interpretation of resources as fixed stocks of environmental materials, shared alike by the authors of *Blueprint*, *Limits to Growth* and *Global 2000*, was increasingly perceived as unrealistic. Furthermore, the experience of the 1980s was that of apparently increasing abundance of food, fuel and minerals. While it is true that famine and malnutrition were not banished, more and more people began to accept that their causes were essentially social, economic and political rather than environmental.

This does not mean that environmental concerns evaporated. On the contrary, major new emphases such as deforestation, loss of biodiversity and land degradation emerged. A major shift occurred from concern about the adequacy of environmental resources to supply human needs of food and fuel to anxiety about the capacity of the environment to absorb the waste products of human activity. The concepts of sustainable development and sustainable use of environmental resources emerged strongly and rapidly, to the extent that by 1990 they were firmly established in the vocabulary of environmental discourse. The apparently wide and enthusiastic adoption of these concepts has led some to suggest that we are now on the brink of the third great revolution of the human race: that following the Agricultural Revolution and the Industrial Revolution there will now be the Sustainability Revolution (e.g. Ruckelhaus, 1989; Harrison, 1993). Time will tell whether this will be the case. At present we need to be careful to distinguish between the rhetoric and the reality of sustainability, and between the slogan and the substance.

The origins of sustainability

The concepts of sustainable development and of the sustainable use of environmental resources are deep-rooted, even if they did not blossom forth until the 1980s. As will be shown later in this chapter, the concepts themselves can be interpreted in different ways, and in particular we need to note that sustainable use of resources and sustainable development are not necessarily synonymous. Because of the vague and overlapping nature of concepts of sustainability, it is unlikely that their origins can ever be explained comprehensively or convincingly. They are simply ideas whose time has apparently come. Nevertheless, a series of steps in their evolution and dissemination can be identified, as can the wellsprings from which they have emanated.

Antecedents

At one level, the concept of the sustainable use of resources is simple and well defined. Principles of sustainable yield management were being applied by German foresters by the nineteenth century and strongly influenced the first conservation movement in the United States by the beginning of the twentieth century (Chapter 3). Sustainable yield management –

cropping at a rate not exceeding the natural replenishment of the forest or fishery – is appealing in its simplicity and apparent logic. It simply 'makes sense', even if it is accepted that there may be circumstances in which the imperatives of economic rationality dictate that different cropping rates are employed. But the fact that the concept did not reach political and environmental agendas until the 1970s and 1980s shows that the availability of technical knowledge on the biological basis of sustainable resource use was not a sufficient condition for its emergence, and in particular for its transmutation into a concept of sustainable development.

At another level, the roots are intertwined with those of environmental movements and their growth as a global phenomenon, especially over the last few decades. Among those who have attempted to trace the origins of the concept is Pierce (1992), who outlines the lineage of the concept back to a diverse set of influences including those of Malthus, nineteenth-century nature philosophers such as Thoreau, and the policies of the Roosevelts during the first two conservation movements (Chapter 3). Less broadly, Adams (1990) identifies a number of overlapping themes that help to define the context of its emergence. The first of these is nature preservation in both developed and tropical countries, and particularly in Africa. From this arose the first international environmental agreements (Chapter 6), and a framework of international environmental organizations began to emerge. This eventually evolved into the International Union for the Protection of Nature (IUPN), established in 1949 (it became the International Union for the Conservation of Nature (IUCN) in 1956). As is indicated subsequently, IUCN played an important role in developing the modern concept of sustainability.

A second theme identified by Adams is the growth of ecological science in development planning in the tropics, and especially in parts of British Africa. The growth of 'ecological managerialism' was manifested, for example, in development plans produced for areas such as Tanganyika and Uganda in the 1940s. Scientists, and especially applied ecologists, had been part of the colonial élites in some oceanic islands and in India since at least the early nineteenth century, and as thus were well placed to provide inputs into development planning.

Adams's other themes focus on the rise of perceptions of a global environmental crisis during the 1960s, and in particular the emergence of the 'Spaceship Earth' concept. This rise was manifested in the growth of the environmental movement (Chapter 3), and culminated in the United Nations Conference on the Human Environment held in Stockholm in 1972.

The Stockholm Conference

The Stockholm Conference is usually regarded as a landmark in the emergence of global environmentalism. It marks the coming of age of the environmental movement and is important because of the legitimacy that it afforded to environmental issues in international relations (Thomas, 1992). That environmental issues were now clearly on the international political agenda is clearly shown by the fact that it was attended by representatives of no fewer than 113 countries and of 19 intergovernmental agencies (McCormick, 1989). The scale of this international gathering itself symbolized the significance of environmental issues as they were now perceived, and a legacy was left which included the setting up of the United Nations Environmental Programme (UNEP). It also gave rise, though not immediately, to the Convention on International Trade in Endangered Species (Chapter 6). Stockholm also helped to strengthen environmental groups and to expand their roles (Chapter 3). Meetings of nongovernmental organizations paralleled those of government representatives, providing both a forum and a means of international networking for these groups. This, in turn, was accompanied by a broadening of their interests away from narrow concerns such as species protection and towards wider issues such as funding policies and indeed the concept of sustainability itself.

In addition, Stockholm helped to promote the development of national environmental policies and the setting up in many countries of environmental agencies and ministries. Furthermore, formal dialogue between developed and developing countries over their environmental concerns now began – although this was not the initial focus of the conference. The driving force behind the decision of the United Nations to hold the conference came primarily from developed-world concerns about the environmental effects of industrialization. Important among these were Swedish concerns about the transnational effects of acid rain, especially on its forests and fresh waters. Although some of these issues had important economic dimensions (such as those relating to forest damage), they did not focus primarily on development issues. Indeed at that time environmental problems and development problems were usually regarded separately, often with the tacit or expressed assumption that environmental conservation and development were mutually opposed. If anyone had seriously believed previously that developing countries would be prepared to forgo economic development in order to achieve environmental goals defined by the North, the conference soon disabused them of that notion. From then on it became apparent that a synthesis would somehow have to be sought between conservation and development, and it was in this context that the concept of sustainability emerged with a new prominence.

Although IUPN/IUCN had been in existence for 20 years prior to the Stockholm Conference, its achievments had been modest. One persistent problem was funding. This was partly alleviated by the setting up of the World Wildlife Fund in 1961, conceived initially as a fund-raising organization to support IUCN but soon developing into a major actor in its own right. Another problem was the narrow focus on nature *protection*, which characterized the organization in its early days. By the early 1960s, it was becoming apparent that such a narrow concern was unrealistic in the face of the realities of post-colonial Africa in particular. IUCN then launched its African Special Project to encourage the new African leaders to embrace environmental objectives. If there were to be a realistic hope of doing so, these would have to be more wide ranging and less preservationist than the traditional approaches of establishing nature reserves and protecting particular species. By the end of the 1960s, IUCN had begun to show a greater general concern for economic development and was considering the notion of a strategic approach to conservation. Dialogue with UNEP developed after the Stockholm Conference, and in 1977 IUCN was commissioned by UNEP to prepare what eventually became the World Conservation Strategy. UNEP and WWF provided financial support as well as contributing to development of its basic themes and structure.

World Conservation Strategy

As might be expected from a parentage involving a number of bodies and differing environmental attitudes and perspectives, the World Conservation Strategy (WCS) went through several drafting stages before it was finally launched – simultaneously in 40 countries – in 1980. As its preface made clear, it reflected compromises, both among conservationists,

and between conservationists and 'practitioners of development'. The published document also acknowledged its limitations in aggregating, generalizing and simplifying complex issues. These, its authors considered, were less important than the need to agree on a set of conservation requirements and priorities. The Strategy was targeted at three main groups of potential users – government policymakers, conservationists and those directly concerned with living resources, and development practitioners, including aid agencies, industry and trade unions.

Conservation was defined as 'the management of human use of the biosphere so that it may yield the greatest sustainable benefit to present generations while maintaining its potential to meet the needs and aspirations of future generations' (IUCN *et al.*, 1980). The authors were careful to emphasize that, as thus defined, conservation was 'positive', embracing 'preservation, maintenance, sustainable utilization, restoration, and enhancement of the natural environment' (p. 1). Three main objectives were identified:

- To maintain essential ecological processes and life-support systems.
- To preserve genetic diversity.
- To ensure the sustainable utilization of species and ecosystems.

The practical benefits that would stem from the preservation of genetic diversity and the sustainable utilization of ecological resources were emphasized. Despite its origins in a body originally concerned with nature preservation, the tone of the Strategy is thoroughly anthropocentric: human benefits would follow from appropriate forms of environmental management. Essentially, WCS attempted to reconcile conservation and development goals by incorporating conservation guidelines into development rather than by opposing development. At this level, strict preservation was not a viable option, while the concept of sustainable development could be accepted by a wide variety of countries and interests. The fact that sustainable development could be interpreted in different ways was more of a help than a hindrance in this respect, and it is a feature of the Strategy that no attempt was made to define the meaning of the concept in detail. At the same time, it is noticeable that distributional issues were openly acknowledged. For example, the opening section highlights the gross disparities of resource consumption in different parts of the world, graphically illustrating how one Swiss person consumed as much as 40 Somalis.

At around the same time as WCS was launched, the concept of ecodevelopment emerged, initially as a planning concept in UNEP (Riddell, 1981). It was initially concerned with conserving renewable resources, with regulating the use of non-renewable resources and with controlling the discharge of waste products, but was soon transformed into a more ideological weapon in the fight against social injustice and exploitation. Some aspects of ecodevelopment were akin to sustainable development, but the concept became associated with radicalism and was soon overshadowed by the burgeoning concept of sustainable development.

World Commission on Environment and Development

If the World Conservation Strategy was the initial launchpad for the concept of sustainable development, a major boost was provided a few years later by the publication of the report of the World Commission on Environment and Development (the Brundtland Report) in 1987 (WCED, 1987). The Commission had been established by the United Nations General Assembly in 1983, as an independent body 'linked to but outside the control of governments and the UN system' (p. 3). Its 23 commissioners came from 22 countries, with a strong majority from the developing world. The Commission was mandated to examine critical environmental and developmental issues and to formulate proposals for dealing with them; to propose new forms of international cooperation; and to raise levels of understanding and commitment to action on the part of individuals, corporations and governments.

Implicitly, Brundtland rejected the 1970s notion of environmental limits to growth. It focused on how development should be achieved, and not on whether it was desirable. The report is notable for the way in which it brought together environmental and developmental issues. Degraded and deteriorating environments were seen to be inimical to continued development. On the other hand, economic growth could help to provide resources for environmental protection. It is especially notable for the emphasis placed on sustainable development, which in the view of the Commission, would seek to meet the needs and aspirations of the present without compromising the ability to meet those of the future. With a large complement of members from the developing world, there was little prospect that a cessation of economic growth would be advocated as a means of solving environmental problems. On the contrary, further growth was seen as essential if problems of poverty and under-development were to be addressed. 'Sustainable development requires meeting the basic needs of all and extending to all the opportunity to fulfil their aspirations for a better life (WCED, 1987: 8). Distributional issues were thus highlighted – including both those relating to the present generation and the marked disparities between the 'haves' and the 'have nots', and those of an inter-generational nature. The industrialized countries, with one-quarter of the world's population, consumed one-third of the world's food calories and around four-fifths of its paper, metals and commercial energy. Growth, while essential, would have to become less demanding of material and energy and be more equitable in its occurrence and impact. It would also, in the view of the Commission, have to be geared to the available stock of capital and not deplete that stock (what is meant by capital is discussed later in the chapter). In short, the Brundtland Report helped to crystallize and publicize the concept of sustainable development. It also proposed that a UN Programme on Sustainable Development should be established, and that a major international conference be convened to review progress and to promote action.

The second World Conservation Strategy

The momentum of sustainability concepts was further boosted with the publication of the second World Conservation Strategy (*Caring for the Earth*) in 1991. Produced by the same bodies as those responsible for the first Strategy in 1980 (i.e. IUCN, UNEP and WWF), it had two aims. One was to promote a new ethic for sustainable living, and to translate its general principles into practice. The other was further to integrate conservation and development. It was now claimed that WCS (1980) had stated a new message, to the effect that conservation was not the opposite of development. *Caring for the Earth* sought to extend and to emphasize this message. A set of principles for building a sustainable society was therefore outlined, each principle being accompanied by a number of actions required for its implementation (Table 12.1).

Criticisms of the first WCS were clearly taken into account in the second version. While the 1980 Strategy was perceived by some as aiming to protect the environment at the expense of basic human needs, *Caring for the Earth* offered a broader view of environment in its developmental context. Com-

Table 12.1 World Conservation Strategy 2: principles of a sustainable society

Principle	Action
Respect and care for the community of life	Develop the world ethic for living sustainably
	Promote world ethic for living sustainably at national level
	Implement the world ethic through action in all sectors of society
	Establish a world organization to monitor implementation of the world ethic and to prevent and combat serious breaches in its observation.
Improving the quality of human life	In lower-income countries, increase economic growth to advance human development
	In upper-income countries, adjust national development policies and strategies to ensure sustainability
	Provide the services that will promote a long and healthy life
	Provide universal primary education and reduce illiteracy
	Develop more meaningful indicators of quality of life
	Enhance security against natural disasters and social strife.
Conserving the Earth's vitality and diversity	Adopt a precautionary approach to pollution
	Cut emissions of greenhouse and other gases
	Prepare for climate change
	Adopt integrated approach to land and water management, using the drainage basin as the unit of management
	Maintain as much as possible of each country's natural and modified ecosystems
	Take pressure off natural and modified ecosystems by protecting the best farmland and managing it in ecologically sound ways.
	Halt net deforestation, protect large areas of old-growth forest, and maintain a permanent estate of modified forest
	Complete and maintain a comprehensive system of protected areas
	Improve conservation of wild plants and animals
	Improve knowledge and understanding of species and ecosystems
	Use a combination of *in situ* and *ex situ* conservation to maintain species and genetic resources
	Harvest wild resources sustainably
	Support management of wild renewable resources by local communities; and increase incentives to conserve biological diversity
Keeping within the Earth's carrying capacity	Increase awareness about the need to stabilize resource consumption and population
	Integrate resource consumption and population issues in national development policies and planning
	Develop, test and adopt resource-efficient methods and technologies
	Tax energy and other resources in high-consumption countries
	Encourage 'green consumer' movements
	Improve maternal and child health care
	Double family planning services
Changing personal attitudes and practices	Ensure that national strategies for sustainability include action to motivate, educate and equip individuals to lead sustainable lives
	Review the status of environmental education and make it an integral part of formal education at all levels
	Determine the training needs for a sustainable society and plan to meet them.
Enabling communities to care for their own environments	Provide secure access to resources and equitable shares in managing them
	Improve exchange of information, skills and technologies
	Enhance participation in conservation and development
	Develop more effective local governments
	Care for the local environment in every community
	Provide financial and technical support to community environmental action.
Providing a national framework for integrating development and conservation	Adopt an integrated approach to environmental policy, with sustainability as the overall goal
	Develop strategies for sustainability and implement them directly through regional and local planning
	Subject proposed development projects, programmes and policies to environmental impact assessment and to economic appraisal

Table 12.1 Continued

Principle	Action
	Establish a commitment to the principles of a sustainable society in constitutional or other fundamental statements of national policy
	Establish a comprehensive system of environmental law and provide for its implementation and enforcement
	Review the adequacy of legal and administrative controls and of implementation and enforcement mechanisms
	Ensure that national policies, development plans, budgets and decisions on investments take full account of their effects on the environment
	Use economic policies to achieve sustainability
	Provide economic incentives for conservation and sustainable use
	Strengthen the knowledge base, and make information on environmental matters more accessible
Creating a global alliance	Strengthen existing international agreements to conserve life-support systems and biological diversity
	Conclude new international agreements to help achieve global sustainability
	Develop a comprehensive and integrated conservation regime for Antarctica and the Southern Ocean
	Prepare and adopt a universal declaration and covenant on sustainability
	Write off official debt of low-income countries and retire enough of their commercial debt to restore economic progress
	Increase capacity of lower-income countries to help themselves
	Increase development assistance and devote it to helping countries develop sustainable societies and economies
	Recognize the value of global and national non-governmental action, and strengthen it
	Strengthen the United Nations system as an effective force for global sustainability

Source: compiled from IUCN/UNEP/WWF (1991) *Caring for the Earth. A strategy for sustainable living.* Gland, Switzerland.

pared with the 1980 Strategy, there is clearly a major emphasis on development and on distributional aspects. Perhaps the most striking feature, however, is the promotion of an ethic. Perhaps the emergence of ethics in this way reflects an acknowledgement that purely technical solutions to environmental problems are unlikely to be effective, whether these stem from economics or from the applied sciences such as agriculture or engineering. At the same time, the ethic that is advocated does not relate to the environment alone, but rather it embraces questions of equity in human (social and political) relationships (Chapter 1).

The Rio Conference

The publication of *Caring for the Earth* was overshadowed by the United Nations Conference on Environment and Development (UNCED), held in Rio in 1992. If any single event marks the coming of age of concepts of sustainability, the Rio Conference did so. This is not, of course, to say that sustainability

has become the dominant paradigm in the management of environmental resources, but the maturing of the concept is reflected in the setting up of an United Nations Commission on Sustainable Development following the conference. This Commission, and the conference itself, originated with the Brundtland Report.

The Rio 'Earth Summit' was the largest conference ever staged by the United Nations. It brought together over 100 heads of state and government, and in all some 178 governments were represented. Official delegates numbered almost 10 000, and many thousands more attended in other capacities. (Many of these were members of environmental groups and other non-governmental organizations. By 1990, over 900 NGOs had established consultative status with the United Nations (Starke, 1990).) At the end of 1989, the UN General Assembly had expressed its concern about environmental trends and resolved to convene the 'Earth Summit'. It accepted the invitation of the Brazilian government to hold it at Rio, the timing in 1992 marking the twentieth anniversary of the Stockholm Conference. Several major agreements were reached at UNCED, including a Convention on Biological Diversity (Chapter 6), a Framework Convention on Climate Change, Forest Principles (Chapter 5), the Rio Declaration of 27 principles (Table 12.2), and *Agenda 21*, which is essentially an action plan for reconciling conservation and development and for achieving sustainable development during the twenty-first century. Consisting of some 40 chapters and 500 pages, *Agenda 21* addresses general development issues, issues relating to specific resources, the roles of the various groups, and means of implementation. It is not a binding treaty (unlike the Conventions), but nevertheless carries much political authority as 'a collection of agreed negotiated wisdom' (Grubb *et al.*, 1993:17). As a major contribution to international 'soft law', it has moral if not legal force (Hughes and Lea, 1993). It is up to each country to translate *Agenda 21* into terms appropriate for its own circumstances. At the global level, the UN General Assembly established the Commission on Sustainable Development to review progress in implementing of *Agenda 21*, drawing on reports from UN organizations, national governments and non-governmental organizations such as environmental groups.

The Rio Declaration differs from the Stockholm Declaration in giving greater emphasis to developmental issues and less to ecological concerns. While in the latter, wildlife and other environmental concerns are highlighted, the former is more socio-political in tone. The prevailing outlook is decidedly anthropocentric (Principle 1, Table 12.2). Behind the sometimes rather bland wording can be heard the echoes of conflicts between the priorities of North and South, and tensions between national sovereignty and global interdependence. One specific aspect of these tensions and conflicts which pervaded the Conference was the question of funding, and especially of funding for the implementation of *Agenda 21*. Estimated costs amount to around $600 billion per year during the 1990s, with an estimated requirement of $125 billion per year in aid or loans to developing countries on favourable terms. This latter figure corresponds roughly with the United Nations target of achieving 0.7 per cent of GNP as aid from rich countries, but is more than twice the current level of assistance (Donlon, 1992) . And in addition to the question of the distribution of costs and benefits, there is that of the overall adequacy of funding. Some commentators regard it as quite inadequate: for example Myers (1993) compares it very unfavourably with the current

level of subsidies to agriculture in the developed world, which he, like many others, regards as quite unsustainable.

The Rio Conference could certainly not be regarded as a complete success. In addition to these general issues about North–South relations, sovereignty and funding, there were specific (and indeed spectacular) failures. For example the concept of a Global Forest Convention, promoted especially by developed countries, met with opposition from many developing countries. Developing nations feared that it would be a mechanism by which they would be prevented from using and managing their forests in ways which they themselves chose. Brazil, for example, completely rejected the principle of international intervention in what it saw as its own affairs. Some countries saw the emphasis on tropical deforestation and moves for the creation of global commons as a latter-day manifestation of imperialism (Thomas, 1992), and resented any move towards preserving developing-world forests as a sink for carbon dioxide emitted in the North. A legally binding Convention was therefore not agreed, although the 'Forest Principles' (Chapter 5) emerged as a compromise. Nevertheless, while the realities of international politics prevented a Forest Convention from being negotiated, the fact remains that environmental issues were now very clearly on the international political agenda. At least the rudiments of an institutional framework for discussing environmental problems were now in existence. And the concept of sustainable development had diffused far and wide.

The meaning of sustainability

Sustainability, and in particular sustainable development, has gained widespread currency and has reached contemporary political agendas despite – and possibly because of – vagueness of meaning. This vagueness may have helped to make it a concept acceptable in a wide range of political settings and to a wide range of people. Most of its formulations are so bland and general to be unexceptionable, and the scope for creativity and originality in its interpretation is enormous. Indeed as Pearce *et al.* (1989:1) suggest, 'There is some truth in the criticism that [sustainable development] has come to mean whatever suits the particular advocacy of the individual concerned'. It is not surprising, therefore, that Mrs Thatcher, when she was UK prime minister, found

Table 12.2 The Rio Declaration on Environment and Development

1. Human beings are at the centre of concerns for sustainable development. They are entitled to a healthy and productive life in harmony with nature.
2. States have the sovereign right to exploit their resources and the responsibility to ensure that such exploitation does not cause damage to the environments of other states.
3. The right to development should equitably meet the needs of present and future generations.
4. In order to achieve sustainable development, environmental protection shall be an integral part of the development process.
5. All States shall cooperate in eradicating poverty.
6. Developing countries, and especially the least developed and most environmentally vulnerable, will be given special priority.
7. States have common but differentiated responsibilities. The developed countries acknowledge their responsibility in view of the pressures they place on the global environment.
8. States should reduce and eliminate unsustainable patterns of production and consumption and promote appropriate demographic policies.
9. States should cooperate to strengthen capacity for sustainable development through exchanges of scientific and technological knowledge.
10. States shall facilitate and encourage public awareness by making environmental information widely available.
11. States shall enact effective environmental legislation.
12. States should cooperate to promote a supportive and open international economic system.
13. States shall develop national law on liability and compensation for victims of pollution and other environmental damage.
14. States should cooperate to discourage or prevent the relocation and transfer to other states of any activities or substances that cause severe environmental degradation or harm to human health.
15. The precautionary principle shall be applied by states.
16. National authorities should promote the internalization of environmental costs and the use of economic instruments. The polluter should, in principle, bear the cost of pollution.
17. Environmental impact assessment shall be undertaken for proposed activities that are likely to have a significant adverse impact on the environment.
18. States shall notify other states of natural disasters or other emergencies likely to produce sudden harmful environmental effects.
19. States shall notify potentially affected states of activities that may have a significant trans-boundary environmental effect.
20. Women have a vital role in environmental management and their participation is therefore essential to achieve sustainable development.
21. The creativity, ideals and courage of youth should be mobilized to forge a global partnership in order to achieve sustainable development.
22. Indigenous people and other local communities have a vital role in environmental management and development.
23. The environment and natural resources of people under oppression, domination and occupation shall be protected.
24. Warfare is inherently destructive of sustainable development. States shall therefore respect international law providing protection for the environment in times of armed conflict.
25. Peace, development and environmental protection are interdependent and indivisible.
26. States shall resolve all their environmental disputes peacefully and by appropriate means in accordance with the Charter of the United Nations.
27. States and people shall cooperate in the fulfilment of the principles embodied in this Declaration and in further development of international law in the field of sustainable development.

Source: United Nations (1993) The global partnership for environment and development: a guide to Agenda 21.
Note: In this table, the principles are expressed in summarized and paraphrased form. See source for original wording.

the phrase 'weak and wimpish' (Pearce, 1994). Fears have also been expressed that the concept becomes so abused as to be meaningless (e.g. O'Riordan, 1988).

Several scores of different definitions have been suggested, with varying degrees of precision and specificity (by 1989 Pezzey had identified over 60 definitions). The Food and Agriculture Organization (FAO) of the United Nations defines sustainable development as 'the management and conservation of the natural resource base, and the orientation of technological and institutional change in such a manner as to ensure the attainment and continued satisfaction of human needs for present and future generations. Such sustainable development ... conserves land, water, plant and animal genetic resources, is environmentally non-degrading, technically appropriate, economically viable and socially acceptable' (FAO, 1992: 144). According to FAO, three basic concepts are involved : externalities, intergenerational equity, and the impact of one country's actions on the environment of other countries. Although there may be consensus on such concepts, FAO nevertheless acknowledges that the operational meaning of sustainability depends on the viewpoint of the observer and in particular on disciplinary and other backgrounds. Similar problems confront most, if not all, definitions of sustainability.

Few if any definitions have become more widely established than that of the Brundtland Commission: 'sustainable development is development that meets the needs of the present without compromising the ability of future generations to meet their own needs' (WCED, 1987: 43). This definition gives centre stage to human *needs*, for food, clothing and shelter. But as Brundtland points out, 'Perceived needs are culturally determined, and sustainable development requires the promotion of values that encourage consumption standards that are within the bounds of the ecological [*sic*] possible and to which all can reasonably aspire' (p. 44). Endless argument could surround reasonable aspirations, but implicit is the view that needs are greater than merely the biological minima required to maintain life. Also implicit is the idea that natural limits exist, although these are 'not absolute limits but limitations imposed by the present state of technology and social organization on environmental resources and by the ability of the biosphere to absorb the effects of human activities' (p. 8). Therefore in the Brundtland view, neither needs nor limits can be defined in absolute terms. Both are, to a greater or lesser degree, flexible. This means that the Brundtland definition of sustainable development (like many others) is difficult to operationalize, however useful it may be as a general guide to the nature and direction of development.

Sustainability and constancy of capital

This vagueness of definition strikes many as unsatisfactory, even if they accept that a trade-off may exist between the specificity and acceptability of the concept. One major distinction is between the essentially ecological or environmental concept of sustainability on the one hand, and of the economic concept of sustainable development on the other. But even with the former there are problems. Sustainability can relatively easily be applied (as a concept)

within the renewable resource sector. It implies that the productivity of land or sea is not reduced by overcropping or misuse. The resource is not exploited beyond the critical zone (Chapter 1), and in theory the resource can be used in perpetuity. In the non-renewables sector, however, the concept of sustainability is more difficult to apply. Any extraction of a resource such as coal or oil means depletion, and therefore the resource cannot be used in perpetuity. While it is true that some minerals such as iron and copper can be recycled, others such as the fossil fuels are irretrievably altered in the course of consumption. Strictly speaking, therefore, the concept of sustainability, in the physical or environmental sense, cannot be applied to stock or fund resources. In addition, since resource perceptions vary through time, it might not even be considered desirable to maintain a constant stock of non-renewable resources. In prehistory, for example, maintaining a constancy of the stock of flint, for instance, would have proved rather pointless when metals began to be used for tools (Chapter 8). Furthermore, there are problems in applying the concept to 'sink' functions of the environment, except in a general way related to the absorptive capacity of the atmosphere or the oceans.

If problems beset the precise definition of sustainability in the purely environmental sense, even greater difficulties can arise when more broadly based interpretations are considered. One initially attractive principle is that sustainable development should mean that the next generation should inherit a stock of wealth no less than that inherited by the present generation. Two intractable questions then arise. First, should 'stock of wealth' mean natural capital alone, or natural capital together with human capital? In other words, can human capital be substituted for natural capital in the quest for constancy of the 'stock of wealth', or should the goal of constancy apply to natural capital or environmental assets alone? Daly (1991) uses the term 'strong sustainability' for the case of constancy of natural capital alone, and 'weak sustainability' for that of natural together with human capital. Second, if natural capital should be considered alone, should the aim be to maintain a constant physical stock alone (in terms, for example, of hectares or tonnes), or a constant economic value? (See Pearce *et al.*, 1990 for further discussion of these issues.)

Neither of these questions can be easily answered. At one level, there are obvious limits to the extent to which human capital can be substituted for natural capital. The loss of a plant or animal species, for example, is irreversible, and no increase in human capital (through, for example, an advance in technology) can be a direct substitute. Nevertheless, some environmental economists contend that species, like paintings, though regarded as priceless can yet be expressed in monetary terms (Chapter 2). To this extent, an increase in human capital could compensate for the loss of natural capital. 'Complete' substitutability of human for natural capital is, however, unrealistic. Because of the uncertainty surrounding some environmental functions and benefits and because of the irreversible nature of their elimination, no trade-off will be acceptable in relation to some environmental elements. These may constitute critical capital, which on the precautionary principle must be safeguarded. Pearce *et al.* (1990) therefore conclude that there may be a rationale for conserving the entire stock of natural capital, at least until a clearer understanding is established of what constitutes *optimal* stock. On the other hand, strict constancy of natural capital, in the sense of maintaining physical stocks, is completely incompatible with *any* use of fund resources. (It is possible, however, to conceive of sustainability in such a context as implying that depletion proceeds at a rate no greater than the rate at which substitutes become available.)

The concept of sustainability, therefore, does not easily stand up to logical analysis, nor is it likely to benefit politically or pragmatically from too close scrutiny. Two main clusters of meaning can be identified. One relates primarily to the ecological dimension, and can be traced back to the concept of sustained yield management. The other embraces economic, social and political dimensions, and focuses as much on how development can be achieved as it does on how the environment can be protected. Many apparent contradictions (as well as uncertainties) in its meaning have been pointed out: see for example (among many others) Redclift (1987, 1992). Dovers and Handmer (1993) consider that some of the contradictions may be insurmountable, including the paradox of technology (whether it is cause or cure); economic growth versus ecological limits; and individual versus collective interests. Even if such contradictions can be resolved, some commentators consider that it may take centuries for an agreed definition (and in particular) an agreed *precise* definition to evolve. Manning (1990), for example, likens the concept to that of social justice – a broad high-order social goal shared by many even if difficult to reduce to a simple and precise definition.

This, however, does not necessarily mean that the concept is untenable or that it lacks utility. The concept for the first time (at least in modern history) draws together environmental, economic and social concerns. At the very least, it provides a setting in which the shared goals of environmentalists and development interests can be identified. In comparison with the entrenched opposition of these interests, it is a welcome advance. It provides an intellectual arena for dialogue between these interests, and it has won political endorsement at both national and international levels in a way in which 'purer' or more ecocentric environmental interests could not have achieved. To be sure, some of those lying towards the 'deep ecology' end of the spectrum of environmental attitudes (Chapter 1) may dismiss it as merely legitimizing development at a time when the only environmental hope lies in the end of growth. It would be a brave politician, however, who would espouse the cause of zero growth, whether he or she were in the developed or developing world. The acid test of the concept of sustainability, however, will not be the way in which it stands up to logical analysis or philosophical acceptablility, but the effectiveness of its application as a guiding principle of the management of environmental resources.

Progress of 'sustainability'

It is still too early to judge on the effectiveness of sustainability as a guiding principle and on whether a transition towards sustainability is now under way. The afterglow of Rio has not yet faded completely, but neither have the new institutions and procedures stemming from it had time to deliver. In some sectors and in some countries, however, signs of a shift towards sustainability (or at least towards more sustainable ways of managing environmental resources) were apparent even before Rio. For example the policies of the World Bank by the late 1980s were reflecting an increasing awareness of the need for environmental protection concurrent with economic development. Compared with its policies of ten years earlier, the World Bank had taken on a greener hue, probably at least partly because of pressures brought to bear on it by environmental groups as well as because of the influence of Brundtland. In the early and mid-1980s the environmental damage resulting from projects it had supported, including ranching in Latin America and large dams, attracted strong criticism (e.g. Rich, 1985). By the second half of the decade, however, it had begun to consider more fully the environmental consequences of its projects and policies, and to attempt to incorporate environmental costs more fully in economic analyses (e.g. World Bank, 1987, 1988). For example by the late 1980s it had adopted a general policy of avoiding the conversion of wildlands through its development projects, and of assisting in their conservation (Ledec and Goodland, 1988).

Examples of progress

Following the World Conservation Strategy, the preparation of national strategies marked the beginning of progress towards sustainability in some countries, and other signs of progress became evident in other lands. Australia, New Zealand and Canada are notable examples of cases where at least some progress, in different forms, has been made towards sustainability. These countries are not necessarily representative of the wider world. They are, for example, well endowed with land and other environmental resources, and they have relatively small, well-educated and prosperous populations. Conditions for the adoption of sustainability objectives, therefore, are more favourable than in many other parts of the world, but at the same time some of them have long traditions of land degradation and other aspects of unsustainable use (e.g. Heathcote and Mabbutt, 1988). While the experience of such countries is unlikely to be shared or followed by all nations, it is nevertheless of interest as one of the few criteria by which the practical progress of sustainability (as opposed to its rhetoric) can be gauged.

Australia

In Australia, the National Conservation Strategy endorsed the three main objectives of the world strategy and added as a fourth the need to maintain and enhance environmental quality. In some sectors such as national parks, the federal government could act directly, and it also requested that each of the states produce its own strategy. The results were variable in content and quality, but in general the concept, if not the practice, of sustainability was well established before the publication of the Brundtland Report. Since then, numerous sectoral initiatives have been undertaken, included a National Soil Conservation Strategy Program and a National Tree Planting Program. Several sites have been placed on the World Heritage list (Chapter 6), and in general Moffatt (1992) concludes that 'Australian governmental departments, non-governmental organisations and individuals [have been] actively involved in promoting sustainable development projects throughout the country' (p. 36). Even a large number of individual projects does not, of course, necessarily reflect the adoption of a paradigm of sustainability, but there is also other evidence of progress, in the form, for example, of the use of market-based mechanisms for alleviating pollution and more generally in seeking to develop mechanisms that bring together economic and environmental considerations. While progress may not have been of the type or rate sought by environmental groups, Moffatt (1992) concludes that there are encouraging signs that the move towards sustainable development is being taken seriously.

Canada

Sustainability has also become a prominent concept in Canada, perhaps not least because of the involvement of Canadians as prominent members of the Brundtland Commission and because of public hearings of the Commission staged in several Canadian cities. As in Australia, the concept was stimulated by the publication of WCS. While a national conservation strategy was not produced for Canada in the wake of WCS, a major self-assessment exercise was undertaken (Pollard and McKenzie, 1986). In relation to the objective of sustainable utilization of species and ecosystems, Canada's achievements were found to be impressive but not always sufficient. More generally, it was concluded that Canada had not fully overcome the difficulty of integrating conservation with development. Since Brundtland, however, further progress has been achieved, and the report's recommendations have been endorsed by both federal and provincial governments and expressed in the federal policy statement *The Green Plan* (Government of Canada, 1990).

Practical evaluation of progress is again difficult. Numerous programmes for individual sectors have been instigated, and some institutional changes have been introduced. For example a Bureau for Environmental Sustainability has been set up in the federal agricultural ministry. Within agriculture in particular, much attention has been directed towards the concept of sustainability. A sectoral definition of the concept has been adopted by both the federal and the provincial departments of agriculture: 'Sustainable agri-food systems are those that are economically viable, and meet society's need for safe and nutritious food while conserving and enhancing Canada's natural resources and the quality of the environment for future generations' (Science Council of Canada, 1992: 15). Such a definition brings together both environmental and economic concerns, and hence reflects something of the distinctive character of the concept of sustainability as a potential societal objective. Yet agreed definitions and shared goals do not in themselves ensure that sustainability becomes the dominant paradigm. Many of the practical measures that have been introduced have been fragmented and in some cases localized. Smith (1993) concludes that the programmes and institutional changes do not amount to a wholehearted promotion of sustainability. He notes that issues of social justice and equity have been neglected, and that there is little evidence of macro-structural change to achieve sustainability.

New Zealand

In New Zealand, the principle of sustainability has been incorporated into legislation in the shape of the Resource Management Act of 1991. This Act brings together most of the country's environmental laws with the aim of achieving the sustainable management of environmental resources. Under the Act, new activities that could have an impact on the environment require a 'resource consent' granted after a public hearing and preparation of an environmental impact assessment. The Act also facilitates the internalizing of environmental costs and better environmental accounting, as well as introducing several new instruments such as taxes and tradable permits (Smith, 1993). The Resource Management Act was introduced at a time of radical change in New Zealand, including, for example, the removal of agricultural subsidies. It is therefore difficult to isolate the effects of the Act itself from those of the broader climate of political economy. It is also too early to draw conclusions about the effectiveness of this legislative approach to promoting sustainability. The fact that the concept has reached the statute book, however, indicates that it is now being incorporated into the conventional wisdom.

Other countries

While some of the clearest indications of progress towards sustainability are to be found in countries such as Canada and New Zealand, they are not restricted to such cases. Elsewhere, national conservation strategies (of varying kinds) have been prepared in many other countries, both in the South and in the North. In Costa Rica, for example, a 'National Conservation Strategy for Sustainable Development' was outlined in 1988 (Calvo, 1990). It contained nine objectives. In addition to the standard WCS objectives were those relating to (among others) equity and social justice, balance between rural development and urban growth, and the design of population and immigration policies geared to resource constraints and acceptable living standards.

With the setting up of the UN Commission on Sustainable Development, activity is likely to accelerate. Many problems remain. For example the view is often taken that much of the apparent activity on sustainability is mere rhetoric or window-dressing, and that little has changed (or is changing) behind the façade. Such a view is expressed about the case of the United Kingdom, for instance. Both its initial response to the World Conservation Strategy and its more recent announcements about sustainability have been dismissed by some as complacent and inadequate. David Pearce, an eminent environmental economist and former government adviser, concludes that the United Kingdom's environmental record is 'pathetic', that the United Kingdom has 'enjoyed an orgy of consumption at the expense of the environment and the quality of life', and that in general sustainable development requires a more radical rethink than government has anticipated (Pearce, 1994: 26). Nevertheless, with the publication of the *This Common Inheritance* in 1990 a formal commitment was made, and annual reviews of progress towards sustainability are now published. Sustainable management has now been formally adopted as an objective of forest management, for example, and a national biodiversity plan was published early in 1994 (Department of Environment, 1994). And as in other countries environmental groups have been provided with a stick with which to beat governments performing, in their view, inadequately in relation to environmental issues.

Markets and ethics

Two major trends underlie the growth of the sustainability concept in recent years. To some extent, they seem contradictory, and perhaps the long-term success of the concept will depend on how well they can be reconciled and how closely they can be integrated. One of these trends concerns the role of market forces and the extent to which economic instruments and market regulation can become the vehicle for sustainability. The other is the growing attention focused on the role of ethics, and in particular of the stewardship ethic, in environmental management.

Markets

Many environmentalists have viewed market forces and the general climate of capitalism as a root cause of environmental problems, but recent years have seen a remarkable reversal of this stance, at least on the part of some. This reversal is perhaps a reflection of political realities. The collapse of the command economies has left capitalism unchallenged, and even before that collapse took place it was apparent that state ownership of resources did not necessarily mean sustainable management. There is little prospect, therefore, of increasing state ownership in the foreseeable future. Furthermore, the ethos of de-regulation became increasingly established in many countries during the 1980s. The ability of the state to ensure sustainable management both on land under its ownership and on other areas within its territory was therefore increasingly questioned. 'Command and control' approaches to environmental manage-ment became unfashionable. With this partial vacuum developing in the field of regulation, it is not surprising that attention should turn increasingly to market forces and economic incentives. Even environmental groups now became supporters of market regulation. At the same time, business leaders concluded that market regulation and economic instruments were more acceptable than direct state regulation. In short, a common ground opened up between environmentalists and developers which could provide an arena for dialogue on the means of achieving sustainability. A wide consensus now exists that economic incentives have an important part to play in the improved management of environmental resources. The OECD has played a major role in developing the use of economic instruments for environmental protection, especially in the field of climatic change and its human consequences (e.g. OECD, 1989a, b, 1991b, c, 1992a, 1993). (It has also been active in seeking closer integration between agricultural and environmental policies, and in investigating market and government failures in the management of environmental resources such as forests and wetlands (OECD, 1992c)).

This consensus reflects on the one hand current political realities, and on the other hand the increasing attention which economists have focused on environmental issues. This has brought with it new methods of valuing the environment and the resources which it contains (Chapter 2). If economic means of achieving environmental management are to be effective and successful, an essential prerequisite is that all environmental functions are fully valued. There has also been a growing awareness of the potentially distorting and damaging effects that can result from government subsidies and mis-pricing in agriculture and other resource sectors. For example, subsidies on pesticides can encourage over-use, while under-pricing of water can have the same effect. The result can be detrimental environmental effects. More generally, protectionism and government support to agriculture in the North have encouraged highly intensive agriculture to develop there, and have impeded trade, to the detriment of potential expor-ters in the South. An important step, therefore, has been the recognition that certain forms of govern-ment intervention in the economic aspects of resource management can have adverse environmental effects. In addition, there has been growing recognition that traditional yardsticks such as gross national product or gross domestic product are inadequate measures of development or of welfare. At the Rio Conference in 1982, it was agreed that existing systems of national accounts should be expanded to incorporate environ-mental and social dimensions. In 1993, the United Nations adopted a revised system of national accounts (SNA), under which countries are encour-aged to prepare 'satellite' accounts, using both physical and monetary units (Chapter 2).

Market-based instruments for achieving sustain-able development are regarded by some as more effective (at least under certain circumstances) than the more traditional instrument of government regulation through laws, consents and standards. For example Pearce *et al.* (1990) consider that measures such as taxes on emissions will result in the price of polluting products being higher than those of cleaner products, encouraging the consumer to choose the latter. The same authors also suggest that such instruments can maintain continuous pressure on the polluter, encouraging the introduc-tion of cleaner technology. But despite advances in valuing previously unvalued resources and in design-ing new market-based instruments, it would take a blind act of faith to accept that the market, even with appropriate reform or refinement, can be the solution to all environmental problems. (Rees, 1992, wryly observes that while 'government failure' has led to a move to reduce the direct role of government in environmental issues, 'market failure' has led to market refinement and increasing use of market mechanisms.)

Equity and ethics

Even if the environment is seen merely as a commodity, and even if problems such as the valuing of clean water or of plant and animal species can be overcome, the market cannot resolve all problems of environmental management, and it cannot replace government intervention in its entirety. One basic issue is that of equity. Poverty underlies processes such as deforestation and land degradation in many parts of the developing world, and it is difficult to see how market forces alone can overcome this problem. A link may exist between prices paid and value if there is a basic ability to pay. If that ability is not present the linkage collapses, taking with it the potential role of the market as a means of resolving environmental problems. In addition there is the broader question of distribution of costs between rich and poor (both within and between countries). In other words, while the market may be in theory a means of allocating resources *efficiently*, it does not necessarily do so *equitably*. Economic instruments and market forces alone will not solve the fundamental problems arising from uneven distribution of wealth and consumption patterns, any more than 'command and control' approaches which ignore market forces will solve all environmental problems. A mix of regulations and economic incentives, geared to an underlying objective or ethic, will be required.

Brundtland and others have therefore presented sustainability as an *ethic* and not just as the end state of the operation of (reformed) market forces. As Chapter 1 indicates, the UK White Paper on the environment (*This Common Inheritance*) began with a moral principle, namely that 'The starting policy of this Government is the ethical imperative of stewardship which must underlie all environmental policies'. While the cynic may dismiss such statements as window-dressing if not hypocrisy, there is little doubt that the concept of stewardship and other ethical concerns have attained a new prominence in recent years. At one level, this may reflect a conclusion that technical solutions to environmental problems are not feasible, any more than are solutions through economic instruments and market forces. At the core

of many problems lies the issue of distribution, and in particular distribution between the developed and developing countries, between rich and poor, and between present and future generations. Perhaps the success of the concept of sustainability in gaining world-wide currency and in reaching the agendas of international politics lies in the fact that it relates to questions of distribution and incorporates the concept of stewardship. While it can function as a purely environmental ethic, relating as it does to attitudes to nature, there is also an acknowledgement that environmental relations and social relations are irretrievably intertwined (Chapter 1), and that there may be common roots to environmental and social problems. It has been suggested recently that the central weakness in geography's response to environmental problems has been its failure to engage with questions of ethics (Reed and Slaymaker, 1993). Perhaps a similar diagnosis could be applied to other disciplines (especially to those with positivist traditions), and indeed to resource management in general. Numerous practical and theoretical difficulties confront the adoption of stewardship and similar ethics as the guiding principles of resource management. These include different religions and attitudes to nature, and different resource regimes and political systems. Whether these difficulties can be overcome will perhaps determine whether a 'sustainability revolution' takes place.

A sustainability revolution?

In a recent review of the environment in international relations, it was concluded that the fundamental environmental problem – that of achieving sustainable development – has scarcely begun to be addressed (Thomas, 1992). We clearly have not yet effected a sustainability revolution, either globally or nationally. Some commentators take a very gloomy view of progress thus far, and Myers (1993), for example, concludes that no country is anywhere near to achieving sustainability.

'Revolution' implies an abrupt and radical overturning of the old ways. Such an overturning has clearly not (yet) occurred. But neither the Agricultural Revolution nor the Industrial Revolution, with which the sustainability revolution is sometimes compared, could be regarded as an abrupt and radical overturning of the old ways. They occurred over long periods, and were partial and gradual. Both these revolutions are regarded by some as responses to earlier resource problems. In the case of the Agricultural Revolution, food shortages may have been a significant factor, while scarcity of fuelwood is identified by some as a driving force behind the Industrial Revolution (Chapter 3). If stress is a necessary factor for such a revolution, it is perhaps present in the modern world in terms of impoverished biodiversity and endangered life-support systems. But if stress is a necessary condition for such a revolution, the numerous famines that have afflicted humankind are a reminder that it is not a sufficient condition. Three requirements are identified by Ruckelhaus (1989) as necessary if a sustainability revolution is to occur: appropriate values, motivations and institutions. Stewardship and related ethics may serve as appropriate values, and motivations may be provided through economic incentives and the working of a modified or reformed market. There is some evidence that institutional adjustment is also taking place, with new supra-national bodies emerging and at least some modifications occurring in agencies and ministries at the national level. It remains to be seen, however, whether a synchronized emergence of appropriate values, motivations and institutions will take place.

There can be little doubt that the concept of sustainability has made great advances in recent years. Its fate, however, will depend not only on environmental problems and trends, but also on trends in the world economy and in international relations. One of the major concerns in the early 1990s has been the liberalization of trade and the dismantling of trade barriers. During the GATT negotiations of the early 1990s, environmental groups vigorously lobbied for the reform of GATT to take greater account of environmental concerns, just as they had done previously in relation to the World Bank (Thomas, 1992). As yet, success in integrating environmental safeguards into agreements about trade liberalization has been limited, and there have been acrimonious disputes about the use of trade instruments to achieve environmental protection. For example, environmentally motivated restrictions on imports of tropical timber products from the developing world are prohibited under GATT, although some attempts have been made to impose them (Chase, 1993). The appropriate use of trade measures for environmental aims is still being debated.

A trend towards the freeing of trade has both dangers and potential benefits. Anderson (1992) concludes that the liberalization of agricultural trade could reduce global environmental damage from farming, by discouraging subsidized production in high-intensity systems. As yet, the overall relationship between trade and environment is unclear. Stevens (1993) concludes that direct effects, both positive and negative, may be limited to specific cases, and that indirect effects, resulting from global prices and market conditions, may be greater. In theory, liberalization of trade should facilitate the operation of comparative advantage, so that different lands are able to specialize in the resource products for which they are optimally suited. While at one level specialization and monoculture are often ecologically unwelcome and in some circumstances may be unsustainable, at other levels benefits could flow as the parts of the world that are optimal in terms of potential agricultural productivity are generally not optimal in terms of biodiversity (e.g. Huston, 1993). At least in theory, therefore, food production could be increased and at the same time biodiversity maintained. For this to be achieved in practice, however, the means of estimating and allocating the costs and benefits of conserving biodiversity would need to be developed more fully than is the case at present. In other words, both ethics (in the form of stewardship and social justice) and environmental economics would need to be developed and applied more fully than as yet. Without a corresponding development in the ethical framework, the liberalization of trade could have damaging environmental effects. Strong opposition from environmentalists in developed countries could drive polluting industries and activities such as intensive agriculture and silviculture to developing countries, where opposition is more easily overcome or ignored. (It is not yet clear whether Principle 14 in the Rio Declaration (Table 12.2) will have any meaningful effect in this regard.)

In more general terms, there is the possibility that some countries will continue to seek national sustainability through unsustainable exploitation of environmental resources in other parts of the world, as they have done, in some cases, for centuries. Whether frameworks such as the UN Commission on Sustainable Development will prove sufficiently robust to cope with such possibilities remains to be seen. But even if it does not, such courses of action are likely to be perceived more and more negatively in the countries concerned. This is partly because of the growing effectiveness and sophistication of environmental groups in monitoring what is happening around the world and in bringing pressure to bear on the governments and corporations in question. But it is also partly because the main resource stresses in the world of the late twentieth century are in areas such as global atmospheric and climatic change and biodiversity. Sustainability in these sectors cannot be 'imported', and perhaps as the concept develops further so also will the notion of global inter-dependence with which it is so closely linked. Again there is the problem of the difference between rhetoric and reality. Ample acknowledgement is made of global ecological interdependence (and increasingly also of economic interdependence), but translating that acknowledgement into effective international action is another matter. There are some promising signs, however, that the 'environmental nationalism' of some countries appears to be weakening, and that national sovereignty and global interdependence can be reconciled to at least some extent. In Brazil, for example, subsidies that encouraged deforestation as recently as the late 1980s have been phased out in the face of global concern about the fate of the Amazonian forest (Hurrell and Kingsbury, 1992). Some countries, including India and Malaysia, initially declined to have the environmental conse-quences of their development programmes judged by the international community, but have since agreed to do this by supplying reports to the UN Commission on Sustainable Development (Hughes and Lea, 1993).

The human population is larger, better fed and generally more prosperous than at any time in history. The shadow of fear of global shortages of food and other resources has lifted, and much of the doom-mongering that characterized the 1970s in particular has been proved wrong. Or perhaps it has rather been misdirected. While resources have perhaps never been more abundant, the problems of mal-distribution of these resources have probably never been greater, nor the indirect threats to life-support systems more real. The emergence of the concept of sustainable development itself reflects the unsustainability that pervades the world in the latter part of the twentieth century. Whether the transition from unsustainability to sustainability takes place depends not on technical advances and solutions to physical problems of resource management, but on the much more difficult questions of ethics and values and the distribution of power and wealth.

Further reading

Grubb, M. *et al.* (1993) *The Earth Summit agreements: a guide and assessment.* London: Royal Institute of International Affairs.

Harrison, P. (1992) *The third revolution: population, environment and a sustainable world.* Harmondsworth: Penguin.

IUCN/UNEP/WWF (1980) *World conservation strategy.* Gland: IUCN/UNEP/WWF.

IUCN/UNEP/WWF (1991) *Caring for the Earth: a strategy for sustainable living.* Gland: IUCN/UNEP/WWFI.

Thomas, C. (1992) *The environment in international relations.* London: Royal Institute of International Affairs.

World Commission on Environment and Development (1987) *Our common future.* Oxford: Oxford University Press.

References

Abel, W. (1980) *Agricultural fluctuations in Europe from the thirteenth to the twentieth centuries*. London: Methuen.

Adams, W. (1990) *Green development*. London: Routledge.

Adams, W.M. (1985) River basin planning in Nigeria. *Applied Geography*, **5**, 297–308.

Adams, W.M. (1991) Large scale irrigation in northern Nigeria. *Transactions, Institute of British Geographers*, **16**, 287–300.

Agarwal, B. (1986) *Cold hearths and barren slopes*. London: Zed.

Aiken, S.R. and Leigh C.H. (1992) *Vanishing rainforests: the ecological transition in Malaysia*. Oxford: Clarendon Press.

Aitchison, J.W. and Heal, D.W. (1987) World patterns of fuel consumption: towards diversity and a low-cost energy future. *Geography*, **72**, 235–9.

Alexander, I. (1988) Western Australia: the resource state. *Australian Geographer*, **19**, 117–30.

Alexandratos, N. (ed.) (1988) *World agriculture towards 2000*. London and New York: Pinter/Belhaven Press.

Amery, H.A. (1993) The Litani River of Lebanon. *Geographical Review*, **83**, 229–37.

Anderson, A.B. (1990) Deforestation in Amazonia: dynamics, causes and alternatives. In Anderson, A.B. (ed.) *Alternatives to deforestation: steps towards sustainable use of the Amazon rainforest*. New York: Columbia University Press, pp. 1–23.

Anderson, E.W. (1988) *Strategic minerals: the geopolitical problem for the United States*. New York: Praeger.

Anderson, E.W. (1991) United States Law of the Sea policy and the strategic minerals supply problem. In Smith, H.D. and Vallega, A. (eds) *The development of integrated sea use management*. London: Routledge, pp. 260–72.

Anderson, K. (1992) Agricultural trade liberalization and the environment: a global perspective. *World Economy*, **15**, 153–71.

Andreae, B. (1981) *Farming: development and space*. Berlin: Gruyter.

Andrews, R.N.L. (1980) Class politics or democratic reform: environmentalism and American political institutions. *Natural Resources Journal*, **20**, 221–42.

Aschmann, H. (1970) The natural history of a mine. *Economic Geography*, **46**, 172–89.

Ashworth, W. (1986) *The late Great Lakes: an environmental history*. New York: Alfred A. Knopf.

Attfield, R. (1983) Christian attitudes to nature. *Journal of the History of Ideas*, **44**, 369–86.

Attfield, R. (1991) *The ethics of environmental concern*. Athens: University of Georgia Press.

Auty, R.M. (1990) *Resource-based industrialization: sowing the oil in eight developing countries*. Oxford: Clarendon Press.

Auty, R.M. (1993) *Sustaining development in mineral economies: the resource curse thesis*. London: Routledge.

Bahn, P. and Flenley, V. (1992) *Easter Island, Earth Island*. London: Thames and Hudson.

Baker, R. (1992) Land rights in the Borroloola area of Australia's Northern Territory. *Applied Geography*, **12**, 162–75.

Barbier, E.B. (1989) *Economics, natural resources scarcity and development: conventional and alternative views*. London: Earthscan.

Barde, J.-P. and Pearce, D.W. (eds) (1991) *Valuing the environment: six case studies*. London: Earthscan.

Barnett, H.J. (1979) Scarcity and growth revisited. In Smith, V.K. (ed.) *Scarcity and growth reconsidered*. Baltimore and London: Johns Hopkins Press for Resources for the Future: pp. 163–217.

Barnett, H.J. and Morse, C. (1963) *Scarcity and growth: the economics of natural resource availability*. Baltimore: Johns Hopkins Press for Resources for the Future.

Barney, G.O. (1980) *The Global 2000 report to the President of the US*. New York: Pergamon.

Batisse, M. (1990) Development and implications of the biosphere reserve concept and applicability to coastal regions. *Environmental Conservation*, **17**, 111–16.

Baumann, D.D. and Dworkin, D. (1978) Water resources for our cities. *Association of American Geographers Resource Paper 78*.

Beaumont, P. (1978) Man's impact on river systems: a world-wide view, *Area*, **10**, 38–42.

Beaumont, P. (1989) *Environmental management and development in drylands*. London: Routledge.

Beckerman, W. (1972) Economists, scientists and the environmental catastrophe. *Oxford Economic Papers*, **24**, 327–44.

Beinart, W. (1989) Introduction: the politics of colonial conservation. *Journal of Southern African Studies*, **15**, 143–62.

Bell, F.C. (1988) The sharing of scarce water resources. *Geoforum*, **19**, 353–66.

Bennett, J.W. (1976) *The ecological transition: cultural anthropology and human adaptions.* New York: Pergamon Press.

Bennett, J.W. and Dahlberg, K.A. (1990) Institutions, social organization and cultural values. In Turner, B.L. *et al.* (eds) *The Earth as transformed by human action.* Cambridge: CUP with Clark University, pp. 69–86.

Berkes, F. (1985) Fishermen and 'the tragedy of the commons'. *Environmental Conservation,* **12**, 199–206.

Berkes, F. (ed.). (1989) *Common property resources: ecology and community-based sustainable development.* London: Belhaven.

Berkes, F. and Farvar, M.T. (1989) Introduction and overview. In Berkes, F. (ed.) *Common property resources: ecology and community-based sustainable development.* London: Belhaven, pp. 7–21.

Berkes, F., Feeny, D., McCay, B.J. and Anderson, J.M. (1989) The benefits of the commons. *Nature,* **340**, 91–3.

Bernstam, M.S. (1991) The wealth of nations and the environment. In Davis, K. and Bernstam, M.S. (eds) *Resources, environment and population: present knowledge, future options.* New York and Oxford: OUP, pp. 333–73.

Berry, R.J. (1990) Identification of the commitments and limits of any environmental ethic, and science, mankind and ethics. In Bourdeau, Ph., Fasella, P.M. and Teller, A. (eds) *Environmental ethics: man's relationship with nature, interactions with science.* Brussels and Luxembourg: Office for Official Publications of EC, pp. 203–20 and 289–306.

Berry, R.J. (ed.) (1992) *Environmental dilemmas: ethics and decisions.* London: Chapman & Hall.

Black, J.N. (1970) *The dominion of man.* Edinburgh: Edinburgh University Press.

Blaikie, P. and Brookfield, H. (1987) *Land degradation and society.* London: Methuen.

Blunden, J. (1985) *Mineral resources and their management.* Harlow: Longman.

Blunden, J. (1991) The environmental impact of mining and mineral processing. In Blunden, J. and Reddish, A. (eds) *Energy, resources and environment.* London: Open University, pp. 79–131.

Bogucka, M. (1978) North European commerce as a solution to resource shortage in the 16th–18th centuries. In Maczak, A. and Parker, W.N. (eds) *Natural resources in European history.* Washington DC: Resources for the Future, pp. 9–42.

Bookchin, M. (1985) *Post-scarcity anarchism.* Montreal: Black Rose.

Boserup, E. (1965) *The conditions of agricultural growth: the economics of agricultural change under population pressure.* London: Allen & Unwin.

Boserup, E. (1976) Environment, population and technology in primitive societies. *Population and Development Review,* **2**, 20–36.

Boserup, E. (1983) The impact of scarcity and plenty on development. *Journal of Interdisciplinary History,* **14**, 185–209.

Boughey, A.S. (1980) Environmental crisis – past and present. In Bilsley, L.J. (ed.) *Historical ecology: essays on environment and social change.* New York: Kenneticut Press, pp. 9–32.

Bourdeau, Ph., Fasella, P.M. and Teller, A. (1990) *Environmental ethics: man's relationship with nature, interactions with science.* Brussels and Luxembourg: Office for Official Publications of EC.

Bowonder, B., Prasal, S.S.R. and Unni, N.V.M. (1987) Deforestation around urban centres in India. *Environmental Conservation,* **14**, 23–8.

Bradbury, J. (1982) Some geographical implications of the restructuring of the iron ore industry 1950–1980. *Tijdshrift voor Economische en Sociale Geografie,* **73**, 295–306.

Brader, L. (1987) Plant protection and land transformation. In Wolman, M.G. and Fournier, F.G.A. (eds) *Land transformation in agriculture.* Chichester: Wiley, pp. 227–48.

Bradley, I. (1990) *God is green: Christianity and the environment.* London: Darton, Longman and Todd.

Braudel, F. (1979) *Civilization and capitalism 15th–18th century: vol 1. The structures of everyday life: the limits of the possible.* London: Collins.

Bromley, D.W. (1990) Arresting renewable resources degradation in the Third World *American Journal of Agricultural Economics,* **75**, 1274–5.

Bromley, D.W. (1991) *Environment and economy: property rights and public policy.* Oxford: Blackwell.

Bunker, S.G. (1984) Modes of extraction: unequal exchange and the progressive underdevelopment of an extreme periphery: the Brazilian Amazon 1600–1980. *American Journal of Sociology,* **89**, 1017–64.

Buringh P (1985) The land resource for agriculture. *Philosophical Transactions of the Royal Society of London B* **310**: *151-9.*

Buringh, P. and Dudal, R. (1987) Agricultural land use in space and time. In Wolman, M.G. and Fournier, F.G.A. (eds) *Land transformation in agriculture.* Chichester: Wiley, pp. 9–44.

Burnett, G.W. and Stilwell, H.B. (1990) National park and equivalent reserve creation in French and British Africa. *Society and Natural Resources,* **3**, 29–41.

Burns, R. (1789) *The complete poetical works of Robert Burns* (n.d.). London and Edinburgh: Nelson.

Butzer, K.W. (1992) The Americas before and after 1492: an introduction to current geographical research *Annals, Association of American Geographers,* **82**, 345–69.

Calvo, J.C. (1990) The Costa Rican national conservation strategy for sustainable development: exploring the possibilities. *Environmental Conservation,* **17**, 355–8.

Campbell, J.L. (1988) *Collapse of an industry: nuclear power and the contradictions of US energy policy.* Ithaca: Cornell University Press.

Carson, R. (1963) *Silent spring.* London: Hamilton.

Central Statistical Office (1992) *Social trends.* London: HMSO.

Chandler, W.V. (1984) *The myth of TVA conservation and development in the Tennessee Valley 1933–1983.* Cambridge (MA): Ballinger.

Chapman, K. (1976) *North Sea oil and gas: a geographical perspective.* Newton Abbott: David and Charles.

Chapman, K. (1981) Issues in environmental impact assessment. *Progress in Human Geography,* **34**, 405–16.

Chapman, K. (1987) Control of resources and the development of the petrochemical industry in Alberta. *Canadian Geographer,* **29**, 310–26.

Chapman, K. and Walker, D.F. (1991) *Industrial location* (2nd edn). Oxford: Blackwell.

Chase, B.F. (1993) Tropical forests and trade policy: the

legality of unilateral attempts to promote sustainable development under the GATT. *Third World Quarterly*, **14**, 749–74.

Chenery, H.B. (1965) Land: the effects of resources for economic growth. In Berrill, K. (ed.) *Economic development with special reference to East Asia*. London: Macmillan, pp. 19–52.

Chern, W. and James, W.E. (1988) Measurements of energy productivity in Asian countries. *Energy Policy*, **16**, 494–505.

Chisholm, M. (1980) The wealth of nations. *Transactions of the Institute of British Geographers*, **5**, 255–76.

Chisholm, M. (1982) *Modern world development: a geographical perspective*. London: Hutchinson.

Chisholm, M. (1990) The increasing separation of production and consumption. In Turner, B.L. *et al.* (eds) *The Earth as transformed by human action*. Cambridge: CUP with Clark University, pp. 87–102.

Ciriacy-Wantrup, S.V. (1952) *Resource conservation, economics and policies*. Berkeley: University of California Press.

Ciriacy-Wantrup, S.V. and Bishop, R.C. (1975) 'Common property' as a concept in natural resources policy. *Natural Resources Journal*, **15**, 713–27.

Clark, B. (1990) 'The range of the mountains is his pasture' – environmental ethics in Israel. In Engel, J.R. and Engel, J.V. (eds) *Ethics of environment and development : global challenge, international response*. London: Belhaven, pp. 183–88.

Clark, T.D. (1984) *The greening of the South: the recovery of land and forest*. Lexington: University of Kentucky Press.

Coddington, A. (1970) The economics of ecology. *New Society*, 9 April, 95–7.

Cohen, M.N. (1977) *The food crisis in prehistory*. New Haven: Yale UP.

Cole, H.S.D., Freeman, C., Johoda, M. and Pavitt, K.C.R. (1973) *Thinking about the future: a critique of the Limits of Growth*. London: Chatto and Windus.

Connell, J. and Howitt, R. (1991) *Mining and indigenous people in Australia*. Sydney: Sydney University Press.

Cook, E. (1971) The flow of energy in an industrial society. In Scientific American, *Energy and power*. San Francisco: W.H. Freeman, pp. 83–94.

Cook, E. (1976) *Man, energy, society*. San Francisco: W.H. Freeman.

Cook, E. (1977) Energy the ultimate resource. *Association of American Geographers (Commission on College Geography) Resource Paper* 77-4, Washington DC.

Coppola, P. (1978) Natural resources and economic development in the Mediterranean Basin. In Maczak, A. and Pouler, W.N. (eds) *Natural resources in European history*. Washington DC: Resources for the Future, pp. 205–20.

Cottrell, F. (1955) *Energy and society*. New York: McGraw Hill.

Coull, J.R. (1975) The big fish pond: a perspective on the contemporary situation in the world's fisheries. *Area*, **7**, 103–7.

Coull, J.R. (1993a) *World fisheries resources*. London: Routledge.

Coull, J.R. (1993b) Will a blue revolution follow the green revolution? The modern upsurge of aquaculture. *Area*, **25**, 350–7.

Cox, T.R., Maxwell, R.S., Thomas, P.D. and Malone, J.J. (1985) *This well-wooded land: Americans and their forests from Colonial times to the present*. Lincoln and London: University of Nebraska Press.

Cribb, R. (1988) Conservation policy and politics in Indonesia 1945–1988. In Dargavel, J. *et al.* (eds) *Changing topical forests*. Canberra: ANU, pp. 341–51.

Cronon, W. (1983) *Changes in the land: Indians, colonists and the ecology of New England*. New York: Hill and Wang.

Crosby, A.W. (1986) *Ecological imperialism 900–1900*. Cambridge: CUP.

Crosson, P.R. (1991) Cropland and soils: past performance and policy challenges. In Frederick, K.D. and Sedjo, R.A. (eds) *American renewable resources: historical trends and current challenges*. Washington DC: Resources for the Future, pp. 119–204.

Crosson, P.R. and Rosenburg, N.J. (1989) Strategies for agriculture. *Scientific American*, **261**, 78–85.

Crowson, P.C.F. (1988) A perspective on worldwide exploration for minerals. In Tilton, J.E., Eggert, R.G. and Landsberg, H.H. (eds) *World mineral exploration : trends and economic issues* . Washington DC: Resources for the Future, pp. 21–104.

Daly, H.E. (1991) Sustainable development: from conceptual theory to operational principles. In Davis, K. and Bernstam, M.S. (eds) *Resources, environment and population: present knowledge, future options*. New York and Oxford: OUP, pp. 25–43.

Dando, W.A. (1980) *The geography of famine*. London: Arnold.

Daniel, J.B.M. (1988) Has the South African environment been respected? *South African Geographer*, **15**, 3–11.

Daniel, P. (1992) Economic policy in mineral exporting countries: what have we learned? In Tilton, J.E. (ed.) *Mineral wealth and economic development*. Washington DC: Resources for the Future, pp. 81–121.

Dargarvel, J. (1992) Incorporating natural forests into the new Pacific economic order: processes and consequences. In Dargarvel, J. and Tucker, R. (eds) *Changing Pacific forests: historical perspectives on the forest economy of the Pacific Basin*. Durham, NC: Forest History Society, pp. 1–18.

Dasgupta, P. (1990) The environment as a commodity. *Oxford Review of Economic Policy*, **6**, 51–67.

Dasmann, R.F. (1988) Towards a biosphere consciousness. In Worster, D. (ed.) *The ends of the earth: perceptions on modern environmental history*. Cambridge: CUP, pp. 177–88.

Davidson, J. (1990) Values and uses: seeing the forest through different eyes. In Webb, L.J. and Kikkawa, J. (eds) *Australian tropical rainforests: science, values and meaning*. Melbourne: CSIRO, pp. 124–32.

Davis, D.W. (1991) Oil in the northern Gulf of Mexico. In Smith, H.D. and Vallega, A. (eds) *The development of integrated sea-use management*. London: Routledge, pp. 139–52.

Davis, K. (1991) Population and resources: fact and interpretation. In Davis, K. and Bernstam, M.S. (eds*). Resources, environment and population: present knowledge, future options*. New York and Oxford: OUP, pp. 1–24.

Davis, S.D. *et al.* (1986) *Plants in danger: what do we know?* Gland and Cambridge: IUCN.

Dearden, P. (1989) Wilderness and our common future. *Natural Resources Journal*, **29**, 205–22.

Deen, M.Y.I. (1990) Islamic environmental ethics: law and society. In Engel, J.R. and Engel, J.G. (eds) *Ethics of environment and development: global challenge, international response*. London: Belhaven, pp. 189–98.

Delcourt, H.R. and Harris, W.F. (1980) Carbon budget of the southeastern US biota: analysis of historical change in trend from source to sink. *Science*, **210**, 321–3.

Denevan, W.M. (1992) The pristine myth: the landscape of the Americans in 1492. *Annals, Associations of American Geographers*, **82**, 369–85.

Department of Environment (UK Govt) (1994) *Biodiversity: the UK action plan*. London: HMSO.

Department of Environment (UK Govt) (1994) *Climatic change – the UK programme*. London: HMSO.

Department of Environment (UK Govt) (1994) *Sustainable forestry: the UK programme*. London: HMSO.

Department of Environment (UK Govt) (1994) *Sustainable development: the UK strategy*. London: HMSO.

Department of Trade and Industry (1993) *Digest of UK energy statistics 1993*. London: HMSO.

de Saussay, C. (1987) Land tenure systems and forest policy *FAO Legislative Study no 41*.

Devall, B. (1980) The Deep Ecology movement. *Natural Resources Journal*, **20**, 299–323.

de Vries, J. (1981) Patterns of urbanization in pre-industrial Europe 1500–1800. In Schmal, H. (ed.) *Patterns of European urbanization since 1500*. London: Croom Helm, pp. 77–110.

Diamond, J.M. (1987) Human use of world resources. *Nature*, **328**, 479–80.

Dodeson, P. (1983) Land is sacred to us. In Howitt, R. and Douglas, J. (eds) *Aboriginals and mining companies in northern Australia* . Chippendale: Alternative Publishing Co-operative Ltd.

Donlon, N. (1992) The Earth Summit. *House of Commons Library 92/6*.

Dovers, S.R. and Handmer, V.W. (1993) Contradictions in sustainability. *Environmental Conservation*, **20**, 217–22.

Dudley, R.L. (1990) A framework for natural resource management. *Natural Resource Journal*, **30**, 107–22.

Dunlap, R.E. and Van Liere, K. (1978) The New Environmental Paradigm: a proposed measuring instrument and preliminary results. *Journal of Environmental Education*, **9**, 10–19.

Durning, A.B. (1990) Apartheid's environmental toll *Worldwatch Paper 95*. New York: Worldwatch Institute.

Earle, C. (1988) The myth of the southern soil miner: macrohistory, agricultural innovation, and environmental change. In Worster, D. (ed.) *The ends of the earth*. Cambridge: CUP, pp. 175–210.

Earney, F.C.F. (1990) *Marine mineral resources*. London: Routledge.

Economic Council of Canada (1985) *Connections: an energy strategy for the future*. Ottawa: Minister of Supply and Services.

Edgell, M.C.R. and Nowell, D.E. (1989) The New Environmental Paradigm scale: wildlife and environmental beliefs in British Columbia. *Society and Natural Resources*, **2**, 285–96.

Edmonds, J. and Reilly, J.M. (1985) *Global energy: assessing the future*. New York: OUP.

Ehrlich, P. (1968) *The population bomb: population control or race oblivion*. New York: Ballantine.

Ehrlich, P.H. and Holdren, J.R. (1971) The impact of population growth, *Science*, **171**, 1212–17.

Eidsvik, H.K. (1980) National parks and other protected areas: some reflections on the past and prescriptions for the future. *Environmental Conservation*, **7**, 185–90.

Eidsvik, H.K. (1989) The status of wilderness: an international review. *Natural Resources Journal*, **29**, 57–82.

Engel, R. (1988) Ethics. In Pitt, D.C. (ed.) *The future of the environment: the social dimension of conservation and ecological alternatives*. London, Routledge, pp. 23–45.

Evans, J. (1982, 1992) *Plantation forestry in the tropics*. Oxford: Clarendon.

Evans, J. (1987) Site and species selection – changing perspectives. *Forest Ecology and Management*, **21**, 299–310.

Eyre, L.A. (1987) Jamaica: test case for tropical deforestation? *Ambio*, **16**, 338–43.

Eyre, S.R. (1978) *The real wealth of nations*. London: Arnold.

Falkenmark, M. (1986) Fresh water – time for a modified approach. *Ambio*, **15**, 192–200.

Faber, M. (1986) States, mining companies and state corporations: a review article. *Institute of Development Studies Bulletin*, **17**(4), 66.

FAO annual *Production yearbooks*.

FAO (1982) *Map of the fuelwood situation in developing countries*. Rome: FAO.

FAO (1984) *Land, food, people*. Rome: FAO.

FAO (1990) *FAO production yearbook*, vol. 44. Rome: FAO.

FAO (1991) *The state of food and agriculture*. Rome: FAO.

FAO (1992) *The state of food and agriculture 1992*. Rome: FAO.

Favre, D.S. (1989) *International trade in endangered species: a guide to CITES* . Dordrecht: Martinus Nijhoff.

Fernie, J. and Pitkethly, A.S. (1985) *Resources: environment and policy*. London: Harper & Row.

Finkel, H.J. (1982) *Handbook of irrigation technology vol. 1*. Boca Raton, Florida: CRC press.

Firey, W.I. (1960) *Man, mind and land: a theory of resource use*. Glencoe Ill: Free Press.

Fisher, A.C., Krutilla, J.V. and Cicchetti, C.V. (1972) Alternative uses of natural environments: the economics of environmental modification. In Krutilla, J.V. (ed.) *Natural environments: studies in theoretical and applied analysis*. Washington DC: Resources for the Future, pp. 18–53.

Forbes, D. (1982) Energy imperialism and a new international division of resources: the case of Indonesia. *Tijdschrift voor Economische en Sociale Geografie*, **73**, 94–108.

Forrester, J.W. (1971) *World dynamics*. Cambridge, Mass.: Wright Allen.

Fowler, C. (1993) Biological diversity in a North–South context. In Bergesen, H.O. and Parmann, G. (eds) *Green Globe Yearbook 1993*, Oxford: OUP, pp. 35–44.

Friedrich, E. (1904) Wesen und geographische Verbreitung de Raubwritschift. *Petermanns Mitteilungen*, **50**, 68–79, 92–95.

Frosch, A. and Gallopoulos, N.E. (1989) Strategies for manufacturing. *Scientific American*, **261**(3), 94–103.

Gadgil, M. and Iyer, P. (1989) On the diversification of common-property resource use by Indian society. In

Berkes, F. (ed.) *Common property resources: ecology and community-based sustainable development*. London: Belhaven, pp. 240–72.

Gardner, B.D. (1991) Rangeland resources: changing uses and productivity. In Frederich, K.D. and Sedjo, R.A. (eds) *America's renewable resources: historical trends and current challenges*. Washington DC: Resources for the Future, pp. 123–69.

Genovese, E.D. (1965) *The political economy of slavery: studies in the economy and society of the slave south*. New York: Pantheon.

GESAMP (1990) *The state of the marine environment*. Oxford: Blackwell.

Gibbs, C.V.N. and Bromley, D.W. (1989) Institutional arrangements for management of rural resources, common property regions. In Berkes, F. (ed.) *Common property resources: ecology and community-based sustainable development*. London: Belhaven, pp. 22–32.

Gillis, M. (1992) Forest concession management and revenue policies. In Sharma, N.P. (ed.) *Managing the world's forests*. Dubuque, Iowa: Kendall/Hunt, pp. 139–76.

Ginsburg, N. (1957) Natural resources and economic development. *Annals, Association of American Geographers*, **47**, 197–212.

Ginsburg, N. Osborn, J. and Blank, G. (1986) Geographic perceptions on the wealth of nations. Department of Geography, University of Chicago, *Research Paper 220*.

Glassner, M.I. (1990) *Neptune's domain: a political geography of the sea*. Boston: Unwin Hyman.

Gleick, P.H. (1987) *Global climatic changes and regional hydrology: impacts and responses*. Internat. Assoc. of Scientific Hydrology Publication 168, pp. 389–402.

Goldsmith, E., Allen, R., Allaby, M., Davoll, J. and Lawrence, S. (editors of *The Ecologist*) (1972) *Blueprint for survival*. Harmondsworth: Penguin.

Golley, F.B. (1992) Environmental attitudes in North America. In Berry, R.J. (ed.) *Environmental dilemmas, ethics and decisions*. London: Chapman & Hall, pp. 20–32.

Gomez-Pompa, A. and Kaus, A. (1990) Traditional management of tropical forests in Mexico. In Anderson, A.B. (ed.) *Alternatives to deforestation: steps towards sustainable use of the Amazon rainforest*. New York: Columbia University Press, pp. 45–64.

Goodland, R., Ledec, G. and Webb, M., (1989) Meeting environmental concerns caused by common property mismanagement in economic development projects. In Berkes, F. (ed.) *Common property resources: ecology and community-based sustainable development*. London: Belhaven, pp. 148–63.

Gordon, H.S. (1954) The economic theory of a common property resource – the fishery. *Journal of Political Economy*, **62**, 124–142.

Government of Canada (1990) *The Green Plan: a framework for discussion on the environment*. Ottawa: Government of Canada.

Graf, W.L. (1992) Science, public policy and western American rivers. *Transactions of Institute of British Geographers*, **17**, 5–19.

Grainger, A. (1993) *Controlling tropical deforestation*. London: Earthscan.

Griffin, K. (1987) *World hunger and the world economy*. London: Macmillan.

Grigg, D. (1981) The historiography of hunger: changing views on the world food problem 1945–1980. *Transactions of Institute of British Geographers*, **6**, 279–92.

Grigg, D. (1982) Counting the hungry: world patterns of undernutrition. *Tijdschrift voor Economische en Sociale Geografie*, **73**, 66–79.

Grigg, D. (1985) *The world food problem*. Oxford: Blackwell.

Grigg, D. (1987) The Industrial Revolution and land transformation, In Wolman, M.G. and Fournier, F.G.A. (eds) *Land transformation in agriculture*. Chichester: Wiley, pp. 79–110.

Grigg, D. (1992) *The transformation of agriculture in the west*. Oxford: Blackwell.

Grove, R. (1988) Conservation and colonialism, the evolution of environmental attitudes and conservation policies on St Helena, Mauritius and in West India 1660–1854. In Dargavel, J., Dixon, K. and Semple, N. (eds) *Changing tropical forests*. Canberra: ANU, pp. 19–45.

Grove, R.H. (1990) Colonial conservation, ecological hegemony and popular resistance: towards a global synthesis. In Mackenzie J.M. (ed.) *Imperialism and the natural world*. Manchester: Manchester UP, pp. 15–50.

Grove, R.H. (1992) Origins of Western environmentalism. *Scientific American*, **267**, 22–27.

Grubb, M. *et al.* (1993) *The Earth Summit agreements: a guide and assessment*. London: Earthscan.

Haglund, D.G. (1989) The new geopolitics of minerals: an inquiry into the changing international significance of strategic minerals. In Haglund, D.G. (ed.) *The new geopolitics of minerals: Canada and international resource trade* . Vancouver: University of British Columbia Press, pp. 3–36.

Haigh, N. (1989) *EEC environmental policy and Britain*. London: Longman.

Hall, D.C. and Hall, V.V. (1984) Concepts and measures of natural resource scarcity, with a summary of recent trends. *Journal of Environmental Economics and Management*, **11**, 363–79.

Hardin, G. (1968) The tragedy of the commons. *Science*, **162**, 1243–8.

Hardin, G. (1972) *Exploring new ethics for survival: the voyage of the Spaceship Beagle*. New York: Viking Press.

Hardin, G. (1974) Living on a lifeboat. *BioScience*, **24**, 10.

Harrington, W. (1991) Wildlife: severe decline and partial recovery. In Frederick, K.D. and Sedjo, R.A. (eds) *America's renewable resources*. Washington: Resources for the Future, pp. 205–28.

Harrison, P. (1993) *The third revolution: population, environment and a sustainable world*. Harmondsworth: Penguin.

Hawkins, K. (1984) *Environment and enforcement : regulation and the social definition of pollution*. Oxford: Oxford University Press.

Hay, A.M. (1976) A simple location theory for mining activity. *Geography*, **61**, 65–76.

Hays, S.P. (1959) *Conservation and the gospel of efficiency: the progressive conservation movement 1890–1920*. Cambridge, MA: Harvard University Press.

Hays, S.P. (1987) *Beauty, health and permanency: environmental politics in the United States 1955–1985*. Cambridge: CUP.

Healey, M.J. and Ilbery, B.W. (1990) *Location and change: perspectives on economic geography*. Oxford: OUP.

Heathcote, R.C. and Mabbutt, J.A. (eds) (1988) *Land, water and people, geographical essays in Australian resource management*. Sydney: Allen & Unwin.

Hecht, S. and Cockburn, A. (1989) *The fate of the forest*. Harmondsworth: Penguin.

Helm, D. and Pearce, D. (1990) Economic policy towards the environment. *Oxford Review of Economic Policy*, **6**, 1–16.

Homer-Dixon, T.F., Boutwell, J.H.H. and Rathjens, G.W. (1993) Environmental change and violent conflict. *Scientific American*, **268**, 16–23.

Hosang, J.B. (1992) Trade with endangered species. In Bergesen, H.O., Nordenhang, M. and Parmann, G. (eds) *Green Globe Yearbook 1992*. Oxford: OUP, pp. 58–70.

Hosier, R.H., Mwandosya, M.J. and Luhanga, M.L. (1993) Future energy development in Tanzania: the energy costs of urbanization. *Energy Policy*, **21**, 524–42.

Hough, J.H. (1988) Obstacles to effective management of conflicts between national parks and surrounding human communities in developing countries. *Environmental Conservation*, **15**, 129–36.

Houghton, R.A. and Skole, D.L. (1990) Carbon. In Turner II, B.L. *et al.* (eds) *The Earth as transformed by human action*. Cambridge: CUP, pp. 393–408.

Houghton, R.A. and Woodwell, G.M. (1989) Global climatic change. *Scientific American*, **260**, 18–27.

Houghton, R.A., Hobbie, J.E., Melillo, J.M., Moore, B., Peterson, B.J., Shaver, G.R. and Woodwell, G.M. (1983) Changes in the carbon content of terrestrial biota and soils between 1860 and 1980: a net release of CO_2 to the atmosphere. *Ecological Monograph*, **53**, 235–62.

Howitt, R. (1992) Weipa: industrialisation and indigenous rights in a remote Australian mining area. *Geography*, **77**, 223–35.

Hubbert, M.K. (1962) *Energy resources*. Washington DC: National Academy of Sciences Publication 1000-D.

Hubbert, M.K. (1971) The energy resources of the Earth. In Scientific American *Energy and power*. San Francisco: W.H. Freeman, pp. 31–40.

Hudson, N.W. (1985) A world view of the development of soil conservation. *Agricultural History*, **59**, 326–39.

Hughes, P. (1993) Biodiversity. *House of Commons Library Research Paper* 93/94.

Hughes, P. and Lea, W. (1993) The Earth Summit: one year on. *House of Commons Library Research Paper* 93/71.

Humphrey, W.S. and Stanislaw, J. (1979) Economic growth and energy consumption in the UK, 1700–1975. *Energy Policy*, **7**, 29–42.

Hurrell, A. and Kingsbury, B. (eds) (1992) *The international politics of the environment*. Oxford: Clarendon.

Huston, M. (1993) Biological diversity, soils and economics. *Science*, **262**, 1676–9.

Ingold, T. (1986) *The appropriation of nature: essays on human ecology and social relations*. Iowa City: University of Iowa Press.

Inman, K. (1993) Fueling expansion in the Third World: population, development, debt and the global decline of forests. *Society and Natural Resources* **6**, 17–40.

Intergovernmental Panel on Climate Change (IPCC) (1990) *Climate change: the IPCC scientific assessment* (edited by Houghton, J., Jenkins, G.J. and Ephraums, J.J.). Cambridge: Cambridge University Press, pp. 207.

IUCN (1985) *United Nations list of national parks and protected areas*. Gland: IUCN.

IUCN with UNEP and WWF (1980) *World Conservation Strategy*. Gland: IUCN.

IUCN, UNEP, WWF (1991) *Caring for the Earth, a strategy for sustainable living*. London: Earthscan.

Jackson, C.I. (1993) The Great Lakes: exploring the ecosystem. In Thomas C. and Howlett D. (eds) *Resource politics: freshwater and regional relations*. Buckingham: Open University, pp. 23–46.

Jeffrey, J.W. (1987) The fatal flaws in the Sizewell report: a review of the economics of Sizewell B. *Energy Policy*, **15**, 456–62.

Jennings, J.S. (1989) The deep offshore – commercial and technical perspectives. (Address by Group Managing Director). London, Shell International Petroleum.

Jevons, W.S. 1886 *The coal question: an inquiry concerning the progress of the nation and the probable exhaustion of our coal mines*. London:, Macmillan.

Joffe, G. (1993) The issue of water in the Middle East and North Africa. In Thomas C. and Howlett D. (eds) *Resource politics: freshwater and regional relations*. Buckingham: Open University, pp. 65–85.

Johnson, C.V. (1990) Ranking countries for mineral exploration. *Natural Resources Forum*, **14**, 178–86.

Johnston, R.J. (1989) *Environmental problems: nature, economy and state*. London: Belhaven.

Johnston, R.J. (1992) Laws, states and superstates, international law and the environment. *Applied Geography*, **12**, 211–28.

Jones, P.D., Wigley, T.M.L. and Wright, P.B. (1986) Global temperature variations between 1861 and 1984. *Nature* **322**, 430–34.

Jones, T. and Wibe, S. (1991) *Forests: market and intervention failures*. London: Earthscan.

Kanowski, P.J. and Savill, P.S. (1992) 'Forest plantations: towards sustainable practice'. In Sargent, C. and Bass, S. (eds) *Plantation politics: forest plantations in development*. London: Earthscan.

Kanowski, P. *et al.* (1992) Plantation forestry. In Sharman, N.P. (ed.) *Managing the world's forests*. Dubuque, Iowa: Kendall/Hunt, pp. 375–402.

Kates, R.W. (1992) Times of hunger. In Wong, S.T. (ed.) *Person, place and thing: interpretive and empirical essays in cultural geography*. Baton Rouge, GeoScience Publications, pp. 275–300.

Kay, J. (1985) Preconditions of natural resource conservation. *Agricultural History*, **59**, 124–35.

Kay, J. and Brown, C.J. (1985) Common beliefs about land and natural resources. *Journal of Historical Geography*, **11**, 253–67.

Kimuyu, P.K. (1993) Urbanization and consumption of petroleum products in Kenya. *Energy Policy*, **21**, 403–7.

Kiss, A. (1985) The protection of the Rhine against pollution. *Natural Resources Journal*, **25**, 613–38.

Kiss, A. and Shelton, D. (1991) *International environmental law*. New York: Ardsley.

Kitabatake, Y. (1992) What can be learned from domestic and international aspects of Japan's forest resource utilization? *Natural Resources Journal*, **32**, 856–81.

Kliot, N. (1993) *Water resources and conflict in the Middle East*. Routledge: London.

Komarov, B. (1980) *The destruction of nature in the Soviet Union*. New York: White Plains.

Kondratieff, N.D. (1935) The long waves in economic life. *Review of Economic Statistics*, **17**, 105–15.

Koppes, C.R. (1988) Efficiency, equity and esthetics: shifting themes in American conservation. In Worster, D. (ed.) *The ends of the Earth*. Cambridge: CUP, pp. 230–51.

Kornai, G. (1987) Historical analysis of international trade in forest products. In Kallin, M., Dykstra, D.P. and Binkley, C.S. (eds) *The global forest sector: an analytical perspective*. Chichester: Wiley, pp. 432–56.

Kummer, D.M. (1992) Deforestation in the postwar Philippines. *Chicago University, Geography Department Research Paper 234*.

Kuzmiak, D.T. (1991) The American environmental movement. *Geographical Journal*, **157**, 265–78.

L'vovich, M.I., White, G.F., Belyaev, A.V., Kindler, J., Koronkevic, N.I., Lee, T.R. and Voropaev, G.V. (1990) Use and transformation of terrestrial water systems. In Turner, B.L. *et al.* (eds) *The Earth as transformed by human action*. Cambridge: CUP, pp. 235–52.

Laarman, J.G. (1988) Export of tropical hardwoods in the twentieth century. In Richards, J.F. and Tucker, R.P. (eds) *World deforestation in the twentieth century*. Durham and London: Duke University Press, pp. 148–63.

Lado, C. (1992) Problems of wildlife management and land use in Kenya. *Land Use Policy*, **9**, 169–84.

Lamb, D. (1990) *Exploiting the tropical rainforest: an account of pulpwood logging in Papua New Guinea*. Paris: Parthenon.

Lamb, D. (1991) Combining traditional and commercial uses of rainforests. *Nature and Resources*, **27**(2), 3–11.

Lamb, H.H. (1977) *Climate: present, past and future vol 2 Climatic history and the future*. London: Methuen.

Lanning, G. and Mueller, M. (1979) *Africa undermined*. Harmondsworth: Penguin.

Lappé, F.M. and Collins, J. (1988) *World hunger: 12 myths*. London: Earthscan.

Larson, E.D., Ross, M.H. and William, R.H. (1986) Beyond the era of materials. *Scientific American*, **254**, 24–31.

Lawson, R.M. (1984) *Economics of fisheries development*. London: Pinter.

Lazenby, J.B.C. and Jones, P.M.S. (1987) Hydroelectricity in West Africa: its future role. *Energy Policy*, **15**, 441–55.

Le Heron, R.B. (1988) The internationalization of New Zealand's forestry companies and the social reappraisal of New Zealand's exotic forest resource. *Environment and Planning A*, **20**, 489–515.

Ledec, G. and Goodland, R. (1988) *Wildlands. Their protection and management in economic development*. Washington DC: World Bank.

Leipziger, D.M. and Mudge, J.L. (1976) *Seabed mineral resources and the economic interests of developing countries*. Cambridge, MA: Ballinger.

Lemma, A. and Malaska, P. (1989) *Africa beyond famine*. London and New York: Tycooly.

Leonard, H. J. (1988) *Pollution and the struggle for the world product*. Cambridge: CUP.

Leopold, A. (1949) *A Sand County almanac: and sketches here and there*. London: OUP.

Lesbirel, S.H. (1990) Implementing nuclear energy policy in Japan. *Energy Policy*, **18**, 267–82.

Lewis, C.S. (1947) *The abolition of man* New York: Macmillan.

Lilienthal, D.E. (1953) *TVA democracy on the march* . New York: Harper.

Linnerooth, J. (1990) The Danube river basin: negotiating settlements to transboundary environmental issues. *Natural Resources Journal*, **30**, 629–60.

Lonergan, S. (1993) Water and security in the Middle East. In Foster, H.D. (ed.) *Advances in resource management* London: Belhaven, pp. 199–226.

Lowe, P. and Goyder, J. (1983) *Environmental groups in politics*. London: Allen & Unwin.

Lucas, R.C. (1989) A look at wilderness use and users in transition. *Natural Resources Journal*, **29**, 41–55.

Machlis, G.E. and Tichnell, D.L. (1987) Economic development and threats to national parks: a preliminary analysis. *Environmental Conservation*, **14**, 151–6.

MacKenzie, J.M. (1988) *The empire of nature: hunting, conservation and British imperialism*. Manchester: Manchester University Press.

Mackinnon, J., Mackinnon, K., Child, G. and Thorsell, V. (1986) *Managing protected areas in the tropics*. Gland: IUCN.

Maddison, A. (1964) *Economic growth in the west*. London: George Allen & Unwin.

Mahar, D. (1989) Deforestation in Brazil's Amazon Region: magnitude, rate and causes. In Schramm, G. and Warford, J.J. (eds) *Environmental management and economic development*. Washington: World Bank, pp. 87–116.

Mandel, R. (1988) *Conflict over the world's resources*. New York: Greenwood Press.

Mangone, G.J. (ed.) (1983) *The future of oil and gas from the sea*. New York: Van Nostrand Reinhold.

Manners, G. (1992) Unresolved conflicts in Australian mineral and energy resource policies. *Geographical Journal*, **158**, 129–44.

Manning, E.W. (1990) Sustainable development: the challenge. *Canadian Geographer*, **34**, 290-302.

Manning, R.E. (1989) The nature of America: vision and revision of wilderness. *Natural Resources Journal*, **29**, 25–40.

Mark, A.F. and McSweeney, G.D. (1990) Pattern of impoverishment in natural communities: case history studies in forest ecosystem – New Zealand. In Woodwell, G.M. (ed.) *The Earth in transition: pattern and processes of biotic impoverishment*. Cambridge: CUP, pp. 151–76.

Martin, H.L. and Lo Sun-Jen (1988) Are ore grades declining? The Canadian experience, 1939–1989. In Tilton, J.E., Eggert, R.G. and Landsberg, H.H. (eds) *World mineral exploration: trends and economic issues*. Washington DC: Resources for the Future, pp. 283–330.

Mather, A.S. (1990) *Global forest resources*. London: Belhaven.

Mather, A.S. (ed.) (1993) *Afforestation, policy, pattern and progress*. London: Belhaven.

Matthews, E. (1983) Global vegetation and land use: new high resolution data bases for climatic studies. *Journal of Climate and Applied Meteorology*, **22**, 474–87

May, R.M. (1988) How many species are there on Earth? *Science*, **241**, 1441–8.

McCloskey, J.M. and Spalding H. (1990) The world's remaining wilderness. *Geographical Magazine*, **62**, 14–18.

McCormick, J. (1989) *The global environmental movement: reclaiming paradise*. London: Belhaven.

McCutcheon, R. (1979) *Limits of a modern world*. London: Butterworth.

McDonald, A.T. and Kay, D. (1988) *Water resources: issues and strategies*. Harlow: Longman.

McEvoy, A.F. (1988) Towards an interactive theory of nature and culture: ecology, production and cognition in the Californian fishing industry. In Worster, D. (ed.) *The ends of the Earth*. Cambridge: CUP, pp. 211–29.

McHarg, I. (1969) *Design with nature*. New York: Natural History Press.

McKean, M.A. (1986) Management of traditional common lands in Japan. In *Proceedings of the Conference on Common Property Resource Management*. Washington DC: National Academy Press, pp. 533–89.

McLaren, D.J. and Skinner, B.J. (eds) (1987) *Resources and world development*. Chichester: Wiley.

McNeely, J. (1988) Protected areas. In Pitt, D.C. (ed.) *The future of the environment: the social dimensions of conservation and ecological alternatives*. London and New York: Routledge, pp. 126–44.

McNeely, J.A. (1989a) Management of protected areas for sustaining society. In Verwey, W.D. (ed.) *Nature management and sustainable development*. Amsterdam: IOS, pp. 235–45.

McNeely, J.A. (1989b) Protected areas and human ecology: how national parks can contribute to sustaining societies of the twenty-first century. In Western, D. and Pearl, M. (eds) *Conservation for the twenty-first century*. New York: OUP, pp. 150–7.

McNeely, J.A. (1990) The future of national parks. *Environment*, **32**, 16–20, 36–41.

McNeely, J.A. and MacKinnon, J.R. (1990) Protected areas: development and land use in the tropics. *Resource Management and Optimization*, **7**, 191–208.

McNeill, J.R. (1992) *The mountains of the Mediterranean world: an environmental history*. Cambridge: CUP.

McNeill, J., Winsemius, P. and Yakushiji, T. (1991) *Beyond interdependence: the meshing of the world's economy and the Earth's ecology*. Oxford: OUP.

Meadows, D.H., Meadows, D.L. and Randers, J. (1992) *Beyond the limits: global collapse or a sustainable future*. London: Earthscan.

Meadows, D.H., Meadows, D.L., Randers, J. and Behrens III, W.W. (1972) *Limits to growth*. London: Pan.

Mercado, J. (1990) Philippines fight to prevent its forests from becoming a wasted heritage. *Ceres*, **22**, 42–6.

Mercer, D.E. and Soussan, J. (1992) Fuelwood problems and solutions. In Sharma, N.P. (ed.) *Managing the world's forests*. Dubuque, Iowa: Kendall/Hunt, pp. 177–214.

Merchant, C. (1992) *Radical ecology: the search for a livable world*. New York and London: Routledge.

Micklin, P.P. (1984) Recent developments in large scale water transfers in the USSR. *Soviet Geography*, **25**, 261–63.

Mikesell, R.F. (1992) *Economic development and the environment*. London: Mansell.

Milbrath, L.W. (1983) Images of scarcity in four nations. In Welch, S. and Miewald, R. (eds) *Scarce national resources: the challenge to public policymaking*. Beverley Hills CA: Sage, pp. 105–24.

Milbrath, L.W. (1985) Culture and the environment in the United States. *Environmental Management*, **9**, 161–72.

Miller, G.T. (1990) *Resource conservation and management*. Belmont: Walsworth.

Millward, H. (1985) A model of coalfield development: six stages exemplified by the Sydney field. *Canadian Geographer*, **29**, 234–48.

Mitchell, B. (1979) *Geography and resource analysis*. New York: Longman.

Mitchell, B. (1989) *Geography and resource analysis* (2nd edn). Harlow: Longman.

Moffat, I. (1992) The evolution of the sustainable development concept: a perspective from Australia. *Australian Geographical Studies*, **30**, 27–42.

Mohai, P. (1992) Men, women and the environment: an examination of the gender gap in environmental concern and activism. *Society and Natural Resources*, **5**, 1–19.

Moncrief, C.W. (1970) The cultural basis for our environment crisis. *Science*, **170**, 508–12.

Moore, R.J. (1990) A new Christian reformation. In Engel, J.R. and Engel, J.G. (eds) *Ethics of environment and development: global challenge, international response*. London: Belhaven, pp. 104–13.

Munasinghe, M. (1992) Biodiversity protection policy: environmental valuation and distribution issues. *Ambio*, **21**, 227–36.

Myers, N. (1993) Population, environment and development. *Environmental Conservation*, **20**, 205–16.

Naess, A. (1990) Sustainable development and deep ecology. In Engel, J.R. and Engel, J.G. (eds) *Ethics of environmental development: global challenge, international response*. London: Belhaven, pp. 87–96.

Nagy, A. (ed.) (1988) *International trade in forest products*. Bicester: AB Academic Publishers.

Nash, R.F. (1989) *The rights of nature: a history of environmental ethics*. Madison: University of Wisconsin Press.

Nef, J.V. (1932) *The rise of the British coal industry*. London: Routledge.

Neil, C., Tykkylainen, M. and Bradbury, J. (eds) (1992) *Coping with closure: an international comparison of mine-town experiences*. London: Routledge.

Nesting, R.M. (1986) What alpine peasants have in common: observations on communal tenure in a Swiss village, *Human Ecology*, **4**, 135–46.

Newson, M. (1992) *Land, water and development*. London: Routledge.

Nordhaus, W.D. (1973) World dynamics: assessment without data. *Economic Journal*, **83**, 1156–83.

North, D.C. (1958) Ocean freight rates and economic development 1750–1913. *Journal of Economic History* **18**, 537–55.

North, D.C. and Thomas, R.P. (1977) The First Economic Revolution. *Economic History Review*, **30**, 229–41.

Norton, G.A. (1984) *Resource economics*. London: Edward Arnold.

Odell, P.R. (1973) The future of oil: a rejoinder. *Geographical Journal*, **139**.

Odell, P.R. (1988) The West European gas market. *Energy Policy*, **16**, 480–93.

Odell, P.R. (1989) Draining the world of energy. In Johnston, R.J. and Taylor, P. (eds) *A world in crisis:*

geographical perspectives (2nd edn). Oxford: Blackwell, pp. 79-100.

Odell, P.R. and Rosing, K.E. (1983) *The future of oil* (2nd edn). London: Kogan Page.

OECD (1989a) *Economic instruments for environmental protection*. Paris: OECD.

OECD (1989b) *Renewable natural resources: economic incentives for improved management*. Paris: OECD.

OECD (1991a) *The state of the environment*. Paris: OECD.

OECD (1991b) *Climate change: evaluating the socio-economic impacts*. Paris: OECD.

OECD (1991c) *Responding to climate change: selected economic issues*. Paris: OECD.

OECD (1992a) *Global energy: the changing outlook*. Paris: OECD

OECD (1992b) *Climate change: designing a tradable present system*. Paris: OECD.

OECD (1992c) *Market and government failures in environmental management: wetlands and forests*. Paris: OECD.

OECD (1993) *International economic instruments and climate change*. Paris: OECD.

Omari, C.K. (1990) Traditional African land ethics. In Engel, J.R. and Engel, J.G. (eds*) Ethics of environment and development: global challenge, international response*. London: Belhaven: pp. 167–75.

Ophuls, W. (1977) *Ecology and the politics of scarcity: prologue to a political theory of the steady state*. San Francisco: W.H. Freeman.

O'Riordan, T. (1971) The third American conservation movement. *Journal of American Studies*, **5**, 155–71.

O'Riordan, T. (1977) Environmental ideologies. *Environment and Planning A*, **9**, 3–15.

O'Riordan, T. (1981) *Environmentalism* (2nd edn) London: Pion.

O'Riordan, T. (1988) The politics of sustainability. In Turner, R.K. (ed.) *Sustainable environmental management*. London: Belhaven, pp. 29–50.

Ostrom, E. (1990) *Governing the commons: the evolution of institutions for collective action*. Cambridge: CUP.

Owen, A.D. (1988) Australia's role as an energy exporter. *Energy Policy*, **16**, 131–51.

Owen, M. (1973) *The Tennessee Valley Authority*. London: Praeger.

Paddock, W. and Paddock, P. (1967) *Famine 1975! America's decision: who will survive?* Boston: Little Brown.

Palmer, J and Synnott, T.J. (1992) The management of natural forests. In Sharma, N.P. (ed.) *Managing the world's forests*. Dubuque Iowa: Kendall/Hunt, pp. 337–73.

Palmer, M. (1990) The encounter of religion and conservation. In Engel, J.R. and Engel, J.G. (eds) *Ethics of environment and development: global challenge: international response*. London: Belhaven, pp. 50–62.

Park, C. (1991) Trans-frontier air pollution: some geographical issues. *Geography*, **76**, 21–35.

Parker, D.J. and Sewell, W.R.D. (1988) Evolving water institutions in England and Wales: an assessment of two decades of experience. *Natural Resources Journal*, **28**, 751–86.

Parry, M.L. (1978) *Climatic change: agriculture and settlement*. Folkestone. Dawson.

Parsons, H.L. (ed.) (1977) *Marx and Engels on ecology*. Westport, Connecticut: Greenwood.

Pasqualetti, M.J. (1989) Introducing the geosocial context of nuclear decommissioning: policy implications in the US and UK. *Geoforum*, **20**, 381–96.

Paterson, J.L. (1989) Religious beliefs, environmental stewardship and industrialising agriculture: the Christian Farmers' Federation of Alberta. In Welch, R. (ed.) *Geography in action. NZ Geographical Society Conference Series* No 15, 64–8.

Pawson, E. and Cant, G. (1972) Land rights in historical and contemporary context. *Applied Geography*, **12**, 95–108.

Pearce, D. (1994) Green Britain wilts in a land of hot air. *The Observer*, 23 January, 26.

Pearce, D., Markandya, A. and Barbier, E.B. (1989) *Blueprint for a green economy*. London: Earthscan.

Pearce, D., Barbier, E. and Markandya, A. (1990) *Sustainable development: economics and environment in the third world*. Aldershot: Edward Elgar.

Pearce, D.W. and Turner, R.K. (1990) *Economics of natural resources and the environment*. Hemel Hempstead: Harvester Wheatsheaf.

Pearse, P.H. (1991) Scarcity of natural resources and the implications for sustainable development. *Natural Resources Forum*, **15**, 74–9.

Peet, R. (1969) The spatial expansion of commercial agriculture in the 19th century: a von Thunian interpretation. *Economic Geography*, **45**, 283–301.

Peluso, N.L. (1993) Coercing conservation? – the politics of state resource control, *Global Environmental Change*, **3**, 199–217.

Pepper, D. (1984) *The roots of modern environmentalism*. London: Croom Helm .

Perelman, M. (1975) Natural resources and agriculture under capitalism: Karl Marx's economic model. *American Journal of Agricultural Economics*, **57**, 701–4.

Persson, R. (1974) *World forest resources: review of the world's forest resources in the early 1970s*. Research Note No. 17, Dept of Forest Survey, Royal College of Forestry, Stockholm.

Peters, C., Gentry, A. and Mendelsohn, O. (1989) Valuation of an Amazonian rainforest. *Nature*, **339**, 65–6.

Peters, E.J. (1992) Protecting the land under modern land claims agreements: the effectiveness of the environmental regime negotiated by the James Bay Cree in the James Bay and Northern Quebec Agreement. *Applied Geography*, **12**, 133–45.

Peters, R.L. and Lovejoy, T.E. (1990) Terrestrial fauna. In Turner II, B.L. *et al. The Earth as transformed by human action*. Cambridge: CUP, pp. 353–70.

Petulla, J. (1977) *American environmental history: the exploitation and conservation of natural resources*. San Francisco: Boyd and Fraser.

Pezzey, J. (1989) Definitions of sustainability. *Discussion Paper 9*. UK Centre for Economic and Environmental Development.

Pierce, J.T. (1990) *The food resource*. Harlow: Longman.

Pierce, J.T. (1992) Progress and the biosphere: the dialectics of sustainable development. *Canadian Geographer*, **36**, 306–19.

Pierce, J.T. and Furuseth, O.J. (1986) Constraints to expanding food production: a North American perspective. *Natural Resources Journal*, **26**, 15–40.

Pollard, D.F.W. and McKenzie, M.R. (1986) *World*

Conservation Strategy – Canada: a report on achievements in conservation. Ottawa: Environment Canada.

Ponting, C. (1991) *A green history of the world*. Harmondsworth: Penguin.

Pope John Paul II (1990) Message of His Holiness Pope John Paul II for the Celebration of World Day of Peace 1 January 1990 (Peace with God the Creator: Peace with all of Creation). *Natural Resources Journal*, **30**,1–8.

Postel, S. and Heise, L. (1988) Reforesting the earth. *Worldwatch Paper No 83*. Washington DC: Worldwatch Institute.

Premm, H.J. (1992) Spanish colonisation and Indian property in Central Mexico 1521–1620. *Annals, Association of American Geographers*, **82**, 444–59.

Prescott, J.R.V. (1975) *The political geography of the oceans*. Newton Abbott: David and Charles.

Prestwich, R. (1975) America's dependence on the world's metal resources: shifts in import emphasis. *Transactions of Institute of British Geographers*, **64**, 97–118.

Pullan, R.A. (1988) Conservation and the development of national parks in the humid tropics of Africa. *Journal of Biogeography*, **15**, 171–83.

Rappaport, R.A. (1971) The flow of energy in an agricultural society. In Scientific American *Energy and power*. San Francisco: W.H. Freeman, pp. 69–80.

Redclift, M. (1987) *Sustainable development: exploring the contradictions*. London: Routledge.

Redclift, M. (1988) Economic models and environmental values: a discourse on theory. In Turner, R.K. (ed.) *Sustainable environmental management*. London, Belhaven, pp. 51–66.

Redclift, M. (1992) The meaning of sustainable development. *Geoforum*, **23**, 395–403.

Reed, M.G. and Slaymaker, O. (1993) Ethics and sustainability: a preliminary perspective. *Environment and Planning A*, **25**, 723–39.

Rees, J. (1985) (1990 2nd edn) *Natural resources, allocation, economics, policy*. London: Methuen.

Rees, J. (1989) Natural resources: economy and society. In Gregory, D. and Walford, R. (eds) *Horizons in human geography*. London: Macmillan, pp. 364–95.

Rees, J. (1991a) Resources and the environment: scarcity and sustainability. In Bennett, R.J. and Estall, R.C. (eds) *Global change and challenge*. London: Routledge, pp. 5–26.

Rees, J. (1991b) Equity and environmental policy. *Geography*, **76**, 292–303.

Rees, J. (1992) Markets – the panacea for environmental regulation? *Geoforum*, **23**, 383–394.

Reid, W.V. (1992) How many species will there be? In Whitmore, T.C. and Sayer. J.A. (eds) *Tropical deforestation and species extinction*. London: Chapman & Hall, pp. 55–73.

Reiger, J.F. (1992) Wildlife, conservation and the first forest resource. In Steen, H.K. (ed.) *The origins of the national forests*. Durham, NC: Forest History Society, pp. 106–21.

Repetto, R. (1989) Economic incentives for sustainable production. In Schramm, G. and Warford, J.J. (eds) *Environmental management and economic development*. Washington DC: World Bank, pp. 69–86.

Repetto, R. (1990) Deforestation in the tropics. *Scientific American*, **262**, 18–25.

Repetto, R. (1992) Accounting for environmental assets. *Scientific American*, **266**, 64–70.

Repetto, R. and Gillis, M. (eds) (1988) *Public policy and the misuse of forest resources*. Cambridge: CUP.

Repetto, R. and Holmes, T. (1983) The role of population in resource depletion. *Population and Development Review*, **9**.

Repetto, R. *et al.* (1989) *Wasting assets: natural resources in the national income amounts*. New York: World Resources Institute.

Rich, B. (1985) Multilateral development banks: their role in destroying the global environment. *The Ecologist*, **15**, 635–50.

Rich, B. (1990) Multilateral development trends and tropical deforestation. In Heal, S. and Heinzman, R. (eds) *Lessons of the rainforest*. San Francisco: Sierra Club Books, pp. 118–30.

Richards, J.F. (1990) Land transformation. In Turner II, B.L. *et al.* (eds) *The Earth as transformed by human action*. Cambridge: CUP, pp. 163–78.

Richards, J.F. and Tucker, R.P. (eds) (1988) *World deforestation in the twentieth century*. Durham and London: Duke University Press.

Richards, J.F., Haymen, E.S. and Hagen, J.R. (1985) Changes in the land and human productivity in Northern India 1870–1970. *Agricultural History*, **59**, 523–48.

Riddell, R. (1981) *Ecodevelopment*. London: Gower.

Robb, G.A. (1994) Environmental consequences of coal mine closure. *Geographical Journal*, **160**, 33–40.

Rolston, H. III (1988) Human values and natural systems. *Society and Natural Resources*, **1**, 271–84.

Rosenzweig, C. and Parry, M.L. (1994) Potential impact of climate change on world food supply. *Nature*, **367**, 133–8.

Ross, W.M. (1971) The management of international common property resources. *Geographical Review*, **61**, 325–37.

Rostow, W.W. (1978) *The world economy: history and prospect*. London: Macmillan.

Rotberg, R.I. and Rabb, T.K. (eds) (1985) *Hunger and history: the impact of changing food production and consumption patterns in society*. Cambridge: CUP.

Rowe, R., Sharma, N.P. and Browder, J. (1992) Deforestation: problems, causes and concerns. In Sharma, N.P. (ed.) *Managing the world's forests: looking for balance between conservation and development*. Dubuque, Iowa: Kendall/Hunt, pp. 33–46.

Rowntree, K. (1990) Political and administrative constraints on integrated river basin development: an evaluation of the Tana and Athi Rivers Development Authority, Kenya, *Applied Geography*, **10**, 21–41.

Ruckelhaus, W.D. (1989) Towards a sustainable world. *Scientific American*, **261**, 114–20.

Runte, A. (1979) *National parks: the American experience*. Lincoln and London: University of Nebraska Press.

Ryan, J. (1992) Conserving biological diversity. In Brown, L.R. (ed.) *State of the world 1992*. London: Earthscan: 9–26.

Sagoff, M. (1988) *The economy of the Earth: philosophy, law and environment*. Cambridge: CUP.

Sax, J.L. (1993) Nature and habitat conservation and protection in the United States. *Ecology Law Quarterly*, **20**, 47–56.

Sayer, J.A., Harcount, C.S. and Collin, N.M. (1992) *The*

conservation atlas of tropical forests, Africa. London: Macmillan.

Schiff, M. and Valdes, A. (1990) Poverty, food intake and malnutrition: implications for food scarcity in developing countries. *American Journal of Agricultural Economics*, **72**, 1318–22.

Schlager, E. and Ostrom, E. (1992) Property-rights regimes and natural resources: a conceptual analysis. *Land Economics*, **68**, 249–62.

Schramm, G. and Warford, J.J. (eds) (1989) *Environmental management and economic development*. Washington: World Bank.

Schwarz, H.E., Emel, J., Dickens, W.J., Rogers, P. and Thompson, J. (1990) Water quality and flows. In Turner II, B.L. *et al.* (eds) *The Earth as transformed by human action*. Cambridge: CUP, pp. 253–70.

Science Council of Canada (1992) *Sustainable agriculture: the research challenge*. Ottawa: Science Council of Canada.

Sedjo, R.A. (1984) An economic assessment of industrial forest plantations. *Forest Ecology and Management*, **9**, 245–58.

Sedjo, R.A. (1987) Forest resources of the world: forest in transition. In Kallio, M., Dyskstra, D.P. and Binkley, C.S. (eds) *The global forest sector: an analytical perspective*. Chichester: Wiley, pp. 7–33.

Sedjo, R.A. (ed.) (1985) *Investments in forestry: resources, land use and public policy*. Boulder: Westview.

Sedjo, R.A. and Clawson, M. (1983) Tropical deforestation: how serious? *Journal of Forestry*, **81**, 792–4.

Sewell, W.R.D., Dearden, P. and Dumbrell, J. (1989) Wilderness decision making and the role of environmental interest groups. *Natural Resources Journal*, **29**, 147–70.

Sharma, N.P., Rowe, R., Opershaw, K. and Jacobson, M. (1992) World forests in perspective. In Sharma, N.P. (ed.) *Managing the world's forests*. Dubuque, Iowa: Kendall Hunt, pp. 17–32.

Shell Briefing Service (1983) *The offshore challenge*. London: Shell International Petroleum Co. Ltd.

Shuval, H.I. (1987) The development of water re-use in Israel. *Ambio*, **16**.

Simmons, I.G. (1987) Transformation of the land in pre-industrial time. In Wolman, M.G. and Fournier, F.G.A. (eds) *Land transformation in agriculture*. Chichester: Wiley, pp. 45–78.

Simmons, I.G. (1989) *Changing the face of the Earth*. Oxford: Blackwell.

Simon, J.L. (1980) Resources, population and environment: an oversupply of false bad news. *Science*, **208**, 1431–37.

Simon, J.L. (1981) *The ultimate resource*. Oxford: Martin Robertson.

Simon, J.L. (1990) *Population matters: people, resources, environment and immigration*. New Brunswick (NJ): Transaction Publishers.

Simpson, R.D. and Sedjo, R.A. (1992) Contracts for transferring rights to indigenous genetic resources. *Resources*, **109**, 1–6.

Singer, M. (1989) *Passage to a human world: the dynamics of creating global wealth*. New Brunswick and Oxford: Transaction Publications.

Skog, K.E. and Watterson, I.A. (1984) Residential fuel-

wood uses in the United States. *Journal of Forestry*, **82**, 742–7.

Smith, A. (1776) *The wealth of nations* (1970 edn). Harmondsworth: Penguin.

Smith, F.D.M., May, R.M., Pellew, R., Johnson, T.H. and Walter, K.S. (1993) Estimating extinction rates. *Nature*, **364**, 494–6.

Smith, K. (1991) *Environmental hazards*. London: Routledge.

Smith, V.K. (1978) Measuring natural resources scarcity: theory and practice. *Journal of Environmental Economics and Management*, **5**, 150–71.

Smith, W. (1993) Sustainable development: the economic and environmental case for policy reforms. *New Zealand Geographer*, **49**, 69–74.

Southgate, D. (1990) The causes of land degradation along spontaneously expanding agricultural frontiers in the Third World. *Land Economics*, **66**, 93-101.

Southgate, D., Sanders, J. and Thin, S. (1990) Resource degradation in Africa and Latin America: population pressure, politics and property arrangements. *American Journal of Agricultural Economics*, **72**, 1259–63.

Spooner, D.J. (1981a) *Mining and regional development*. Oxford: OUP.

Spooner, D.J. (1981b) The geography of coal's second coming. *Geography*, **66**, 29–41.

Stankey, G.H. (1989) Linking parks to people: the key to effective management. *Society and Natural Resources*, **2**, 245–50.

Starke, L. (1990) *Signs of hope: working towards our common future*. Oxford: OUP.

Starr, J.R. (1991) Water wars. *Foreign Policy*, **82**: 17–36.

Stevens, C. (1993) The environmental effects of trade. *World Economy*, **16**, 439–51.

Stevenson, G.G. (1991) *Common property economics: a general theory and land use applications*. Cambridge: CUP.

Stretton, H. (1976) *Capitalism, socialism and the environment*. Cambridge: CUP.

Suarez, C.E. (1990) Long term evolution of oil prices. *Energy Policy*, **18**, 170–4.

Suliman, M. (1993) Civil war in the Sudan: from ethnic to ecological conflict. *The Ecologist*, **23**, 104–9.

Sullivan, K.M. (1989) Conflict in the management of a northwest Atlantic transboundary cod stock. *Marine Policy*, **13**, 118–36.

Swanson, T.M. and Barbier, E.B. (1992) *Economics for the wilds: wildlife, wildlands, diversity and development*. London: Earthscan.

Sweet, C. (1990) Does nuclear power have a future? *Energy Policy*, **18**, 406–22.

Symes, D. (1991) UK demersal fisheries and the North Sea: problems in renewable resources management. *Geography*, **76**, 131–42.

Tanzer, M. (1980) *The race for resources: continuing struggles over minerals and fuels*. New York: Monthly Review Press.

Tarrant, J.R. (1987) Variability in world cereal yields. *Transactions, Institute of British Geographers*, **12**, 315–26.

Teclaf, L.A. and Teclaf, E. (1985) Transboundary toxic pollution and the drainage basin concept. *Natural Resources Journal*, **25**, 589-612.

Thapa, G.B. and Weber, K.E. (1990) Actions and factors of

deforestation in tropical Asia. *Environmental Conservation*, **17**, 19–27.

Thomas, C. (1992) *The environment in international relations*. London: Royal Institute of International Affairs.

Thomas, K. (1983) *Man and the natural world: changing attitudes in England 1500–1800*. London: Allen Lane.

Tilton, V.E. and Skinner, B.J. (1987) The meaning of resources. In Tilton, V.E. and Skinner, B.J. (eds) *Resources and world development*. Chichester: Wiley, pp. 13–27.

Tisdell, C.A. (1990) *Natural resources: growth and development, economics, ecology and resource scarcity*. New York: Praeger.

Tolba, M.K. and El-Kholy, O.A. (1992) *The world environment 1972–1992: two decades of challenge*. London: Chapman & Hall.

Toynbee, A. (1972) The religious background of the present environmental crisis. *International Journal of Environmental Studies*, **3**, 141–46.

Trimble, S.W. (1985) Perceptions of the history of soil erosion control in the eastern United States. *Agricultural History*, **59**, 162–80.

Tuan, Y.F. (1968) Discrepancies between environmental attitude and behaviour: examples from Europe and China. *Canadian Geographer*, **12**, 176–81.

Tuan, Y.F. (1974) *Topophila: a study of environmental perception, attitudes and values*. Englewood Cliffs: Prentice Hall.

Tucker, R.P. and Richards, J.F. (eds) (1983) *Global deforestation and the nineteenth century world economy*. Durham, NC: Duke University Press.

Turner, B.L. II, Clark, W.C., Kates, R.W., Richards, J.F., Mathews, J.T. and Meyer, W.B. (eds) (1990) *The Earth as transformed by human action*. Cambridge: CUP with Clark University.

Ullman, E.L. (1954) Amenities as a factor in regional growth. *Geographical Review*, **44**, 119–32.

UNCTAD (1993) *Handbook of international trade and development statistics 1992*. New York: UN.

UNECE and FAO (1985) *The forest resources of the ECE Region* (Europe, Soviet Union, North America). Geneva: ELE/FAO.

UN Environmental Programme (1989/90) *Environmental data report*. Oxford: Blackwell.

UN Environmental Programme (1991/92) *Environmental data report*. Oxford: Blackwell.

United Nations (1993) *Report of the United Nations Conference on Environment and Development*. (3 vols). New York: United Nations.

US Bureau of the Census (1960) *Historical statistics of the United States: colonial times to 1957*. Washington DC: Department of Commerce.

US Bureau of Mines (annual) *Minerals yearbook*. Washington DC: US Bureau of Mines.

US Department of Energy (1993) *Annual energy review 1992*. Washington DC: Energy Information Administration.

US Department of Energy (1994) *Energy use and carbon emissions: some international comparisons*. Washington DC: Energy Information Administration.

Usher, P.J., Tough, F.J. and Galois, R.M. (1992) Reclaiming the land, aboriginal title, trading rights and land claims in Canada. *Applied Geography*, **12**, 109–32.

Utterström, G. (1955) Climatic fluctuations and population problems in Early Modern history. *Scandinavian Economic History Review*, **3**, 3–47.

Vitousek, P.M. *et al.* (1986) Human appropriation of the products of photosynthesis. *BioScience*, **36**, 368–73.

Walker, L.L. and Hoesada, J.A. (1986) Indonesia: forestry by degree. *Journal of Forestry*, **84**, 38–43.

Wallerstein, I. (1974) *The modern world systems: I Capitalist agriculture and the origins of the European world economy in the sixteenth century*. New York: Academic Press.

Warford, J.J. (1987) Natural resources and economic policy in developing countries. *Annals of Regional Science*, **21**, 3–17.

Warford, J.J. (1989) Economic development and environmental protection, *Natural Resources Forum* **13**, 238–41.

Warman, H.R. (1972) The future of oil. *Geographical Journal*, **138**, 287–97.

Warren, K. (1973) *Mineral resources*. Newton Abbot: David and Charles.

Watson, R.B. and Muraoka, D.D. (1992) The northern spotted owl controversy. *Society and Natural Resources*, **5**, 85–90.

Wells, H.G. (1914) *The world set free*. London: Macmillan.

Westoby, J. (1989) *Introduction to world forestry*. Oxford: Blackwell.

Whitaker, J.R. (1940) World view of destruction and conservation of natural resources. *Annals, Association of American Geographers*, **30**, 143–62.

Whitaker, J.R. (1941) Sequence and equilibrium in destruction and conservation of natural resources. *Annals, Association of American Geographers*, **31**, 129–44.

White, L. (1967) The historical roots of our ecological crisis. *Science*, **155**, 1203–7.

Whitmore, T.C. and Sayer, J.Q. (eds) (1992) *Tropical deforestation and species estimates*. London: Chapman & Hall.

Whitmore, T.M. *et al.* (1990) Long-term population change. In Turner II, B.L. *et al.* (eds) *The Earth as transformed by human action*. Cambridge: CUP, pp. 25–40.

Wilkerson, O.L. and Edgell, M.C.R. (1993) The role and limitations of paradigm research in environmental management. In Foster, H.D. (ed.) *Advances in Resource Management*. London: Belhaven, pp. 53–82.

Wilkinson, L. (ed.) (1980) *Earth keeping: Christian stewardship of natural resources*. Grand Rapids: Eerdmans.

Wilkinson, R.G. (1973) *Poverty and progress: an ecological perspective on economic development*. London: Methuen.

Williams, M. (1989a) Deforestation: past and present. *Progress in Human Geography*, **13**, 176–208.

Williams, M. (1989b) *Americans and their forest: a historical geography*. Cambridge: CUP.

Willis, K. and Garrod, G. (1991) Landscape values: a contingent valuation approach and case study of the Yorkshire Dales National Park. *Countryside Change Working Paper Series No. 21*. University of Newcastle upon Tyne.

Willis, K.G., Benson, J.F. and Saunders, C.M. (1988) Agricultural policy and nature conservation. *Land Economics*, **64**, 147–57.

Willis, K.G. and Benson, J.F. (1989) Recreational values of forests. *Forestry*, **62**, 93–110.

Wilson, E.O. (1989) Threats to biodiversity. *Scientific American*, **261**, 60–6.

Wilson, M.G.A. (1968) Changing patterns of pit location in the New South Wales coalfield. *Annals, Association of American Geographers*, **58**, 78–90.

Winid, B. (1981) Comments on the development of the Awash Valley, Ethiopia. In Saha, S.K. and Barrow, C.J. (eds) *River basin planning: theory and practice*. Chichester: Wiley, pp. 147–65.

Wolf, A. and Ross, J. (1992) The impact of scarce water resources on the Arab–Israeli conflict. *Natural Resources Journal*, **32**, 919–58.

Woodwell, G. (1992) The role of forests in climate change. In Sharma, N.P. *et al.* (eds) *Managing the world's forests*. Dubuque, Iowa: Kendall/Hunt, pp. 75–92.

World Bank (1987) *Environment, growth and development*. Washington DC: World Bank.

World Bank (1988) *The World Bank and the environment*. Washington DC: World Bank.

World Commission on Environment and Development (1987) *Our common future*. Oxford: OUP.

World Energy Council (1993) *Renewable energy resources: opportunities and constraints 1990–2020; energy for tomorrow's world*. London: World Energy Council.

World Resources Institute (1986, 1988, 1990, 1992) *World resources*. New York: OUP.

Worster, D. (1977) *Nature's economy: a history of ecological ideas*. Cambridge: CUP.

Worster, D. (1979) *Dust bowl: the Southern Plains in the 1930s*. New York: OUP.

Wrigley, E.A. (1962) The supply of raw materials in the Industrial Revolution. *Economic History Review*, **15**, 1–16.

Wrigley, E.A. (1986) Elegance and experience: Malthus at the bar of history. In Coleman, D. and Schofield, R. (eds) *The state of population theory: forward from Malthus*. Oxford: Blackwell, pp. 46–64.

Young, J. (1990) *Post environmentalism*. London: Belhaven.

Young, O.R. (1981) *Natural resources and the state: the political economy of resource management*. Berkeley: University of California Press.

Young, O.R. (1982) *Resource regimes: natural resources and social institutions*. Berkeley: University of California Press.

Young, O.R. (1989) *International cooperation: building regimes for natural resources and the environment*. Ithaca and London: Cornell UP.

Zimmerman, E.W. (1951) *World resources and industries*. New York: Harper and Row.

Zobler, L. (1962) An economic-historical view of natural resource use and conservation. *Economic Geography*, **38**, 189–94.

Zube, E.H. (1986) Local and extra-local perceptions of national parks and protected areas, *Landscape and Urban Planning*, **13**, 11–17.

Index